AN INTRODUCTION TO LTE

AN INTRODUCTION TO LTE

LTE, LTE-ADVANCED, SAE AND 4G MOBILE COMMUNICATIONS

Christopher Cox
Director, Chris Cox Communications Ltd, UK

A John Wiley & Sons, Ltd., Publication

This edition first published 2012
© 2012 John Wiley & Sons Ltd

Registered office
John Wiley & Sons Ltd, The Atrium, Southern Gate, Chichester, West Sussex, PO19 8SQ, United Kingdom

For details of our global editorial offices, for customer services and for information about how to apply for permission to reuse the copyright material in this book please see our website at www.wiley.com.

The right of the author to be identified as the author of this work has been asserted in accordance with the Copyright, Designs and Patents Act 1988.

All rights reserved. No part of this publication may be reproduced, stored in a retrieval system, or transmitted, in any form or by any means, electronic, mechanical, photocopying, recording or otherwise, except as permitted by the UK Copyright, Designs and Patents Act 1988, without the prior permission of the publisher.

Wiley also publishes its books in a variety of electronic formats. Some content that appears in print may not be available in electronic books.

Designations used by companies to distinguish their products are often claimed as trademarks. All brand names and product names used in this book are trade names, service marks, trademarks or registered trademarks of their respective owners. The publisher is not associated with any product or vendor mentioned in this book. This publication is designed to provide accurate and authoritative information in regard to the subject matter covered. It is sold on the understanding that the publisher is not engaged in rendering professional services. If professional advice or other expert assistance is required, the services of a competent professional should be sought.

Library of Congress Cataloging-in-Publication Data

Cox, Christopher (Christopher Ian), 1965-
 An introduction to LTE : LTE, LTE-advanced, SAE and 4G mobile communications / Christopher Cox.
   p. cm.
 Includes bibliographical references and index.
 ISBN 978-1-119-97038-5 (cloth)
 1. Long-Term Evolution (Telecommunications) 2. Mobile communication systems – Standards. I. Title.
 TK5103.48325.C693 2012
 621.3845'6 – dc23
                                                                                    2011047216

A catalogue record for this book is available from the British Library.

Print ISBN: 9781119970385

Set in 10/12 Times by Laserwords Private Limited, Chennai, India
Printed and bound in Great Britain by CPI Group (UK) Ltd, Croydon, CR0 4YY

To my nieces, Louise and Zoe

Contents

Preface   xvii

Acknowledgements   xix

List of Abbreviations   xxi

1 Introduction   1
1.1 Architectural Review of UMTS and GSM   1
    1.1.1 High Level Architecture   1
    1.1.2 Architecture of the Radio Access Network   2
    1.1.3 Architecture of the Core Network   4
    1.1.4 Communication Protocols   4
1.2 History of Mobile Telecommunication Systems   6
    1.2.1 From 1G to 3G   6
    1.2.2 Third Generation Systems   7
1.3 The Need for LTE   8
    1.3.1 The Growth of Mobile Data   8
    1.3.2 Capacity of a Mobile Telecommunication System   10
    1.3.3 Increasing the System Capacity   11
    1.3.4 Additional Motivations   11
1.4 From UMTS to LTE   11
    1.4.1 High Level Architecture of LTE   11
    1.4.2 Long Term Evolution   12
    1.4.3 System Architecture Evolution   13
1.5 From LTE to LTE-Advanced   15
    1.5.1 The ITU Requirements for 4G   15
    1.5.2 Requirements of LTE-Advanced   15
    1.5.3 4G Communication Systems   15
    1.5.4 The Meaning of 4G   16
1.6 The 3GPP Specifications for LTE   16
    References   18

2 System Architecture Evolution   21
2.1 Architecture of LTE   21
    2.1.1 High Level Architecture   21
    2.1.2 User Equipment   21
    2.1.3 Evolved UMTS Terrestrial Radio Access Network   23

	2.1.4	Evolved Packet Core	24
	2.1.5	Roaming Architecture	25
	2.1.6	Network Areas	26
	2.1.7	Numbering, Addressing and Identification	27
2.2	Communication Protocols		28
	2.2.1	Protocol Model	28
	2.2.2	Air Interface Transport Protocols	29
	2.2.3	Fixed Network Transport Protocols	30
	2.2.4	User Plane Protocols	31
	2.2.5	Signalling Protocols	31
2.3	Example Information Flows		33
	2.3.1	Access Stratum Signalling	33
	2.3.2	Non Access Stratum Signalling	33
	2.3.3	Data Transport	35
2.4	Bearer Management		36
	2.4.1	The EPS Bearer	36
	2.4.2	Tunnelling Using GTP	37
	2.4.3	Tunnelling Using GRE and PMIP	38
	2.4.4	Signalling Radio Bearers	39
2.5	State Diagrams		40
	2.5.1	EPS Mobility Management	40
	2.5.2	EPS Connection Management	40
	2.5.3	Radio Resource Control	41
2.6	Spectrum Allocation		42
	References		44
3	**Digital Wireless Communications**		**47**
3.1	Radio Transmission and Reception		47
	3.1.1	Signal Transmission	47
	3.1.2	Signal Reception	49
	3.1.3	Channel Estimation	50
	3.1.4	Multiple Access Techniques	51
	3.1.5	FDD and TDD Modes	52
3.2	Multipath, Fading and Inter-Symbol Interference		53
	3.2.1	Multipath and Fading	53
	3.2.2	Inter-Symbol Interference	55
3.3	Error Management		56
	3.3.1	Forward Error Correction	56
	3.3.2	Automatic Repeat Request	57
	3.3.3	Hybrid ARQ	58
	References		60
4	**Orthogonal Frequency Division Multiple Access**		**61**
4.1	Orthogonal Frequency Division Multiplexing		61
	4.1.1	Reduction of Inter-Symbol Interference using OFDM	61
	4.1.2	The OFDM Transmitter	62
	4.1.3	Initial Block Diagram	65

4.2		OFDMA in a Mobile Cellular Network	66
	4.2.1	Multiple Access	66
	4.2.2	Fractional Frequency Re-Use	67
	4.2.3	Channel Estimation	68
	4.2.4	Cyclic Prefix Insertion	69
	4.2.5	Use of the Frequency Domain	70
	4.2.6	Choice of Sub-Carrier Spacing	72
4.3		Single Carrier Frequency Division Multiple Access	72
	4.3.1	Power Variations from OFDMA	72
	4.3.2	Block Diagram of SC-FDMA	73
		References	76
5		**Multiple Antenna Techniques**	**77**
5.1		Diversity Processing	77
	5.1.1	Receive Diversity	77
	5.1.2	Closed Loop Transmit Diversity	78
	5.1.3	Open Loop Transmit Diversity	79
5.2		Spatial Multiplexing	80
	5.2.1	Principles of Operation	80
	5.2.2	Open Loop Spatial Multiplexing	82
	5.2.3	Closed Loop Spatial Multiplexing	84
	5.2.4	Matrix Representation	85
	5.2.5	Implementation Issues	88
	5.2.6	Multiple User MIMO	88
5.3		Beamforming	90
	5.3.1	Principles of Operation	90
	5.3.2	Beam Steering	91
	5.3.3	Dual Layer Beamforming	92
	5.3.4	Downlink Multiple User MIMO Revisited	93
		References	94
6		**Architecture of the LTE Air Interface**	**95**
6.1		Air Interface Protocol Stack	95
6.2		Logical, Transport and Physical Channels	97
	6.2.1	Logical Channels	97
	6.2.2	Transport Channels	97
	6.2.3	Physical Data Channels	98
	6.2.4	Control Information	99
	6.2.5	Physical Control Channels	100
	6.2.6	Physical Signals	100
	6.2.7	Information Flows	101
6.3		The Resource Grid	101
	6.3.1	Slot Structure	101
	6.3.2	Frame Structure	103
	6.3.3	Uplink Timing Advance	105
	6.3.4	Resource Grid Structure	106
	6.3.5	Bandwidth Options	106

6.4	Multiple Antenna Transmission	107
	6.4.1 Downlink Antenna Ports	107
	6.4.2 Downlink Transmission Modes	107
6.5	Resource Element Mapping	109
	6.5.1 Downlink Resource Element Mapping	109
	6.5.2 Uplink Resource Element Mapping	109
	References	112

7 Cell Acquisition — 113
7.1	Acquisition Procedure	113
7.2	Synchronization Signals	114
	7.2.1 Physical Cell Identity	114
	7.2.2 Primary Synchronization Signal	115
	7.2.3 Secondary Synchronization Signal	116
7.3	Downlink Reference Signals	116
7.4	Physical Broadcast Channel	117
7.5	Physical Control Format Indicator Channel	118
7.6	System Information	119
	7.6.1 Organization of the System Information	119
	7.6.2 Transmission and Reception of the System Information	121
7.7	Procedures After Acquisition	121
	References	122

8 Data Transmission and Reception — 123
8.1	Data Transmission Procedures	123
	8.1.1 Downlink Transmission and Reception	123
	8.1.2 Uplink Transmission and Reception	125
	8.1.3 Semi Persistent Scheduling	126
8.2	Transmission of Scheduling Messages on the PDCCH	127
	8.2.1 Downlink Control Information	127
	8.2.2 Resource Allocation	128
	8.2.3 Example: DCI Format 1	129
	8.2.4 Radio Network Temporary Identifiers	130
	8.2.5 Transmission and Reception of the PDCCH	131
8.3	Data Transmission on the PDSCH and PUSCH	132
	8.3.1 Transport Channel Processing	132
	8.3.2 Physical Channel Processing	133
8.4	Transmission of Hybrid ARQ Indicators on the PHICH	135
	8.4.1 Introduction	135
	8.4.2 Resource Element Mapping of the PHICH	136
	8.4.3 Physical Channel Processing of the PHICH	136
8.5	Uplink Control Information	137
	8.5.1 Hybrid ARQ Acknowledgements	137
	8.5.2 Channel Quality Indicator	137
	8.5.3 Rank Indication	138
	8.5.4 Precoding Matrix Indicator	139

	8.5.5	Channel State Reporting Mechanisms	139
	8.5.6	Scheduling Requests	140
8.6	Transmission of Uplink Control Information on the PUCCH		140
	8.6.1	PUCCH Formats	140
	8.6.2	PUCCH Resources	142
	8.6.3	Physical Channel Processing of the PUCCH	142
8.7	Uplink Reference Signals		143
	8.7.1	Demodulation Reference Signal	143
	8.7.2	Sounding Reference Signal	144
8.8	Uplink Power Control		145
	8.8.1	Uplink Power Calculation	145
	8.8.2	Uplink Power Control Commands	146
8.9	Discontinuous Reception		146
	8.9.1	Discontinuous Reception and Paging in RRC_IDLE	146
	8.9.2	Discontinuous Reception in RRC_CONNECTED	147
	References		148
9	**Random Access**		**151**
9.1	Transmission of Random Access Preambles on the PRACH		151
	9.1.1	Resource Element Mapping	151
	9.1.2	Preamble Sequence Generation	153
	9.1.3	Signal Transmission	153
9.2	Non Contention Based Procedure		154
9.3	Contention Based Procedure		155
	References		156
10	**Air Interface Layer 2**		**159**
10.1	Medium Access Control Protocol		159
	10.1.1	Protocol Architecture	159
	10.1.2	Timing Advance Commands	160
	10.1.3	Buffer Status Reporting	161
	10.1.4	Power Headroom Reporting	162
	10.1.5	Multiplexing and De-Multiplexing	162
	10.1.6	Logical Channel Prioritization	163
	10.1.7	Scheduling of Transmissions on the Air Interface	163
10.2	Radio Link Control Protocol		164
	10.2.1	Protocol Architecture	164
	10.2.2	Transparent Mode	165
	10.2.3	Unacknowledged Mode	165
	10.2.4	Acknowledged Mode	166
10.3	Packet Data Convergence Protocol		167
	10.3.1	Protocol Architecture	167
	10.3.2	Header Compression	168
	10.3.3	Prevention of Packet Loss During Handover	169
	References		170

11	**Power-On and Power-Off Procedures**	**173**
11.1	Power-On Sequence	173
11.2	Network and Cell Selection	175
	11.2.1 Network Selection	175
	11.2.2 Closed Subscriber Group Selection	175
	11.2.3 Cell Selection	176
11.3	RRC Connection Establishment	177
	11.3.1 Basic Procedure	177
	11.3.2 Relationship with Other Procedures	178
11.4	Attach Procedure	179
	11.4.1 IP Address Allocation	179
	11.4.2 Overview of the Attach Procedure	180
	11.4.3 Attach Request	180
	11.4.4 Identification and Security Procedures	182
	11.4.5 Location Update	183
	11.4.6 Default Bearer Creation	184
	11.4.7 Attach Accept	185
	11.4.8 Default Bearer Update	186
11.5	Detach Procedure	187
	References	188

12	**Security Procedures**	**191**
12.1	Network Access Security	191
	12.1.1 Security Architecture	191
	12.1.2 Key Hierarchy	192
	12.1.3 Authentication and Key Agreement	193
	12.1.4 Security Activation	195
	12.1.5 Ciphering	196
	12.1.6 Integrity Protection	197
12.2	Network Domain Security	197
	12.2.1 Security Protocols	197
	12.2.2 Security in the Evolved Packet Core	198
	12.2.3 Security in the Radio Access Network	199
	References	200

13	**Quality of Service, Policy and Charging**	**201**
13.1	Policy and Charging Control	201
	13.1.1 Introduction	201
	13.1.2 Quality of Service Parameters	202
	13.1.3 Policy Control Architecture	203
13.2	Session Management Procedures	205
	13.2.1 IP-CAN Session Establishment	205
	13.2.2 Mobile Originated QoS Request	206
	13.2.3 Server Originated QoS Request	207
	13.2.4 Dedicated Bearer Establishment	208
	13.2.5 Other Session Management Procedures	210

| 13.3 | Charging and Billing | 210 |

13.3.1 High Level Architecture 210
13.3.2 Offline Charging 211
13.3.3 Online Charging 212
References 212

14 Mobility Management 215
14.1 Transitions Between Mobility Management States 215
    14.1.1 S1 Release Procedure 215
    14.1.2 Paging Procedure 216
    14.1.3 Service Request Procedure 217
14.2 Cell Reselection in RRC_IDLE 219
    14.2.1 Objectives 219
    14.2.2 Cell Reselection on the Same LTE Frequency 219
    14.2.3 Cell Reselection to a Different LTE Frequency 220
    14.2.4 Fast Moving Mobiles 222
    14.2.5 Tracking Area Update Procedure 222
    14.2.6 Network Reselection 224
14.3 Measurements in RRC_CONNECTED 224
    14.3.1 Objectives 224
    14.3.2 Measurement Procedure 224
    14.3.3 Measurement Reporting 225
    14.3.4 Measurement Gaps 226
14.4 Handover in RRC_CONNECTED 227
    14.4.1 X2 Based Handover Procedure 227
    14.4.2 Handover Variations 229
    References 230

15 Inter-System Operation 231
15.1 Inter-Operation with UMTS and GSM 231
    15.1.1 S3-Based Architecture 231
    15.1.2 Gn/Gp-Based Architecture 233
    15.1.3 Bearer Management 233
    15.1.4 Power-On Procedures 234
    15.1.5 Cell Reselection in RRC_IDLE 234
    15.1.6 Idle Mode Signalling Reduction 235
    15.1.7 Measurements in RRC_CONNECTED 235
    15.1.8 Handover in RRC_CONNECTED 236
15.2 Inter-Operation with Generic Non 3GPP Technologies 239
    15.2.1 Network Based Mobility Architecture 239
    15.2.2 Host Based Mobility Architecture 241
    15.2.3 Attach Procedure 241
    15.2.4 Cell Reselection and Handover 243
15.3 Inter-Operation with cdma2000 HRPD 244
    15.3.1 System Architecture 244
    15.3.2 Preregistration with cdma2000 244

		15.3.3	Cell Reselection in RRC_IDLE	246
		15.3.4	Measurements and Handover in RRC_CONNECTED	246
		References		249
16		**Delivery of Voice and Text Messages over LTE**		**251**
16.1		The Market for Voice and SMS		251
16.2		Third Party Voice over IP		252
16.3		The IP Multimedia Subsystem		253
		16.3.1	IMS Architecture	253
		16.3.2	IMS Procedures	255
		16.3.3	SMS over the IMS	256
16.4		Circuit Switched Fallback		256
		16.4.1	Architecture	256
		16.4.2	Combined EPS/IMSI Attach Procedure	257
		16.4.3	Voice Call Setup	258
		16.4.4	SMS over SGs	260
		16.4.5	Circuit Switched Fallback to cdma2000 1xRTT	261
16.5		VoLGA		262
		References		262
17		**Enhancements in Release 9**		**265**
17.1		Multimedia Broadcast/Multicast Service		265
		17.1.1	Introduction	265
		17.1.2	Multicast/Broadcast over a Single Frequency Network	266
		17.1.3	Implementation of MBSFN in LTE	266
		17.1.4	Architecture of MBMS	268
		17.1.5	Operation of MBMS	269
17.2		Location Services		270
		17.2.1	Introduction	270
		17.2.2	Positioning Techniques	270
		17.2.3	Location Service Architecture	271
		17.2.4	Location Service Procedures	271
17.3		Other Enhancements in Release 9		273
		17.3.1	Dual Layer Beamforming	273
		17.3.2	Commercial Mobile Alert System	273
		17.3.3	Enhancements to Earlier Features of LTE	274
		References		274
18		**LTE-Advanced and Release 10**		**277**
18.1		Carrier Aggregation		277
		18.1.1	Principles of Operation	277
		18.1.2	UE Capabilities	279
		18.1.3	Scheduling	279
		18.1.4	Data Transmission and Reception	280
		18.1.5	Uplink and Downlink Feedback	280

		18.1.6 Other Physical Layer and MAC Procedures	281
		18.1.7 RRC Procedures	281
	18.2	Enhanced Downlink MIMO	281
		18.2.1 Objectives	281
		18.2.2 Downlink Reference Signals	282
		18.2.3 Downlink Transmission and Feedback	283
	18.3	Enhanced Uplink MIMO	283
		18.3.1 Objectives	283
		18.3.2 Implementation	284
	18.4	Relays	284
		18.4.1 Principles of Operation	284
		18.4.2 Relaying Architecture	285
		18.4.3 Enhancements to the Air Interface	286
	18.5	Release 11 and Beyond	287
		18.5.1 Coordinated Multipoint Transmission and Reception	287
		18.5.2 Enhanced Carrier Aggregation	287
		References	288
19		**Self Optimizing Networks**	**291**
	19.1	Self Optimizing Networks in Release 8	291
		19.1.1 Self Configuration of an eNB	291
		19.1.2 Automatic Neighbour Relations	292
		19.1.3 Interference Coordination	293
		19.1.4 Mobility Load Balancing	294
	19.2	New Features in Release 9	295
		19.2.1 Mobility Robustness Optimization	295
		19.2.2 Random Access Channel Optimization	297
		19.2.3 Energy Saving	297
	19.3	Drive Test Minimization in Release 10	298
		References	298
20		**Performance of LTE and LTE-Advanced**	**301**
	20.1	Coverage Estimation	301
	20.2	Peak Data Rates of LTE and LTE-Advanced	302
		20.2.1 Increase of the Peak Data Rate	302
		20.2.2 Limitations on the Peak Data Rate	304
	20.3	Typical Data Rates of LTE and LTE-Advanced	304
		20.3.1 Total Cell Capacity	304
		20.3.2 Data Rate at the Cell Edge	306
		References	307
Bibliography			**309**
Index			**311**

Preface

This book is about the world's dominant 4G mobile telecommunication system, LTE.

In writing the book, my aim has been to give the reader a concise, system level introduction to the technology that LTE uses. The book covers the whole of the system, both the techniques used for radio communication between the base station and the mobile phone, and the techniques used to transfer data and signalling messages across the network. I have avoided going into excessive detail, which is more appropriate for specialized treatments of individual topics and for the LTE specifications themselves. Instead, I hope that the reader will come away from this book with a sound understanding of the system and of the way in which its different components interact. The reader will then be able to tackle the more advanced books and the specifications with confidence.

The target audience is twofold. Firstly, I hope that the book will be valuable for engineers who are working on LTE, notably those who are transferring from other technologies such as GSM, UMTS and cdma2000, those who are experts in one part of LTE but who want to understand the system as a whole and those who are new to mobile telecommunications altogether. Secondly, the book should give a valuable overview to those who are working in non technical roles, such as project managers, marketing executives and intellectual property consultants.

Structurally, the book has four parts of five chapters each. The first part lays out the foundations that the reader will need in the remainder of the book. Chapter 1 is an introduction, which relates LTE to earlier mobile telecommunication systems and lays out its requirements and key technical features. Chapter 2 covers the architecture of the system, notably the hardware components and communication protocols that it contains and its use of radio spectrum. Chapter 3 reviews the radio transmission techniques that LTE has inherited from earlier mobile telecommunication systems, while Chapters 4 and 5 describe the more recent techniques of orthogonal frequency division multiple access and multiple input multiple output antennas.

The second part of the book covers the air interface of LTE. Chapter 6 is a high level description of the air interface, while Chapter 7 relates the low level procedures that a mobile phone uses when it switches on, to discover the LTE base stations that are nearby. Chapter 8 covers the low level procedures that the base station and mobile phone use to transmit and receive information, while Chapter 9 covers a specific procedure, random access, by which the mobile phone can contact a base station without prior scheduling. Chapter 10 covers the higher level parts of the air interface, namely the medium access control, radio link control and packet data convergence protocols.

The third part covers the signalling procedures that govern how a mobile phone behaves. In Chapter 11, we describe the high level procedures that a mobile phone uses when it switches on, to register itself with the network and establish communications with the outside world. Chapter 12 covers the security procedures used by LTE, while Chapter 13 covers the procedures that manage the quality of service and charging characteristics of a data stream. Chapter 14 describes the mobility management procedures that the network uses to keep track of the mobile's location, while Chapter 15 describes how LTE inter-operates with other systems such as GSM, UMTS and cdma2000.

The final part covers more specialized topics. Chapter 16 describes how operators can implement voice and messaging applications across LTE networks. Chapters 17 and 18 describe the enhancements that have been made to LTE in later releases of the specifications, while Chapter 19 covers the self optimization features that straddle the different releases. Finally, Chapter 20 reviews the performance of LTE, and provides estimates of the peak and typical data rates that a network operator can achieve.

LTE has a large number of acronyms, and it is hard to talk about the subject without using them. However, they can make the material appear unnecessarily impenetrable to a newcomer, so I have aimed to keep the use of acronyms to a reasonable minimum, often preferring the full name or a colloquial one. There is a full list of abbreviations in the introductory material and new terms are highlighted using italics throughout the text.

I have also endeavoured to keep the book's mathematical content to the minimum needed to understand the system. The LTE air interface makes extensive use of complex numbers, Fourier transforms and matrix algebra, but the reader will not require any prior knowledge of these in order to understand the book. We do use matrix algebra in one of the subsections of Chapter 5, to cover the more advanced aspects of multiple antennas, but readers can skip this material without detracting from their overall appreciation of the subject.

Acknowledgements

Many people have given me assistance, support and advice during the creation of this book. I am especially grateful to Susan Barclay, Sophia Travis, Sandra Grayson, Mark Hammond and the rest of the publishing team at John Wiley & Sons, Ltd for the expert knowledge and gentle encouragement that they have supplied throughout the production process.

I am indebted to Michael Salmon and Geoff Varrall, for encouraging me to write this book, and to Michael Salmon and Julian Nolan, for taking time from busy schedules to review a draft copy of the manuscript and for offering me invaluable advice on how it might be improved. I would also like to extend my thanks to the delegates who have attended my training courses on LTE. Their questions and corrections have extended my knowledge of the subject, while their feedback has regularly suggested ways to explain topics more effectively.

Several diagrams in this book have been reproduced from the technical specifications for LTE, with permission from the European Telecommunications Standards Institute (ETSI), © 2009, 2010, 2011. 3GPP™ TSs and TRs are the property of ARIB, ATIS, CCSA, ETSI, TTA and TTC who jointly own the copyright for them. They are subject to further modifications and are therefore provided to you 'as is' for information purposes only. Further use is strictly prohibited.

The measurements of network traffic in Figure 1.5 are reproduced by kind permission of Ericsson, © 2011. I am grateful to Svante Bergqvist and Elin Pettersson for making the diagram available for use in this book. Analysys Mason Limited kindly supplied the market research data underlying the illustrations of network traffic and operator revenue in Figures 1.6 and 16.1. I would like to extend my appreciation to Morgan Mullooly, Terry Norman and James Allen, for making this information available.

Nevertheless, the responsibility for any errors or omissions in the text, and for any lack of clarity in the explanations, is entirely my own.

List of Abbreviations

16-QAM	16 quadrature amplitude modulation
1G	First generation
1xRTT	1x radio transmission technology
2G	Second generation
3G	Third generation
3GPP	Third Generation Partnership Project
3GPP2	Third Generation Partnership Project 2
4G	Fourth generation
64-QAM	64 quadrature amplitude modulation
AAA	Authentication, authorization and accounting
ABMF	Account balance management function
ACK	Positive acknowledgement
AES	Advanced Encryption Standard
AF	Application function
AKA	Authentication and key agreement
AM	Acknowledged mode
AMBR	Aggregate maximum bit rate
AMR	Adaptive multi rate
APN	Access point name
APN-AMBR	Per APN aggregate maximum bit rate
ARIB	Association of Radio Industries and Businesses
ARP	Allocation and retention priority
ARQ	Automatic repeat request
AS	Access stratum/Application server
ASME	Access security management entity
ATIS	Alliance for Telecommunications Industry Solutions
AuC	Authentication centre
BBERF	Bearer binding and event reporting function
BCCH	Broadcast control channel
BCH	Broadcast channel
BD	Billing domain
BM-SC	Broadcast/multicast service centre
BPSK	Binary phase shift keying

BSC	Base station controller
BSR	Buffer status report
BTS	Base transceiver station
CA	Carrier aggregation
CBC	Cell broadcast centre
CBS	Cell broadcast service
CC	Component carrier
CCCH	Common control channel
CCE	Control channel element
CCSA	China Communications Standards Association
CDF	Charging data function
CDMA	Code division multiple access
CDR	Charging data record
CFI	Control format indicator
CGF	Charging gateway function
CIF	Carrier indicator field
CM	Connection management
CMAS	Commercial mobile alert system
CoMP	Coordinated multi point
COST	European Cooperation in Science and Technology
CP	Cyclic prefix
CQI	Channel quality indicator
CRC	Cyclic redundancy check
C-RNTI	Cell radio network temporary identifier
CS	Circuit switched
CS/CB	Coordinated scheduling and beamforming
CSCF	Call session control function
CSG	Closed subscriber group
CSI	Channel state information
CTF	Charging trigger function
dB	Decibel
dBm	Decibels relative to one milliwatt
DCCH	Dedicated control channel
DCI	Downlink control information
DeNB	Donor evolved Node B
DFT	Discrete Fourier transform
DFT-S-OFDMA	Discrete Fourier transform spread OFDMA
DHCP	Dynamic host configuration protocol
DL	Downlink
DL-SCH	Downlink shared channel
DRS	Demodulation reference signal
DRX	Discontinuous reception
DSMIP	Dual stack mobile IP
DTCH	Dedicated traffic channel

eAN	Evolved access network
ECGI	E-UTRAN cell global identifier
ECI	E-UTRAN cell identity
ECM	EPS connection management
EDGE	Enhanced Data Rates for GSM Evolution
EEA	EPS encryption algorithm
eHRPD	Evolved high rate packet data
EIA	EPS integrity algorithm
EIR	Equipment identity register
EMM	EPS mobility management
eNB	Evolved Node B
EPC	Evolved packet core
ePCF	Evolved packet control function
ePDG	Evolved packet data gateway
EPS	Evolved packet system
E-RAB	Evolved radio access bearer
ESM	EPS session management
E-SMLC	Evolved serving mobile location centre
ESP	Encapsulating security payload
ETSI	European Telecommunications Standards Institute
ETWS	Earthquake and tsunami warning system
E-UTRAN	Evolved UMTS terrestrial radio access network
EV-DO	Evolution data optimized
FCC	Federal Communications Commission
FDD	Frequency division duplex
FDMA	Frequency division multiple access
FFT	Fast Fourier transform
FTP	File transfer protocol
GBR	Guaranteed bit rate
GERAN	GSM EDGE radio access network
GGSN	Gateway GPRS support node
GMLC	Gateway mobile location centre
GNSS	Global navigation satellite system
GP	Guard period
GPRS	General Packet Radio Service
GPS	Global Positioning System
GRE	Generic routing encapsulation
GSM	Global System for Mobile Communications
GSMA	GSM Association
GTP	GPRS tunnelling protocol
GTP-C	GPRS tunnelling protocol control part
GTP-U	GPRS tunnelling protocol user part
GUMMEI	Globally unique MME identifier
GUTI	Globally unique temporary identity

HARQ	Hybrid ARQ
HeNB	Home evolved Node B
HI	Hybrid ARQ indicator
HLR	Home location register
H-PCRF	Home policy and charging rules function
HRPD	High rate packet data
HSDPA	High speed downlink packet access
HSGW	HRPD serving gateway
HSPA	High speed packet access
HSS	Home subscriber server
HSUPA	High speed uplink packet access
HTTP	Hypertext transfer protocol
I	In phase
I-CSCF	Interrogating call session control function
IEEE	Institute of Electrical and Electronics Engineers
IETF	Internet Engineering Task Force
IKE	Internet key exchange
IMEI	International mobile equipment identity
IM-MGW	IMS media gateway
IMS	IP multimedia subsystem
IMSI	International mobile subscriber identity
IMT	International Mobile Telecommunications
IP	Internet protocol
IP-CAN	IP connectivity access network
IPSec	IP security
IP-SM-GW	IP short message gateway
IPv4	Internet protocol version 4
IPv6	Internet protocol version 6
ISI	Inter symbol interference
ISIM	IP multimedia services identity module
ISR	Idle mode signalling reduction
ITU	International Telecommunication Union
JP	Joint processing
LBS	Location based services
LCS	Location services
LCS-AP	LCS application protocol
LPP	LTE positioning protocol
LTE	Long term evolution
LTE-A	LTE-Advanced
MAC	Medium access control
MAP	Mobile application part
MBMS	Multimedia broadcast/multicast service
MBMS-GW	MBMS gateway

MBR	Maximum bit rate
MBSFN	Multicast/broadcast over a single frequency network
MCC	Mobile country code
MCCH	Multicast control channel
MCE	Multicell/multicast coordination entity
MCH	Multicast channel
MDT	Minimization of drive tests
ME	Mobile equipment
MGCF	Media gateway control function
MGL	Measurement gap length
MGRP	Measurement gap repetition period
MGW	Media gateway
MIB	Master information block
MIMO	Multiple input multiple output
MIP	Mobile IP
MM	Mobility management
MME	Mobility management entity
MMEC	MME code
MMEGI	MME group identity
MMEI	MME identifier
MMSE	Minimum mean square error
MNC	Mobile network code
M-RNTI	MBMS radio network temporary identifier
MSC	Mobile switching centre
MT	Mobile termination
MTCH	Multicast traffic channel
M-TMSI	M temporary mobile subscriber identity
MU-MIMO	Multiple user MIMO
NACK	Negative acknowledgement
NAS	Non access stratum
NH	Next hop
OCF	Online charging function
OCS	Online charging system
OFCS	Offline charging system
OFDM	Orthogonal frequency division multiplexing
OFDMA	Orthogonal frequency division multiple access
OSA	Open service architecture
OSI	Open systems interconnection
OTDOA	Observed time difference of arrival
PBCH	Physical broadcast channel
PBR	Prioritized bit rate
PCC	Policy and charging control
PCCH	Paging control channel
PCEF	Policy and charging enforcement function

PCell	Primary cell
PCFICH	Physical control format indicator channel
PCH	Paging channel
PCRF	Policy and charging rules function
P-CSCF	Proxy call session control function
PDCCH	Physical downlink control channel
PDCP	Packet data convergence protocol
PDN	Packet data network
PDP	Packet data protocol
PDSCH	Physical downlink shared channel
PDU	Protocol data unit
P-GW	Packet data network gateway
PHICH	Physical hybrid ARQ indicator channel
PL	Path loss/Propagation loss
PLMN	Public land mobile network
PLMN-ID	Public land mobile network identity
PMCH	Physical multicast channel
PMD	Pseudonym mediation device
PMI	Precoding matrix indicator
PMIP	Proxy mobile IP
PPR	Privacy profile register
PRACH	Physical random access channel
PRB	Physical resource block
P-RNTI	Paging radio network temporary identifier
PS	Packet switched
PSS	Primary synchronization signal
PSTN	Public switched telephone network
PUCCH	Physical uplink control channel
PUSCH	Physical uplink shared channel
PWS	Public warning system
Q	Quadrature
QAM	Quadrature amplitude modulation
QCI	QoS class identifier
QoS	Quality of service
QPSK	Quadrature phase shift keying
RACH	Random access channel
RADIUS	Remote authentication dial in user service
RANAP	Radio access network application part
RA-RNTI	Random access radio network temporary identifier
RB	Resource block
RBG	Resource block group
RE	Resource element
REG	Resource element group
RF	Radio frequency/Rating function
RI	Rank indication

RIM	Radio access network information management
RLC	Radio link control
RLF	Radio link failure
RN	Relay node
RNC	Radio network controller
RNTI	Radio network temporary identifier
ROHC	Robust header compression
R-PDCCH	Relay physical downlink control channel
RRC	Radio resource control
RS	Reference signal
RSCP	Received signal code power
RSRP	Reference signal received power
RSRQ	Reference signal received quality
RSSI	Received signal strength indicator
RTP	Real time protocol
S1-AP	S1 application protocol
SAE	System architecture evolution
SC	Service centre
SCell	Secondary cell
SC-FDMA	Single carrier frequency division multiple access
S-CSCF	Serving call session control function
SCTP	Stream control transmission protocol
SDP	Session description protocol
SDU	Service data unit
SEG	Secure gateway
SFN	System frame number
SGsAP	SGs application protocol
SGSN	Serving GPRS support node
S-GW	Serving gateway
SIB	System information block
SIM	Subscriber identity module
SINR	Signal to interference plus noise ratio
SIP	Session initiation protocol
SI-RNTI	System information radio network temporary identifier
SMS	Short message service
SMS-GMSC	SMS gateway MSC
SMS-IWMSC	SMS interworking MSC
SMTP	Simple mail transfer protocol
SON	Self optimizing network/Self organizing network
SPR	Subscription profile repository
SPS	Semi persistent scheduling
SR	Scheduling request
SRB	Signalling radio bearer
SRS	Sounding reference signal
SRVCC	Single radio voice call continuity
SS	Supplementary service

SSS	Secondary synchronization signal
S-TMSI	S temporary mobile subscriber identity
SU-MIMO	Single user MIMO
TA	Timing advance/Tracking area
TAC	Tracking area code
TAI	Tracking area identity
TCP	Transmission control protocol
TDD	Time division duplex
TDMA	Time division multiple access
TD-SCDMA	Time division synchronous code division multiple access
TE	Terminal equipment
TEID	Tunnel endpoint identifier
TFT	Traffic flow template
TM	Transparent mode
TMSI	Temporary mobile subscriber identity
TPC	Transmit power control
TR	Technical report
TS	Technical specification
TTA	Telecommunications Technology Association
TTC	Telecommunication Technology Committee
TTI	Transmission time interval
UCI	Uplink control information
UDP	User datagram protocol
UE	User equipment
UE-AMBR	Per UE aggregate maximum bit rate
UICC	Universal integrated circuit card
UL	Uplink
UL-SCH	Uplink shared channel
UM	Unacknowledged mode
UMB	Ultra Mobile Broadband
UMTS	Universal Mobile Telecommunication System
USIM	Universal subscriber identity module
UTRAN	UMTS terrestrial radio access network
VANC	VoLGA access network controller
VLR	Visitor location register
VoIP	Voice over IP
VoLGA	Voice over LTE via generic access
VoLTE	Voice over LTE
V-PCRF	Visited policy and charging rules function
VRB	Virtual resource block
WCDMA	Wideband code division multiple access
WiMAX	Worldwide Interoperability for Microwave Access
WINNER	Wireless World Initiative New Radio
X2-AP	X2 application protocol

1

Introduction

Our first chapter puts LTE into its historical context, and lays out its requirements and key technical features. We begin by reviewing the architectures of UMTS and GSM, and by introducing some of the terminology that the two systems use. We then summarize the history of mobile telecommunication systems, discuss the issues that have driven the development of LTE, and show how UMTS has evolved first into LTE and then into an enhanced version known as LTE-Advanced. The chapter closes by reviewing the standardization process for LTE.

1.1 Architectural Review of UMTS and GSM

1.1.1 High Level Architecture

LTE was designed by a collaboration of national and regional telecommunications standards bodies known as the *Third Generation Partnership Project* (3GPP) [1] and is known in full as 3GPP *Long Term Evolution*. LTE evolved from an earlier 3GPP system known as the *Universal Mobile Telecommunication System* (UMTS), which in turn evolved from the *Global System for Mobile Communications* (GSM). To put LTE into context, we will begin by reviewing the architectures of UMTS and GSM and by introducing some of the important terminology.

A mobile phone network is officially known as a *public land mobile network* (PLMN), and is run by a *network operator* such as Vodafone or Verizon. UMTS and GSM share a common network architecture, which is shown in Figure 1.1. There are three main components, namely the core network, the radio access network and the mobile phone.

The *core network* contains two domains. The *circuit switched* (CS) domain transports phone calls across the geographical region that the network operator is covering, in the same way as a traditional fixed-line telecommunication system. It communicates with the *public switched telephone network* (PSTN) so that users can make calls to land lines and with the circuit switched domains of other network operators. The *packet switched* (PS) domain transports data streams, such as web pages and emails, between the user and external *packet data networks* (PDNs) such as the internet.

An Introduction to LTE: LTE, LTE-Advanced, SAE and 4G Mobile Communications, First Edition. Christopher Cox.
© 2012 John Wiley & Sons, Ltd. Published 2012 by John Wiley & Sons, Ltd.

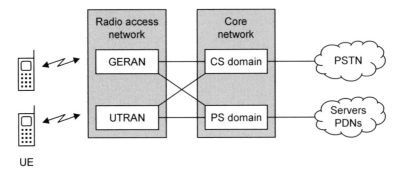

Figure 1.1 High level architecture of UMTS and GSM.

The two domains transport their information in very different ways. The CS domain uses a technique known as *circuit switching*, in which it sets aside a dedicated two-way connection for each individual phone call so that it can transport the information with a constant data rate and minimal delay. This technique is effective, but is rather inefficient: the connection has enough capacity to handle the worst-case scenario in which both users are speaking at the same time, but is usually over-dimensioned. Furthermore, it is inappropriate for data transfers, in which the data rate can vary widely.

To deal with the problem, the PS domain uses a different technique, known as *packet switching*. In this technique, a data stream is divided into packets, each of which is labelled with the address of the required destination device. Within the network, *routers* read the address labels of the incoming data packets and forward them towards the corresponding destinations. The network's resources are shared amongst all the users, so the technique is more efficient than circuit switching. However, delays can result if too many devices try to transmit at the same time, a situation that is familiar from the operation of the internet.

The *radio access network* handles the core network's radio communications with the user. In Figure 1.1, there are actually two separate radio access networks, namely the *GSM EDGE radio access network* (GERAN) and the *UMTS terrestrial radio access network* (UTRAN). These use the different radio communication techniques of GSM and UMTS, but share a common core network between them.

The user's device is known officially as the *user equipment* (UE) and colloquially as the *mobile*. It communicates with the radio access network over the *air interface*, also known as the *radio interface*. The direction from network to mobile is known as the *downlink* (DL) or *forward link* and the direction from mobile to network is known as the *uplink* (UL) or *reverse link*.

A mobile can work outside the coverage area of its network operator by using the resources from two public land mobile networks: the *visited network*, where the mobile is located, and the operator's *home network*. This situation is known as *roaming*.

1.1.2 Architecture of the Radio Access Network

Figure 1.2 shows the radio access network of UMTS. The most important component is the *base station*, which in UMTS is officially known as the *Node B*. Each base station has one

Introduction

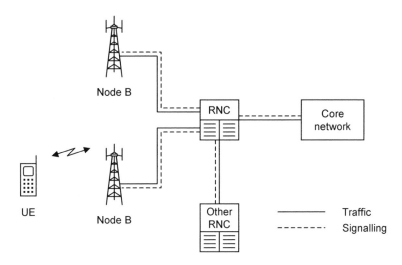

Figure 1.2 Architecture of the UMTS terrestrial radio access network.

or more sets of antennas, through which it communicates with the mobiles in one or more *sectors*. As shown in the diagram, a typical base station uses three sets of antennas to control three sectors, each of which spans an arc of 120°. In a medium-sized country like the UK, a typical mobile phone network might contain several thousand base stations altogether.

The word *cell* can be used in two different ways [2]. In Europe, a cell is usually the same thing as a sector, but in the USA, it usually means the group of sectors that a single base station controls. We will stick with the European convention throughout this book, so that the words cell and sector mean the same thing.

Each cell has a limited size, which is determined by the maximum range at which the receiver can successfully hear the transmitter. It also has a limited capacity, which is the maximum combined data rate of all the mobiles in the cell. These limits lead to the existence of several types of cell. *Macrocells* provide wide-area coverage in rural areas or suburbs and have a size of a few kilometres. *Microcells* have a size of a few hundred metres and provide a greater collective capacity that is suitable for densely populated urban areas. *Picocells* are used in large indoor environments such as offices or shopping centres and are a few tens of metres across. Finally, subscribers can buy *home base stations* to install in their own homes. These control *femtocells*, which are a few metres across.

Looking more closely at the air interface, each mobile and base station transmits on a certain radio frequency, which is known as the *carrier frequency*. Around that carrier frequency, it occupies a certain amount of frequency spectrum, known as the *bandwidth*. For example, a mobile might transmit with a carrier frequency of 1960 MHz and a bandwidth of 10 MHz, in which case its transmissions would occupy a frequency range from 1955 to 1965 MHz.

The air interface has to segregate the base stations' transmissions from those of the mobiles, to ensure that they do not interfere. UMTS can do this in two ways. When using *frequency division duplex* (FDD), the base stations transmit on one carrier frequency, and the mobiles on another. When using *time division duplex* (TDD), the base stations and mobiles transmit on the same carrier frequency, but at different times. The air interface also has to segregate the different base stations and mobiles from each other. We will see the techniques that it uses in Chapters 3 and 4.

When a mobile moves from one part of the network to another, it has to stop communicating with one cell and start communicating with the next cell along. Depending on the circumstances, this process can be carried out using two different techniques, known as *handover* and *cell reselection*. In UMTS, a mobile can actually communicate with more than one cell at a time, in a state known as *soft handover*.

The base stations are grouped together by devices known as *radio network controllers* (RNCs). These have two main tasks. Firstly, they pass the user's voice information and data packets between the base stations and the core network. Secondly, they control a mobile's radio communications by means of signalling messages that are invisible to the user, for example by telling a mobile to hand over from one cell to another. A typical network might contain a few tens of radio network controllers, each of which controls a few hundred base stations.

The GSM radio access network has a similar design, although the base station is known as a *base transceiver station* (BTS) and the controller is known as a *base station controller* (BSC). If a mobile supports both GSM and UMTS, then the network can hand it over between the two radio access networks, in a process known as an *inter-system handover*. This can be invaluable if a mobile moves outside the coverage area of UMTS, and into a region that is covered by GSM alone.

In Figure 1.2, we have shown the user's traffic in solid lines and the network's signalling messages in dashed lines. We will stick with this convention throughout the book.

1.1.3 Architecture of the Core Network

Figure 1.3 shows the internal architecture of the core network. In the circuit switched domain, *media gateways* (MGWs) route phone calls from one part of the network to another, while *mobile switching centre* (MSC) *servers* handle the signalling messages that set up, manage and tear down the phone calls. They respectively handle the traffic and signalling functions of two earlier devices, known as the mobile switching centre and the *visitor location register* (VLR). A typical network might just contain a few of each device.

In the packet switched domain, *gateway GPRS support nodes* (GGSNs) act as interfaces to servers and packet data networks in the outside world. *Serving GPRS support nodes* (SGSNs) route data between the base stations and the GGSNs, and handle the signalling messages that set up, manage and tear down the data streams. Once again, a typical network might just contain a few of each device.

The *home subscriber server* (HSS) is a central database that contains information about all the network operator's subscribers and is shared between the two network domains. It amalgamates the functions of two earlier components, which were known as the *home location register* (HLR) and the *authentication centre* (AuC).

1.1.4 Communication Protocols

In common with other communication systems, UMTS and GSM transfer information using hardware and software *protocols*. The best way to illustrate these is actually through

Introduction 5

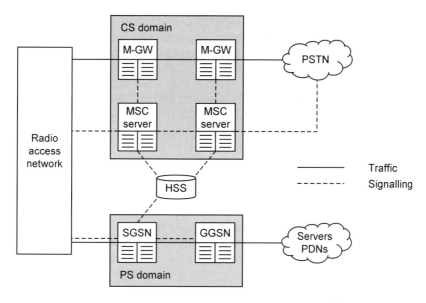

Figure 1.3 Architecture of the core networks of UMTS and GSM.

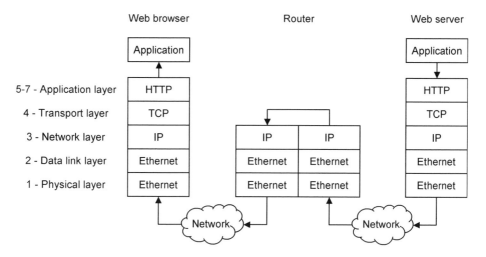

Figure 1.4 Examples of the communication protocols used by the internet, showing their mapping onto the layers of the OSI model.

the protocols used by the internet. These protocols are designed by the *Internet Engineering Task Force* (IETF) and are grouped into various numbered *layers*, each of which handles one aspect of the transmission and reception process. The usual grouping follows a seven layer model known as the *Open Systems Interconnection* (OSI) model.

As an example (see Figure 1.4), let us suppose that a web server is sending information to a user's browser. In the first step, an *application layer* protocol, in this case the *hypertext transfer protocol* (HTTP), receives information from the server's application software, and passes it to the next layer down by representing it in a way that the user's application layer will eventually be able to understand. Other application layer protocols include the *simple mail transfer protocol* (SMTP) and the *file transfer protocol* (FTP).

The *transport layer* manages the end-to-end data transmission. There are two main protocols. The *transmission control protocol* (TCP) re-transmits a packet from end to end if it does not arrive correctly, and is suitable for data such as web pages and emails that have to be received reliably. The *user datagram protocol* (UDP) sends the packet without any re-transmission and is suitable for data such as real time voice or video for which timely arrival is more important.

In the *network layer*, the *internet protocol* (IP) sends packets on the correct route from source to destination, using the IP address of the destination device. The process is handled by the intervening routers, which inspect the destination IP addresses by implementing just the lowest three layers of the protocol stack. The *data link layer* manages the transmission of packets from one device to the next, for example by re-transmitting a packet across a single interface if it does not arrive correctly. Finally, the *physical layer* deals with the actual transmission details; for example, by setting the voltage of the transmitted signal. The internet can use any suitable protocols for the data link and physical layers, such as *Ethernet*.

At each level of the transmitter's stack, a protocol receives a data packet from the protocol above in the form of a *service data unit* (SDU). It processes the packet, adds a header to describe the processing it has carried out, and outputs the result as a *protocol data unit* (PDU). This immediately becomes the incoming service data unit of the next protocol down. The process continues until the packet reaches the bottom of the protocol stack, at which point it is transmitted. The receiver reverses the process, using the headers to help it undo the effect of the transmitter's processing.

This technique is used throughout the radio access and core networks of UMTS and GSM. We will not consider their protocols in any detail; instead, we will go straight to the protocols used by LTE as part of Chapter 2.

1.2 History of Mobile Telecommunication Systems

1.2.1 From 1G to 3G

Mobile telecommunication systems were first introduced in the early 1980s. The *first generation* (1G) systems used analogue communication techniques, which were similar to those used by a traditional analogue radio. The individual cells were large and the systems did not use the available radio spectrum efficiently, so their capacity was by today's standards very small. The mobile devices were large and expensive and were marketed almost exclusively at business users.

Mobile telecommunications took off as a consumer product with the introduction of *second generation* (2G) systems in the early 1990s. These systems were the first to use digital technology, which permitted a more efficient use of the radio spectrum and the introduction of smaller, cheaper devices. They were originally designed just for voice, but were later enhanced to support instant messaging through the *Short Message Service* (SMS). The most popular 2G system was the Global System for Mobile Communications (GSM), which was originally designed as a pan-European technology, but which later became popular throughout the world. Also notable was *IS-95*, otherwise known as *cdmaOne*, which was designed by Qualcomm, and which became the dominant 2G system in the USA.

The success of 2G communication systems came at the same time as the early growth of the internet. It was natural for network operators to bring the two concepts together, by allowing users to download data onto mobile devices. To do this, so-called 2.5G systems built on the original ideas from 2G, by introducing the core network's packet switched domain and by modifying the air interface so that it could handle data as well as voice. The *General Packet Radio Service* (GPRS) incorporated these techniques into GSM, while IS-95 was developed into a system known as *IS-95B*.

At the same time, the data rates available over the internet were progressively increasing. To mirror this, designers first improved the performance of 2G systems using techniques such as *Enhanced Data Rates for GSM Evolution* (EDGE) and then introduced more powerful *third generation* (3G) systems in the years after 2000. 3G systems use different techniques for radio transmission and reception from their 2G predecessors, which increases the peak data rates that they can handle and which makes still more efficient use of the available radio spectrum.

Unfortunately, early 3G systems were excessively hyped and their performance did not at first live up to expectations. Because of this, 3G only took off properly after the introduction of 3.5G systems around 2005. In these systems, the air interface includes extra optimizations that are targeted at data applications, which increase the average rate at which a user can upload or download information, at the expense of introducing greater variability into the data rate and the arrival time.

1.2.2 Third Generation Systems

The world's dominant 3G system is the Universal Mobile Telecommunication System (UMTS). UMTS was developed from GSM by completely changing the technology used on the air interface, while keeping the core network almost unchanged. The system was later enhanced for data applications, by introducing the 3.5G technologies of *high speed downlink packet access* (HSDPA) and *high speed uplink packet access* (HSUPA), which are collectively known as *high speed packet access* (HSPA).

The UMTS air interface has two slightly different implementations. *Wideband code division multiple access* (WCDMA) is the version that was originally specified, and the one that is currently used through most of the world. *Time division synchronous code division multiple access* (TD-SCDMA) is a derivative of WCDMA, which is also known as the low chip rate option of UMTS TDD mode. TD-SCDMA was developed in China, to minimize the country's dependence on Western technology and on royalty payments to Western companies. It is deployed by one of China's three 3G operators, China Mobile.

There are two main technical differences between these implementations. Firstly, WCDMA usually segregates the base stations' and mobiles' transmissions by means of frequency division duplex, while TD-SCDMA uses time division duplex. Secondly, WCDMA uses a wide bandwidth of 5 MHz, while TD-SCDMA uses a smaller value of 1.6 MHz.

cdma2000 was developed from IS-95 and is mainly used in North America. The original 3G technology was known as cdma2000 *1x radio transmission technology* (1xRTT). It was subsequently enhanced to a 3.5G system with two alternative names, cdma2000 *high rate packet data* (HRPD) or *evolution data optimized* (EV-DO), which uses similar techniques to high speed packet access. The specifications for IS-95 and cdma2000 are produced by a similar collaboration to 3GPP, which is known as the *Third Generation Partnership Project 2* (3GPP2) [3].

There are three main technical differences between the air interfaces of cdma2000 and UMTS. Firstly, cdma2000 uses a bandwidth of 1.25 MHz. Secondly, cdma2000 is backwards compatible with IS-95, in the sense that IS-95 mobiles can communicate with cdma2000 base stations and vice versa, whereas UMTS is not backwards compatible with GSM. Thirdly, cdma2000 segregates voice and optimized data onto different carrier frequencies, whereas UMTS allows them to share the same one. The first two issues hindered the penetration of WCDMA into the North American market, where there were few allocations of bandwidths as wide as 5 MHz and there were a large number of legacy IS-95 devices.

The final 3G technology is *Worldwide Interoperability for Microwave Access* (WiMAX). This was developed by the *Institute of Electrical and Electronics Engineers* under IEEE standard 802.16 and has a very different history from other 3G systems. The original specification (IEEE 802.16–2001) was for a system that delivered data over point-to-point microwave links instead of fixed cables. A later revision, known as *fixed WiMAX* (IEEE 802.16–2004), supported point-to-multipoint communications between an omni-directional base station and a number of fixed devices. A further amendment, known as *mobile WiMAX* (IEEE 802.16e), allowed the devices to move and to hand over their communications from one base station to another. Once these capabilities were all in place, WiMAX started to look like any other 3G communication system, albeit one that had been optimized for data from the very beginning.

1.3 The Need for LTE

1.3.1 The Growth of Mobile Data

For many years, voice calls dominated the traffic in mobile telecommunication networks. The growth of mobile data was initially slow, but in the years leading up to 2010 its use started to increase dramatically. To illustrate this, Figure 1.5 shows measurements by Ericsson of the total traffic being handled by networks throughout the world, in petabytes (million gigabytes) per month [4]. The figure covers the period from January 2007 to July 2011, during which time the amount of data traffic increased by a factor of over 100.

This trend is set to continue. For example, Figure 1.6 shows forecasts by Analysys Mason of the growth of mobile traffic in the period from 2011 to 2016. Note the difference in the vertical scales of the two diagrams.

In part, this growth was driven by the increased availability of 3.5G communication technologies. More important, however, was the introduction of the Apple iPhone in 2007, followed by devices based on Google's Android operating system from 2008. These smartphones were more attractive and user-friendly than their predecessors and were designed to support the creation of applications by third party developers. The result was an explosion in the number and use of mobile applications, which is reflected in the diagrams. As a contributory factor, network operators had previously tried to encourage

Introduction

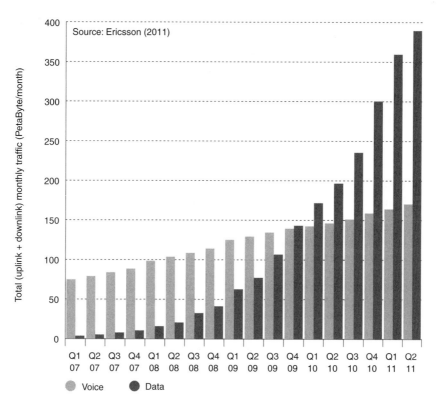

Figure 1.5 Measurements of voice and data traffic in worldwide mobile telecommunication networks, in the period from January 2007 to July 2011. Reproduced by permission of Ericsson.

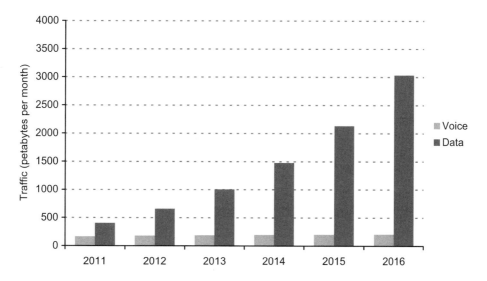

Figure 1.6 Forecasts of voice and data traffic in worldwide mobile telecommunication networks, in the period from 2011 to 2016. Data supplied by Analysys Mason.

the growth of mobile data by the introduction of flat rate charging schemes that permitted unlimited data downloads. That led to a situation where neither developers nor users were motivated to limit their data consumption.

As a result of these issues, 2G and 3G networks started to become congested in the years around 2010, leading to a requirement to increase network capacity. In the next section, we review the limits on the capacity of a mobile communication system and show how such capacity growth can be achieved.

1.3.2 Capacity of a Mobile Telecommunication System

In 1948, Claude Shannon discovered a theoretical limit on the data rate that can be achieved from any communication system [5]. We will write it in its simplest form, as follows:

$$C = B \log_2 (1 + \text{SINR}) \qquad (1.1)$$

Here, SINR is the *signal to interference plus noise ratio*, in other words the power at the receiver due to the required signal, divided by the power due to noise and interference. B is the bandwidth of the communication system in Hz, and C is the *channel capacity* in bits s^{-1}. It is theoretically possible for a communication system to send data from a transmitter to a receiver without any errors at all, provided that the data rate is less than the channel capacity. In a mobile communication system, C is the maximum data rate that one cell can handle and equals the combined data rate of all the mobiles in the cell.

The results are shown in Figure 1.7, using bandwidths of 5, 10 and 20 MHz. The vertical axis shows the channel capacity in million bits per second (Mbps), while the horizontal axis shows the signal to interference plus noise ratio in decibels (dB):

$$\text{SINR(dB)} = 10 \log_{10} (\text{SINR}) \qquad (1.2)$$

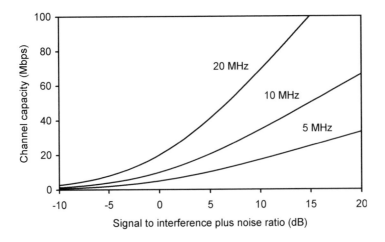

Figure 1.7 Shannon capacity of a communication system, in bandwidths of 5, 10 and 20 MHz.

1.3.3 Increasing the System Capacity

There are three main ways to increase the capacity of a mobile communication system, which we can understand by inspection of Equation (1.1) and Figure 1.7. The first, and the most important, is the use of smaller cells. In a cellular network, the channel capacity is the maximum data rate that a single cell can handle. By building extra base stations and reducing the size of each cell, we can increase the capacity of a network, essentially by using many duplicate copies of Equation (1.1).

The second technique is to increase the bandwidth. Radio spectrum is managed by the *International Telecommunication Union* (ITU) and by regional and national regulators, and the increasing use of mobile telecommunications has led to the increasing allocation of spectrum to 2G and 3G systems. However, there is only a finite amount of radio spectrum available and it is also required by applications as diverse as military communications and radio astronomy. There are therefore limits as to how far this process can go.

The third technique is to improve the communication technology that we are using. This brings several benefits: it lets us approach ever closer to the theoretical channel capacity, and it lets us exploit the higher SINR and greater bandwidth that are made available by the other changes above. This progressive improvement in communication technology has been an ongoing theme in the development of mobile telecommunications and is the main reason for the introduction of LTE.

1.3.4 Additional Motivations

Three other issues are driving the move to LTE. Firstly, a 2G or 3G operator has to maintain two core networks: the circuit switched domain for voice, and the packet switched domain for data. Provided that the network is not too congested, however, it is also possible to transport voice calls over packet switched networks using techniques such as *voice over IP* (VoIP). By doing this, operators can move everything to the packet switched domain, and can reduce both their capital and operational expenditure.

In a related issue, 3G networks introduce delays of the order of 100 milliseconds for data applications, in transferring data packets between network elements and across the air interface. This is barely acceptable for voice and causes great difficulties for more demanding applications such as real-time interactive games. Thus a second driver is the wish to reduce the end-to-end delay, or *latency*, in the network.

Thirdly, the specifications for UMTS and GSM have become increasingly complex over the years, due to the need to add new features to the system while maintaining backwards compatibility with earlier devices. A fresh start aids the task of the designers, by letting them improve the performance of the system without the need to support legacy devices.

1.4 From UMTS to LTE

1.4.1 High Level Architecture of LTE

In 2004, 3GPP began a study into the long term evolution of UMTS. The aim was to keep 3GPP's mobile communication systems competitive over timescales of 10 years and beyond, by delivering the high data rates and low latencies that future users would

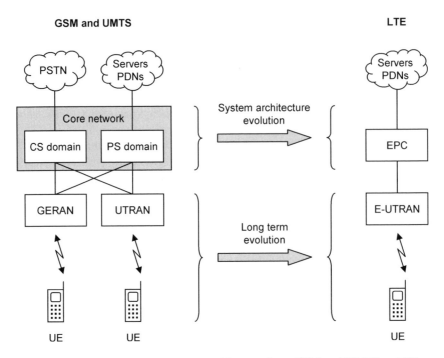

Figure 1.8 Evolution of the system architecture from GSM and UMTS to LTE.

require. Figure 1.8 shows the resulting architecture and the way in which that architecture developed from that of UMTS.

In the new architecture, the *evolved packet core* (EPC) is a direct replacement for the packet switched domain of UMTS and GSM. It distributes all types of information to the user, voice as well as data, using the packet switching technologies that have traditionally been used for data alone. There is no equivalent to the circuit switched domain: instead, voice calls are transported using voice over IP. The *evolved UMTS terrestrial radio access network* (E-UTRAN) handles the EPC's radio communications with the mobile, so is a direct replacement for the UTRAN. The mobile is still known as the user equipment, though its internal operation is very different from before.

The new architecture was designed as part of two 3GPP work items, namely *system architecture evolution* (SAE), which covered the core network, and *long term evolution* (LTE), which covered the radio access network, air interface and mobile. Officially, the whole system is known as the *evolved packet system* (EPS), while the acronym LTE refers only to the evolution of the air interface. Despite this official usage, LTE has become a colloquial name for the whole system, and is regularly used in this way by 3GPP. We will use LTE in this colloquial way throughout the book.

1.4.2 Long Term Evolution

The main output of the study into long-term evolution was a requirements specification for the air interface [6], in which the most important requirements were as follows.

LTE was required to deliver a peak data rate of 100 Mbps in the downlink and 50 Mbps in the uplink. This requirement was exceeded in the eventual system, which delivers peak data rates of 300 Mbps and 75 Mbps respectively. For comparison, the peak data rate of WCDMA, in Release 6 of the 3GPP specifications, is 14 Mbps in the downlink and 5.7 Mbps in the uplink. (We will discuss the different specification releases at the end of the chapter.)

It cannot be stressed too strongly, however, that these peak data rates can only be reached in idealized conditions, and are wholly unachievable in any realistic scenario. A better measure is the *spectral efficiency*, which expresses the typical capacity of one cell per unit bandwidth. LTE was required to support a spectral efficiency three to four times greater than that of Release 6 WCDMA in the downlink and two to three times greater in the uplink.

Latency is another important issue, particularly for time-critical applications such as voice and interactive games. There are two aspects to this. Firstly, the requirements state that the time taken for data to travel between the mobile phone and the fixed network should be less than five milliseconds, provided that the air interface is uncongested. Secondly, we will see in Chapter 2 that mobile phones can operate in two states: an active state in which they are communicating with the network and a low-power standby state. The requirements state that a phone should switch from standby to the active state, after an intervention from the user, in less than 100 milliseconds.

There are also requirements on coverage and mobility. LTE is optimized for cell sizes up to 5 km, works with degraded performance up to 30 km and supports cell sizes of up to 100 km. It is also optimized for mobile speeds up to 15 km hr^{-1}, works with high performance up to 120 km hr^{-1} and supports speeds of up to 350 km hr^{-1}. Finally, LTE is designed to work with a variety of different bandwidths, which range from 1.4 MHz up to a maximum of 20 MHz.

The requirements specification ultimately led to a detailed design for the LTE air interface, which we will cover in Chapters 3 to 10. For the benefit of those familiar with other systems, Table 1.1 summarizes its key technical features, and compares them with those of WCDMA.

1.4.3 System Architecture Evolution

The main output of the study into system architecture evolution was a requirements specification for the fixed network [7], in which the most important requirements were as follows.

Table 1.1 Key features of the air interfaces of WCDMA and LTE

Feature	WCDMA	LTE	Chapter
Multiple access scheme	WCDMA	OFDMA and SC-FDMA	4
Frequency re-use	100%	Flexible	4
Use of MIMO antennas	From Release 7	Yes	5
Bandwidth	5 MHz	1.4, 3, 5, 10, 15 or 20 MHz	6
Frame duration	10 ms	10 ms	6
Transmission time interval	2 or 10 ms	1 ms	6
Modes of operation	FDD and TDD	FDD and TDD	6
Uplink timing advance	Not required	Required	6
Transport channels	Dedicated and shared	Shared	6
Uplink power control	Fast	Slow	8

The evolved packet core routes packets using the Internet Protocol (IP) and supports devices that are using IP version 4, IP version 6, or dual stack IP version 4/version 6. In addition, the EPC provides users with always-on connectivity to the outside world, by setting up a basic IP connection for a device when it switches on and maintaining that connection until it switches off. This is different from the behaviour of UMTS and GSM, in which the network only sets up an IP connection on request and tears that connection down when it is no longer required.

The EPC is designed as a data pipe that simply transports information to and from the user: it is not concerned with the information content or with the application. This is similar to the behaviour of the internet, which transports packets that originate from any application software, but is different from that of a traditional telecommunication system, in which the voice application is an integral part of the system. Because of this, voice applications do not form part of LTE: instead, voice calls are controlled by some external entity such as the *IP multimedia subsystem* (IMS). The EPC simply transports the voice packets in the same way as any other data stream.

Unlike the internet, the EPC contains mechanisms to specify and control the data rate, error rate and delay that a data stream will receive. There is no explicit requirement on the maximum time required for data to travel across the EPC, but the relevant specification suggests a user plane latency of 10 milliseconds for a non roaming mobile, increasing to 50 milliseconds in a typical roaming scenario [8]. To calculate the total delay, we have to add the earlier figure for the delay across the air interface, giving a typical delay in a non roaming scenario of around 20 milliseconds.

Table 1.2 Key features of the radio access networks of UMTS and LTE

Feature	UMTS	LTE	Chapter
Radio access network components	Node B, RNC	eNB	2
RRC protocol states	CELL_DCH, CELL_FACH, CELL_PCH, URA_PCH, RRC_IDLE	RRC_CONNECTED, RRC_IDLE	2
Handovers	Soft and hard	Hard	14
Neighbour lists	Always required	Not required	14

Table 1.3 Key features of the core networks of UMTS and LTE

Feature	UMTS	LTE	Chapter
IP version support	IPv4 and IPv6	IPv4 and IPv6	2
USIM version support	Release 99 USIM onwards	Release 99 USIM onwards	2
Transport mechanisms	Circuit & packet switching	Packet switching	2
CS domain components	MSC server, MGW	n/a	2
PS domain components	SGSN, GGSN	MME, S-GW, P-GW	2
IP connectivity	After registration	During registration	11
Voice and SMS	Included	External	16

Introduction

The EPC is also required to support inter-system handovers between LTE and earlier 2G and 3G technologies. These cover not only UMTS and GSM, but also non 3GPP systems such as cdma2000 and WiMAX.

Tables 1.2 and 1.3 summarize the key features of the radio access network and the evolved packet core, and compare them with the corresponding features of UMTS. We will cover the architectural aspects of the fixed network in Chapter 2 and the operational aspects in Chapters 11 to 15.

1.5 From LTE to LTE-Advanced

1.5.1 The ITU Requirements for 4G

The design of LTE took place at the same time as an initiative by the International Telecommunication Union. In the late 1990s, the ITU had helped to drive the development of 3G technologies by publishing a set of requirements for a 3G mobile communication system, under the name *International Mobile Telecommunications* (IMT) *2000*. The 3G systems noted earlier are the main ones currently accepted by the ITU as meeting the requirements for IMT-2000.

The ITU launched a similar process in 2008, by publishing a set of requirements for a *fourth generation* (4G) communication system under the name *IMT-Advanced* [9–11]. According to these requirements, the peak data rate of a compatible system should be at least 600 Mbps on the downlink and 270 Mbps on the uplink, in a bandwidth of 40 MHz. We can see right away that these figures exceed the capabilities of LTE.

1.5.2 Requirements of LTE-Advanced

Driven by the ITU's requirements for IMT-Advanced, 3GPP started to study how to enhance the capabilities of LTE. The main output from the study was a specification for a system known as *LTE-Advanced* [12], in which the main requirements were as follows.

LTE-Advanced was required to deliver a peak data rate of 1000 Mbps in the downlink, and 500 Mbps in the uplink. In practice, the system has been designed so that it can eventually deliver peak data rates of 3000 and 1500 Mbps respectively, using a total bandwidth of 100 MHz that is made from five separate components of 20 MHz each. Note, as before, that these figures are unachievable in any realistic scenario.

The specification also includes targets for the spectrum efficiency in certain test scenarios. Comparison with the corresponding figures for WCDMA [13] implies a spectral efficiency 4.5 to 7 times greater than that of Release 6 WCDMA on the downlink, and 3.5 to 6 times greater on the uplink. Finally, LTE-Advanced is designed to be backwards compatible with LTE, in the sense that an LTE mobile can communicate with a base station that is operating LTE-Advanced and vice-versa.

1.5.3 4G Communication Systems

Following the submission and evaluation of proposals, the ITU announced in October 2010 that two systems met the requirements of IMT-Advanced [14]. One system was

LTE-Advanced, while the other was an enhanced version of WiMAX under IEEE specification 802.16m, known as mobile WiMAX 2.0.

Qualcomm had originally intended to develop a 4G successor to cdma2000 under the name *Ultra Mobile Broadband* (UMB). However, this system did not possess two of the advantages that its predecessor had done. Firstly, it was not backwards compatible with cdma2000, in the way that cdma2000 had been with IS-95. Secondly, it was no longer the only system that could operate in the narrow bandwidths that dominated North America, due to the flexible bandwidth support of LTE. Without any pressing reason to do so, no network operator ever announced plans to adopt the technology and the project was dropped in 2008. Instead, most cdma2000 operators decided to switch to LTE.

That left a situation where there were two remaining routes to 4G mobile communications: LTE and WiMAX. Of these, LTE has by far the greater support amongst network operators and equipment manufacturers and is likely to be the world's dominant mobile communication technology for some years to come.

1.5.4 The Meaning of 4G

Originally, the ITU intended that the term 4G should only be used for systems that met the requirements of IMT-Advanced. LTE did not do so and neither did mobile WiMAX 1.0 (IEEE 802.16e). Because of this, the engineering community came to describe these systems as 3.9G. These considerations did not, however, stop the marketing community from describing LTE and mobile WiMAX 1.0 as 4G technologies. Although that description was unwarranted from a performance viewpoint, there was actually some sound logic to it: there is a clear technical transition in the move from UMTS to LTE, which does not exist in the move from LTE to LTE-Advanced.

It was not long before the ITU admitted defeat. In December 2010, the ITU gave its blessing to the use of 4G to describe not only LTE and mobile WiMAX 1.0, but also any other technology with substantially better performance than the early 3G systems [15]. They did not define the words 'substantially better', but that is not an issue for this book: we just need to know that LTE is a 4G mobile communication system.

1.6 The 3GPP Specifications for LTE

The specifications for LTE are produced by the Third Generation Partnership Project, in the same way as the specifications for UMTS and GSM. They are organized into *releases* [16], each of which contains a stable and clearly defined set of features. The use of releases allows equipment manufacturers to build devices using some or all of the features of earlier releases, while 3GPP continues to add new features to the system in a later release. Within each release, the specifications progress through a number of different versions. New functionality can be added to successive versions until the date when the release is frozen, after which the only changes involve refinement of the technical details, corrections and clarifications.

Table 1.4 lists the releases that 3GPP have used since the introduction of UMTS, together with the most important features of each release. Note that the numbering scheme was changed after Release 99, so that later releases are numbered from 4 through to 11.

Table 1.4 3GPP specification releases for UMTS and LTE

Release	Date frozen	New features
R99	March 2000	WCDMA air interface
R4	March 2001	TD-SCDMA air interface
R5	June 2002	HSDPA, IP multimedia subsystem
R6	March 2005	HSUPA
R7	December 2007	Enhancements to HSPA
R8	December 2008	LTE, SAE
R9	December 2009	Enhancements to LTE and SAE
R10	March 2011	LTE-Advanced
R11	September 2012	Enhancements to LTE-Advanced

LTE was first introduced in Release 8, which was frozen in December 2008. This release contains most of the important features of LTE and we will focus on it throughout the early chapters of the book. In specifying Release 8, however, 3GPP omitted some of the less important features of the system. These features were eventually included in Release 9, which we will cover in Chapter 17. Release 10 includes the extra capabilities that are required for LTE-Advanced and will be covered in Chapter 18 along with a brief introduction to Release 11. 3GPP have also continued to add new features to UMTS throughout Releases 8 to 11. This process allows network operators who stick with UMTS to remain competitive, even while other operators move over to LTE.

The specifications are also organized into several *series*, each of which covers a particular component of the system. Table 1.5 summarizes the contents of series 21 to 37,

Table 1.5 3GPP specification series used by UMTS and LTE

Series	Scope
21	High level requirements
22	Stage 1 service specifications
23	Stage 2 service and architecture specifications
24	Non access stratum protocols
25	WCDMA and TD-SCDMA air interfaces and radio access network
26	Codecs
27	Data terminal equipment
28	Tandem free operation of speech codecs
29	Core network protocols
30	Programme management
31	UICC and USIM
32	Operations, administration, maintenance, provisioning and charging
33	Security
34	UE test specifications
35	Security algorithms
36	LTE air interface and radio access network
37	Multiple radio access technologies

which contain all the specifications for LTE and UMTS, as well as specifications that are common to LTE, UMTS and GSM. (Some other series numbers are used exclusively for GSM.) Within these series, the breakdown among the different systems varies widely. The 36 series is devoted to the techniques that are used for radio transmission and reception in LTE and is an important source of information for this book. In the other series, some specifications are applicable to UMTS alone, some to LTE alone and some to both, so it can be tricky to establish which specifications are the relevant ones. To help deal with this issue, the book contains references to all the important specifications that we will use.

When written out in full, an example specification number is TS 23.401 v 8.13.0. Here, TS stands for *technical specification*, 23 is the series number and 401 is the number of the specification within that series. 8 is the release number, 13 is the technical version number within that release and the final 0 is an editorial version number that is occasionally incremented for non-technical changes. 3GPP also produces *technical reports*, denoted TR, which are purely informative and have three-digit specification numbers beginning with an 8 or 9.

In a final division, each specification belongs to one of three *stages*. Stage 1 specifications define the service from the user's point of view and lie exclusively in the 22 series. Stage 2 specifications define the system's high-level architecture and operation, and lie mainly (but not exclusively) in the 23 series. Finally, stage 3 specifications define all the functional details. The stage 2 specifications are especially useful for achieving a high-level understanding of the system. The most useful ones for LTE are TS 23.401 [17] and TS 36.300 [18], which respectively cover the evolved packet core and the air interface. There is, however, an important note of caution: these specifications are superseded later on and cannot be relied upon for complete accuracy. Instead, the details should be checked if necessary in the relevant stage 3 specifications.

The individual specifications can be downloaded from 3GPP's specification numbering web page [19] or from their FTP server [20]. The 3GPP website also has summaries of the features that are covered by each individual release [21].

References

1. 3rd Generation Partnership Project (2011) *3GPP*. Available at: http://www.3gpp.org (accessed 12 December, 2011).
2. 4G Americas (May 2010) *MIMO and Smart Antennas for 3G and 4G Wireless Systems*, section 2.
3. 3rd Generation Partnership Project 2 (2010) *Welcome to the 3GPP2 Homepage*. Available at: http://www.3gpp2.org (accessed 12 December, 2011).
4. Ericsson (November 2011) *Traffic and Market Data Report*.
5. Shannon, C. E. (1948) A mathematical theory of communication, *The Bell System Technical Journal*, 27, 379–428 and 623–656.
6. 3GPP TS 25.913 (January 2009) *Requirements for Evolved UTRA (E-UTRA) and Evolved UTRAN (E-UTRAN)*, Release 8.
7. 3GPP TS 22.278 (December 2009) *Service Requirements for the Evolved Packet System (EPS)*, Release 8.
8. 3GPP TS 23.203 (June 2011) *Policy and Charging Control Architecture*, Release 10, section 6.1.7.2.
9. International Telecommunication Union (2008) *Requirements, Evaluation Criteria and Submission Templates for the Development of IMT-Advanced*, ITU report ITU-R M.2133.
10. International Telecommunication Union (2008) *Requirements Related to Technical Performance for IMT-Advanced Radio Interface(s)*, ITU report ITU-R M.2134.
11. International Telecommunication Union (2008) *Guidelines for Evaluation of Radio Interface Technologies for IMT-Advanced*, ITU report ITU-R M.2135.

12. 3GPP TS 36.913 (April 2011) *Requirements for Further Advancements for Evolved Universal Terrestrial Radio Access (E-UTRA) (LTE-Advanced)*, Release 10.
13. 3GPP TS 25.912 (April 2011) *Feasibility Study for Evolved Universal Terrestrial Radio Access (UTRA) and Universal Terrestrial Radio Access Network (UTRAN)*, Release 10, section 13.5.
14. International Telecommunication Union (2010) *ITU Paves Way for Next-Generation 4G Mobile Technologies*. Available at: http://www.itu.int/net/pressoffice/press_releases/2010/40.aspx (accessed 12 December, 2011).
15. International Telecommunication Union (2010) *ITU World Radiocommunication Seminar Highlights Future Communication Technologies*. Available at: http://www.itu.int/net/pressoffice/press_releases/2010/48.aspx (accessed 12 December, 2011).
16. 3rd Generation Partnership Project (2011) *3GPP – Releases*. Available at: http://www.3gpp.org/releases (accessed 12 December, 2011).
17. 3GPP TS 23.401 (September 2011) *General Packet Radio Service (GPRS) Enhancements for Evolved Universal Terrestrial Radio Access Network (E-UTRAN) Access*, Release 10.
18. 3GPP TS 36.300 (October 2011) *Evolved Universal Terrestrial Radio Access (E-UTRA) and Evolved Universal Terrestrial Radio Access Network (E-UTRAN); Overall Description; Stage 2*, Release 10.
19. 3rd Generation Partnership Project (2011) *3GPP - Specification Numbering*, Available at: http://www.3gpp.org/specification-numbering (accessed 12 December, 2011).
20. 3rd Generation Partnership Project (2011) *FTP Directory*. Available at: ftp://ftp.3gpp.org/specs/latest/ (accessed 12 December, 2011).
21. 3rd Generation Partnership Project (2011) *FTP Directory*. Available at: ftp://ftp.3gpp.org/Information/WORK_PLAN/Description_Releases/ (accessed 12 December, 2011).

2

System Architecture Evolution

This chapter covers the high-level architecture of LTE. We begin by describing the hardware components in an LTE network and by reviewing the software protocols that those components use to communicate. We then look in more detail at the techniques used for data transport in LTE, before discussing the state diagrams and the use of radio spectrum. We will leave some more specialized architectural issues until later chapters, notably those related to quality of service, charging and inter-system operation.

Several specifications are relevant to this chapter. TS 23.401 [1] and TS 36.300 [2] are stage 2 specifications that include descriptions of the system architecture, while the relevant stage 3 specifications [3, 4] contain the architectural details. We will also note some other important specifications as we go along.

2.1 Architecture of LTE

2.1.1 High Level Architecture

Figure 2.1 reviews the high-level architecture of the evolved packet system (EPS). There are three main components, namely the user equipment (UE), the evolved UMTS terrestrial radio access network (E-UTRAN) and the evolved packet core (EPC). In turn, the evolved packet core communicates with packet data networks in the outside world such as the internet, private corporate networks or the IP multimedia subsystem. The interfaces between the different parts of the system are denoted Uu, S1 and SGi.

The UE, E-UTRAN and EPC each have their own internal architectures and we will now discuss these one by one.

2.1.2 User Equipment

Figure 2.2 shows the internal architecture of the user equipment [5]. The architecture is identical to the one used by UMTS and GSM.

The actual communication device is known as the *mobile equipment* (ME). In the case of a voice mobile or a smartphone, this is just a single device. However, the mobile

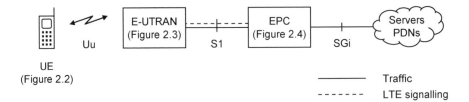

Figure 2.1 High level architecture of LTE.

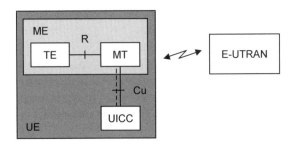

Figure 2.2 Internal architecture of the UE. Reproduced by permission of ETSI.

equipment can also be divided into two components, namely the *mobile termination* (MT), which handles all the communication functions, and the *terminal equipment* (TE), which terminates the data streams. The mobile termination might be a plug-in LTE card for a laptop, for example, in which case the terminal equipment would be the laptop itself.

The *universal integrated circuit card* (UICC) is a smart card, colloquially known as the SIM card. It runs an application known as the *universal subscriber identity module* (USIM) [6], which stores user-specific data such as the user's phone number and home network identity. The USIM also carries out various security-related calculations, using secure keys that the smart card stores. LTE supports mobiles that are using a USIM from Release 99 or later, but it does not support the *subscriber identity module* (SIM) that was used by earlier releases of GSM.

In addition, LTE supports mobiles that are using IP version 4 (IPv4), IP version 6 (IPv6), or dual stack IP version 4/version 6. A mobile receives one IP address for every packet data network that it is communicating with, for example one for the internet and one for any private corporate network. Alternatively, the mobile can receive an IPv4 address as well as an IPv6 address, if the mobile and network both support the two versions of the protocol.

Mobiles can have a wide variety of radio capabilities [7], which cover issues such as the maximum data rate that they can handle, the different types of radio access technology that they support and the carrier frequencies on which they can transmit and receive. Mobiles pass these capabilities to the radio access network by means of signalling messages, so that the E-UTRAN knows how to control them correctly. The most important capabilities are grouped together into the *UE category*. As shown in Table 2.1, the UE category mainly covers the maximum data rate with which the mobile can transmit and receive. It also covers some technical issues that are listed in the last three columns of the table, which we will cover in Chapters 3 and 5.

System Architecture Evolution

Table 2.1 UE categories. Reproduced by permission of ETSI

UE category	Release	Maximum # DL bits per ms	Maximum # UL bits per ms	Maximum # DL layers	Maximum # UL layers	Support of UL 64-QAM?
1	R8	10 296	5 160	1	1	No
2	R8	51 024	25 456	2	1	No
3	R8	102 048	51 024	2	1	No
4	R8	150 752	51 024	2	1	No
5	R8	299 552	75 376	4	1	Yes
6	R10	301 504	51 024	4	1	No
7	R10	301 504	102 048	4	2	No
8	R10	2 998 560	1 497 760	8	4	Yes

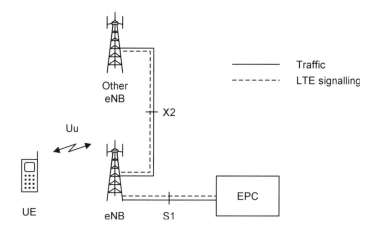

Figure 2.3 Architecture of the evolved UMTS terrestrial radio access network.

2.1.3 Evolved UMTS Terrestrial Radio Access Network

The evolved UMTS terrestrial radio access network (E-UTRAN) [8] is illustrated in Figure 2.3. The E-UTRAN handles the radio communications between the mobile and the evolved packet core and just has one component, the *evolved Node B* (eNB).

Each eNB is a base station that controls the mobiles in one or more cells. A mobile communicates with just one base station and one cell at a time, so there is no equivalent of the soft handover state from UMTS. The base station that is communicating with a mobile is known as its *serving eNB*.

The eNB has two main functions. Firstly, the eNB sends radio transmissions to all its mobiles on the downlink and receives transmissions from them on the uplink, using the analogue and digital signal processing functions of the LTE air interface. Secondly, the eNB controls the low-level operation of all its mobiles, by sending them signalling messages such as handover commands that relate to those radio transmissions. In carrying out these functions, the eNB combines the earlier functions of the Node B and the radio network controller, to reduce the latency that arises when the mobile exchanges information with the network.

Each base station is connected to the EPC by means of the S1 interface. It can also be connected to nearby base stations by the X2 interface, which is mainly used for signalling and packet forwarding during handover. The X2 interface is optional, in that the S1 interface can also handle all the functions of X2, albeit indirectly and more slowly. Usually, the S1 and X2 interfaces are not direct physical connections: instead, the information is routed across an underlying IP based transport network in the manner shown in Figure 1.4. The same issue will apply to the EPC's interfaces below.

A *home eNB* (HeNB) is a base station that has been purchased by a user to provide femtocell coverage within the home [9]. A home eNB belongs to a *closed subscriber group* (CSG) and can only be accessed by mobiles with a USIM that also belongs to the closed subscriber group. From an architectural point of view, a home eNB can be connected directly to the evolved packet core in the same way as any other base station, or can be connected by way of an intermediate device known as a *home eNB gateway* that collects the information from several home eNBs. Home eNBs only control one cell, and do not support the X2 interface until Release 10.

2.1.4 Evolved Packet Core

Figure 2.4 shows the main components of the evolved packet core [10, 11]. We have already seen one component, the home subscriber server (HSS), which is a central database that contains information about all the network operator's subscribers. This is one of the few components of LTE that has been carried forward from UMTS and GSM.

The *packet data network* (PDN) *gateway* (P-GW) is the EPC's point of contact with the outside world. Through the SGi interface, each PDN gateway exchanges data with one or more external devices or packet data networks, such as the network operator's servers, the internet or the IP multimedia subsystem. Each packet data network is identified by an *access point name* (APN) [12]. A network operator typically uses a handful of different APNs, for example one for its own servers and one for the internet.

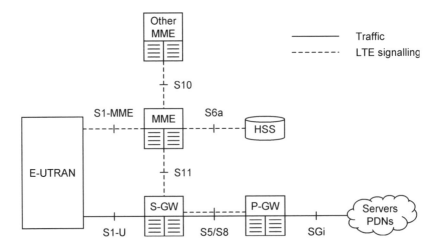

Figure 2.4 Main components of the evolved packet core.

Each mobile is assigned to a default PDN gateway when it first switches on, to give it always-on connectivity to a default packet data network such as the internet. Later on, a mobile may be assigned to one or more additional PDN gateways, if it wishes to connect to additional packet data networks such as private corporate networks. Each PDN gateway stays the same throughout the lifetime of the data connection.

The *serving gateway* (S-GW) acts as a router, and forwards data between the base station and the PDN gateway. A typical network might contain a handful of serving gateways, each of which looks after the mobiles in a certain geographical region. Each mobile is assigned to a single serving gateway, but the serving gateway can be changed if the mobile moves sufficiently far.

The *mobility management entity* (MME) controls the high-level operation of the mobile, by sending it signalling messages about issues such as security and the management of data streams that are unrelated to radio communications. As with the serving gateway, a typical network might contain a handful of MMEs, each of which looks after a certain geographical region. Each mobile is assigned to a single MME, which is known as its *serving MME*, but that can be changed if the mobile moves sufficiently far. The MME also controls the other elements of the network, by means of signalling messages that are internal to the EPC.

Comparison with UMTS and GSM shows that the PDN gateway has the same role as the gateway GPRS support node (GGSN), while the serving gateway and MME handle the data routing and signalling functions of the serving GPRS support node (SGSN). Splitting the SGSN in two makes it easier for an operator to scale the network in response to an increased load: the operator can add more serving gateways as the traffic increases, while adding more MMEs to handle an increase in the number of mobiles. To support this split, the S1 interface has two components: the S1-U interface carries traffic for the serving gateway, while the S1-MME interface carries signalling messages for the MME.

The EPC has some other components that were not shown in Figure 2.4. Firstly, the *cell broadcast centre* (CBC) was previously used by UMTS for the rarely implemented *cell broadcast service* (CBS). In LTE, the equipment is re-used for a service known as the *earthquake and tsunami warning system* (ETWS) [13]. Secondly, the *equipment identity register* (EIR) was also inherited from UMTS, and lists the details of lost or stolen mobiles. We will introduce further components later in the book, when we consider the management of quality of service, and the inter-operation between LTE and other mobile communication systems.

2.1.5 Roaming Architecture

Roaming allows users to move outside their network operators' coverage area by using the resources from two different networks. It relies on the existence of a *roaming agreement*, which defines how the operators will share the resulting revenue. There are two possible architectures [14], which are shown in Figure 2.5.

If a user is roaming, then the home subscriber server is always in the home network, while the mobile, E-UTRAN, MME and serving gateway are always in the visited network. The PDN gateway, however, can be in two places. In the usual situation of *home routed traffic*, the PDN gateway lies in the home network, through which all the user's traffic is all routed. This architecture allows the home network operator to see all the traffic

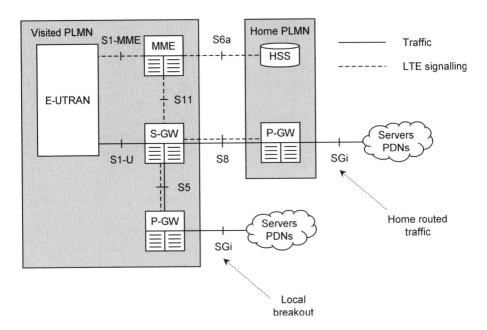

Figure 2.5 Architecture of LTE for a roaming mobile.

and to charge the user for it directly, but can be inefficient if the user is travelling overseas, particularly during a voice call with another user nearby. To deal with this situation, the specifications also support *local breakout*, in which the PDN gateway is located in the visited network. The HSS indicates whether or not the home network will permit local breakout, for each combination of user and APN [15].

The interface between the serving and PDN gateways is known as S5/S8. This has two slightly different implementations, namely S5 if the two devices are in the same network, and S8 if they are in different networks. For mobiles that are not roaming, the serving and PDN gateways can be integrated into a single device, so that the S5/S8 interface vanishes altogether. This can be useful because of the associated reduction in latency.

2.1.6 Network Areas

The EPC is divided into three different types of geographical area [16], which are illustrated in Figure 2.6.

An *MME pool area* is an area through which the mobile can move without a change of serving MME. Every pool area is controlled by one or more MMEs, while every base station is connected to all the MMEs in a pool area by means of the S1-MME interface. Pool areas can also overlap. Typically, a network operator might configure a pool area to cover a large region of the network such as a major city and might add MMEs to the pool as the signalling load in that city increases.

Similarly, an *S-GW service area* is an area served by one or more serving gateways, through which the mobile can move without a change of serving gateway. Every base

System Architecture Evolution   27

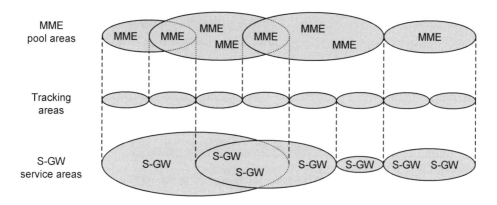

Figure 2.6 Relationship between tracking areas, MME pool areas and S-GW service areas.

station is connected to all the serving gateways in a service area by means of the S1-U interface. S-GW service areas do not necessarily correspond to MME pool areas.

MME pool areas and S-GW service areas are both made from smaller, non-overlapping units known as *tracking areas* (TAs). These are used to track the locations of mobiles that are on standby and are similar to the location and routing areas from UMTS and GSM.

2.1.7 Numbering, Addressing and Identification

The components of the network are associated with several different identities [17]. As in previous systems, each network is associated with a *public land mobile network identity* (PLMN-ID). This comprises a three digit *mobile country code* (MCC) and a two or three digit *mobile network code* (MNC). For example, the mobile country code for the UK is 234, while Vodafone's UK network uses a mobile network code of 15.

Each MME has three main identities, which are shown as the shaded parts of Figure 2.7. The 8 bit *MME code* (MMEC) uniquely identifies the MME within all the pool areas that it belongs to. By combining this with a 16 bit *MME group identity* (MMEGI), we arrive at a 24 bit *MME identifier* (MMEI), which uniquely identifies the MME within a particular

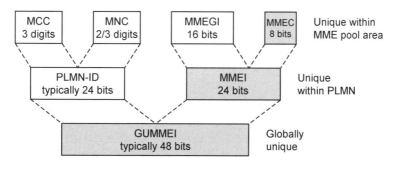

Figure 2.7 Identities used by the MME.

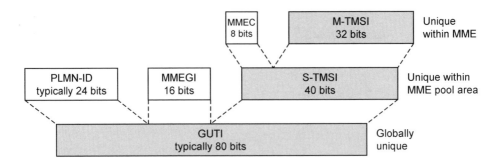

Figure 2.8 Temporary identities used by the mobile.

network. By bringing in the network identity, we arrive at the *globally unique MME identifier* (GUMMEI), which identifies an MME anywhere in the world.

Similarly, each tracking area has two main identities. The 16 bit *tracking area code* (TAC) identifies a tracking area within a particular network. Combining this with the network identity gives the globally unique *tracking area identity* (TAI).

Cells have three types of identity. The 28 bit *E-UTRAN cell identity* (ECI) identifies a cell within a particular network, while the *E-UTRAN cell global identifier* (ECGI) identifies a cell anywhere in the world. Also important for the air interface is the *physical cell identity*, which is a number from 0 to 503 that distinguishes a cell from its immediate neighbours.

A mobile is also associated with several different identities. The most important are the *international mobile equipment identity* (IMEI), which is a unique identity for the mobile equipment, and the *international mobile subscriber identity* (IMSI), which is a unique identity for the UICC and the USIM.

The IMSI is one of the quantities that an intruder needs to clone a mobile, so we avoid transmitting it across the air interface wherever possible. Instead, a serving MME identifies each mobile using temporary identities, which it updates at regular intervals. Three types of temporary identity are important, and are shown as the shaded parts of Figure 2.8. The 32 bit *M temporary mobile subscriber identity* (M-TMSI) identifies a mobile to its serving MME. Adding the MME code results in the 40 bit *S temporary mobile subscriber identity* (S-TMSI), which identifies the mobile within an MME pool area. Finally, adding the MME group identity and the PLMN identity results in the most important quantity, the *globally unique temporary identity* (GUTI).

2.2 Communication Protocols

2.2.1 Protocol Model

Each of the interfaces from the previous section is associated with a protocol stack, which the network elements use to exchange data and signalling messages. Figure 2.9 shows the high-level structure of those protocol stacks.

The protocol stack has two planes. Protocols in the *user plane* handle data that are of interest to the user, while protocols in the *control plane* handle signalling messages that

System Architecture Evolution 29

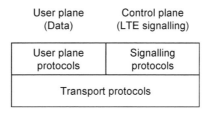

Figure 2.9 High level protocol architecture of LTE.

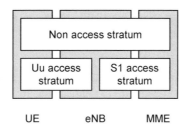

Figure 2.10 Relationship between the access stratum and the non access stratum on the air interface.

are only of interest to the network elements themselves. The protocol stack also has two main layers. The upper layer manipulates information in a way that is specific to LTE, while the lower layer transports information from one point to another. In the E-UTRAN, they are known as the *radio network layer* and *transport network layer* respectively.

There are then three types of protocol. *Signalling protocols* define a language by which two devices can exchange signalling messages with each other. *User plane protocols* manipulate the data in the user plane, most often to help route the data within the network. Finally, the underlying *transport protocols* transfer data and signalling messages from one point to another.

On the air interface, there is an extra level of complexity, which is shown in Figure 2.10 [18]. As noted earlier, the MME controls the high-level behaviour of the mobile by sending it signalling messages. However, there is no direct path between the MME and the mobile, through which those messages can be transported. To handle this, the air interface is divided into two levels, known as the *access stratum* (AS) and the *non access stratum* (NAS). The high-level signalling messages lie in the non access stratum and are transported using the access stratum protocols of the S1 and Uu interfaces.

2.2.2 Air Interface Transport Protocols

The air interface, officially known as the Uu interface, lies between the mobile and the base station. Figure 2.11 shows the air interface's transport protocols. Starting at the bottom, the *air interface physical layer* contains the digital and analogue signal processing functions that the mobile and base station use to send and receive information. The physical layer is described in several specifications that are listed in Chapter 6: the figure shows the most important.

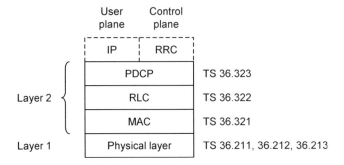

Figure 2.11 Transport protocols used on the air interface. Reproduced by permission of ETSI.

The next three protocols make up the data link layer, layer 2 of the OSI model. The *medium access control* (MAC) protocol [19] carries out low-level control of the physical layer, particularly by scheduling data transmissions between the mobile and the base station. The *radio link control* (RLC) protocol [20] maintains the data link between the two devices, for example by ensuring reliable delivery for data streams that need to arrive correctly. Finally, the *packet data convergence protocol* (PDCP) [21] carries out higher-level transport functions that are related to header compression and security.

2.2.3 Fixed Network Transport Protocols

Each interface in the fixed network uses standard IETF transport protocols, which are shown in Figure 2.12. Unlike the air interface, these interfaces use protocols from layers 1 to 4 of the usual OSI model. At the bottom of the protocol stack, the transport network can use any suitable protocols for layers 1 and 2, such as Ethernet. Every network element is then associated with an IP address, and the fixed network uses the internet protocol (IP) to route information from one element to another across the underlying transport network. LTE supports both IP version 4 [22] and IP version 6 [23] for this task. In

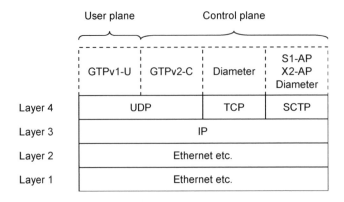

Figure 2.12 Transport protocols used by the fixed network.

the evolved packet core, support of IP version 4 is mandatory and support of version 6 is recommended [24], while the radio access network can use either or both of the two protocols [25, 26].

Above IP, there is a transport layer protocol across the interface between each individual pair of network elements. Three transport protocols are used. The user datagram protocol (UDP) [27] just sends data packets from one network element to another, while the transmission control protocol (TCP) [28] re-transmits packets if they arrive incorrectly. The *stream control transmission protocol* (SCTP) [29] is based on TCP, but includes extra features that make it more suitable for the delivery of signalling messages. The user plane always uses UDP as its transport protocol, to avoid delaying the data. The control plane's choice depends on the overlying signalling protocol, in the manner shown.

2.2.4 User Plane Protocols

The LTE user plane contains mechanisms to forward data correctly between the mobile and the PDN gateway, and to respond quickly to changes in the mobile's location. These mechanisms are implemented by the user plane protocols shown in Figure 2.13. Most of the user plane interfaces use a 3GPP protocol known as the *GPRS tunnelling protocol user part* (GTP-U) [30]. To be precise, LTE uses version 1 of the protocol, denoted GTPv1-U, along with the 2G and 3G packet switched domains from Release 99. Earlier 2G networks used version 0, which is denoted GTPv0-U. Between the serving gateway and the PDN gateway, the S5/S8 user plane has an alternative implementation. This is based on a standard IETF protocol known as *generic routing encapsulation* (GRE) [31].

GTP-U and GRE forward packets from one network element to another using a technique known as *tunnelling*. The two protocols implement tunnelling in slightly different ways, which we will cover as part of Section 2.4.

2.2.5 Signalling Protocols

LTE uses a large number of signalling protocols, which are shown in Figure 2.14.

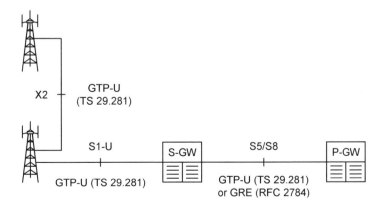

Figure 2.13 User plane protocols used by LTE.

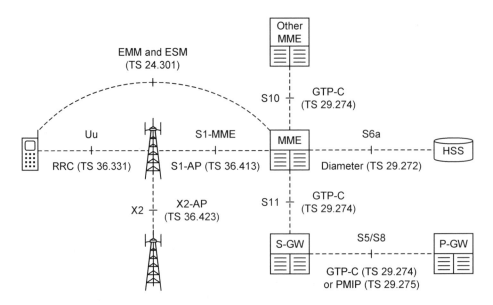

Figure 2.14 Signalling protocols used by LTE.

On the air interface, the base station controls a mobile's radio communications by means of signalling messages that are written using the *radio resource control* (RRC) protocol [32]. In the radio access network, an MME controls the base stations within its pool area using the *S1 application protocol* (S1-AP) [33], while two base stations can communicate using the *X2 application protocol* (X2-AP) [34].

At the same time, the MME controls a mobile's high-level behaviour using two protocols that lie in the air interface's non access stratum [35]. These protocols are *EPS session management* (ESM), which controls the data streams through which a mobile communicates with the outside world, and *EPS mobility management* (EMM), which handles internal bookkeeping within the EPC. The network transports EMM and ESM messages by embedding them into lower-level RRC and S1-AP messages and then by using the transport mechanisms of the Uu and S1 interfaces.

Inside the EPC, the HSS and MME communicate using a protocol based on *Diameter*. The basic Diameter protocol [36] is a standard IETF protocol for authentication, authorization and accounting, which was based on an older protocol known as *Remote Authentication Dial In User Service* (RADIUS) [37, 38]. The basic Diameter protocol can be enhanced for use in specific applications: the implementation of Diameter on the S6a interface [39] is one such application.

Most of the other EPC interfaces use a 3GPP protocol known as the *GPRS tunnelling protocol control part* (GTP-C) [40]. This protocol includes procedures for peer-to-peer communications between the different elements of the EPC, and for managing the GTP-U tunnels that we introduced above. LTE uses version 2 of the protocol, which is denoted GTPv2-C. The 2G and 3G packet switched domains used version 1 of the protocol, GTPv1-C, from Release 99 onwards, while earlier 2G networks implemented GTPv0-C. If the S5/S8 user plane is using GRE, then its control plane uses a signalling protocol known as *proxy mobile IP version 6* (PMIPv6) [41, 42]. PMIPv6 is a standard IETF protocol for the management of packet forwarding, in support of mobile devices such as laptops.

The question then arises of which protocol option to choose on the S5/S8 interface. Operators of legacy 3GPP networks are likely to prefer GTP-U and GTP-C, for consistency with their previous systems and with the other signalling interfaces in the evolved packet core. TS 23.401 describes the system architecture and high level operation of LTE, under the assumption that the S5/S8 interface is using those protocols. We will generally make the same assumption in this book. Operators of non 3GPP networks may prefer GRE and PMIP, which are standard IETF protocols, and which are also used for inter-operation between LTE and non 3GPP technologies. TS 23.402 [43] is a companion specification to TS 23.401, which describes the differences in architecture and operation for a network that uses those protocols.

2.3 Example Information Flows

2.3.1 Access Stratum Signalling

Now that we have introduced the network elements and protocol stacks, it is useful to show a few examples of how the different components fit together. Let us first consider an exchange of RRC signalling messages between the mobile and the base station. Figure 2.15 is the message sequence for an RRC procedure known as *UE Capability Transfer* [44]. Here, the serving eNB wishes to find out the mobile's radio access capabilities, such as the maximum data rate it can handle and the specification release that it conforms to. To do this, the RRC protocol composes a message called *UE Capability Enquiry*, and sends it to the mobile. The mobile responds with an RRC message called *UE Capability Information*, in which it lists the capabilities required.

The corresponding protocol stacks are shown in Figure 2.16. The base station composes its capability enquiry using the RRC protocol, processes it using the PDCP, RLC and MAC and transmits it using the air interface physical layer. The mobile receives the base station's transmission and processes the information by passing it through the same sequence of protocols in reverse. It then reads the enclosed message and composes its reply, which is transmitted and received in exactly the same way.

2.3.2 Non Access Stratum Signalling

The next signalling example is slightly more complex. Figure 2.17(a) shows the message sequence for an EMM procedure known as a *GUTI reallocation* [45]. Using an EMM

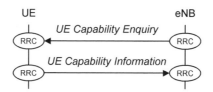

Figure 2.15 UE capability transfer procedure. Reproduced by permission of ETSI.

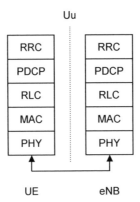

Figure 2.16 Protocol stacks used to exchange RRC signalling messages between the mobile and the base station. Reproduced by permission of ETSI.

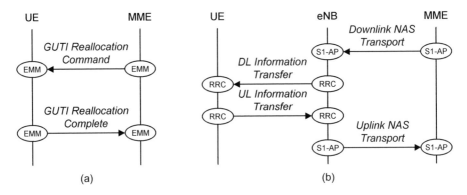

Figure 2.17 GUTI reallocation procedure. (a) Non access stratum messages. (b) Message transport using the access stratum.

GUTI Reallocation Command, the MME can give the mobile a new globally unique temporary identity. In response, the mobile sends the MME an acknowledgement using an EMM *GUTI Reallocation Complete*.

LTE transports these messages by embedding them into S1-AP and RRC messages, as shown in Figure 2.17(b). The usual S1-AP messages are known as *Uplink NAS Transport* and *Downlink NAS Transport* [46], while the usual RRC messages are known as *UL Information Transfer* and *DL Information Transfer* [47]. Their sole function is to transport EMM and ESM messages like the ones shown here. However, the network can also transport non access stratum messages by embedding them into other S1-AP and RRC messages, which can have additional access stratum functions of their own. We will see a few examples later in the book.

Figure 2.18 shows the protocol stacks for this message sequence. The MME writes the *GUTI Reallocation Command* using its EMM protocol, embeds it in the S1-AP *Downlink NAS Transport* message and sends it to the base station using the transport mechanisms of the S1 interface. The base station unwraps the EMM message, embeds it into an RRC *DL Information Transfer* and sends it to the mobile using the air interface protocols that

System Architecture Evolution

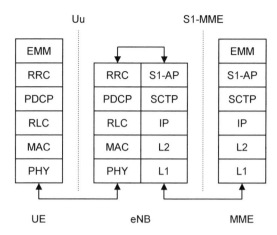

Figure 2.18 Protocol stacks used to exchange non access stratum signalling messages between the mobile and the MME. Reproduced by permission of ETSI.

we covered earlier. The mobile reads the message, updates its GUTI and sends an acknowledgement using the same protocol stacks in reverse.

2.3.3 Data Transport

As a final example, Figure 2.19 shows the protocol stacks that are used to exchange data between the mobile and a server in the outside world. In the figure, we have assumed that the S5/S8 interface is based on GTP rather than PMIP.

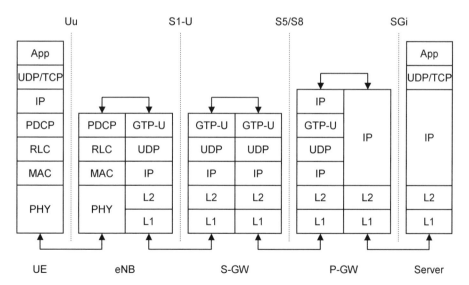

Figure 2.19 Protocol stacks used to exchange data between the mobile and an external server, when using an S5/S8 interface based on GTP. Reproduced by permission of ETSI.

Let us consider the downlink path, from the server to the mobile. The mobile's IP address lies in the address space of the PDN gateway, so the internet routes each of the mobile's data packets towards that device. Using the tunnelling mechanisms described below, the PDN gateway identifies the serving gateway that is looking after the mobile, wraps the incoming packet up inside a second IP packet and sends that packet to the serving gateway's IP address. In turn, the serving gateway unwraps the incoming packet, and repeats the process on the S1 interface towards the base station. Finally, the base station uses the transport mechanisms of the air interface to deliver the packet to the mobile.

2.4 Bearer Management

2.4.1 The EPS Bearer

At a high level, LTE transports data from one part of the system to another using *bearers*. The implementation of bearers depends on whether the S5/S8 interface is based on GTP or PMIP. We will start by describing what happens when using GTP [48, 49], and cover the differences in the case of PMIP later on.

The most important bearer is the *EPS bearer*. This is best thought of as a bi-directional data pipe, which transfers information between the mobile and the PDN gateway with a specific *quality of service* (QoS). In turn, the QoS defines how the data will be transferred, using parameters such as the data rate, error rate and delay. The GTP-U and GTP-C protocols include mechanisms to set up, modify and tear down EPS bearers, and to specify and implement their quality of service.

The information carried by an EPS bearer comprises one or more *service data flows*, each of which carries packets for a particular service such as a streaming video application. In turn, each service data flow comprises one or more *packet flows*, such as the audio and video streams which make up that service. LTE gives the same quality of service to all the packet flows within a particular EPS bearer.

EPS bearers can be classified in two ways. In the first classification, a *GBR bearer* is associated with a *guaranteed bit rate*, which is a long term average data rate that the mobile can expect to receive. GBR bearers are suitable for real-time services such as voice. A *non GBR bearer* receives no such guarantees, so is suitable for non real-time services such as web browsing.

In the second classification, the EPC sets up one EPS bearer, known as a *default bearer*, whenever a mobile connects to a packet data network. A default bearer is always a non GBR bearer. As shown in Figure 2.20, a mobile receives one default bearer as soon as it registers with the EPC, to provide it with always-on connectivity to a default packet data network such as the internet. At the same time, the mobile receives an IP address for it to use when communicating with that network, or possibly an IPv4 address and an IPv6 address. Later on, the mobile can establish connections with other packet data networks, for example private company networks. If it does so, then it receives an additional default bearer for every network that it connects to, together with an additional IP address.

After connecting to a packet data network and establishing a default bearer, a mobile can also receive one or more *dedicated bearers* that connect it to the same network. This does not lead to the allocation of any new IP addresses: instead, each dedicated bearer shares an IP address with its parent default bearer. A dedicated bearer typically has a

System Architecture Evolution

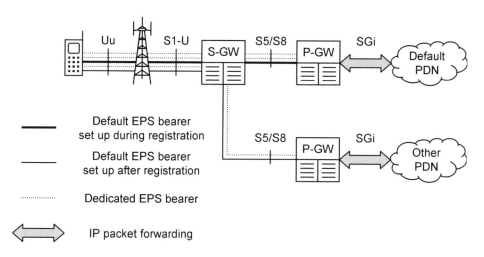

Figure 2.20 Default and dedicated EPS bearers, when using an S5/S8 interface based on GTP.

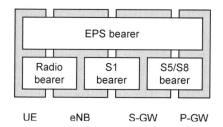

Figure 2.21 Bearer architecture of LTE, when using an S5/S8 interface based on GTP. Reproduced by permission of ETSI.

better quality of service than the default bearer can provide and in particular can have a guaranteed bit rate. A mobile can have a maximum of 11 EPS bearers [50], to give it connectivity to several networks using several different qualities of service.

The EPS bearer spans three different interfaces, so it cannot be implemented directly. To deal with this problem (Figure 2.21), the EPS bearer is broken down into three lower-level bearers, namely the *radio bearer*, the *S1 bearer* and the *S5/S8 bearer*. Each of these is also associated with a set of QoS parameters, and receives a share of the EPS bearer's error rate and delay. The radio bearer is then implemented by a suitable configuration of the air interface protocols, while the S1 and S5/S8 bearers are implemented using GTP-U tunnels in the manner described below. The combination of a radio bearer and an S1 bearer is sometimes known as an *evolved radio access bearer* (E-RAB).

2.4.2 Tunnelling Using GTP

The GTP-U protocol carries out a mapping between the S1 and S5/S8 bearers and the fixed network's transport protocols, by associating each bearer with a bi-directional GTP-U *tunnel*. In turn, each tunnel is associated with two *tunnel endpoint identifiers* (TEIDs),

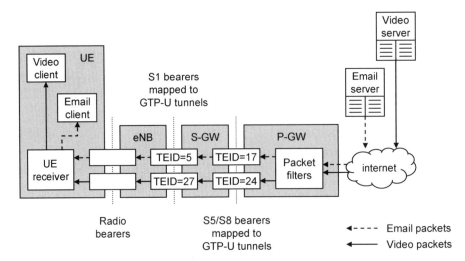

Figure 2.22 Implementation of tunnelling in the downlink, when using an S5/S8 interface based on GTP. Reproduced by permission of ETSI.

one for the uplink and one for the downlink. These identifiers are set up using GTP-C signalling messages, and are stored by the network elements at both ends of the tunnel.

To illustrate how the tunnels are used, let us consider the flow of data packets on the downlink. In Figure 2.22, a mobile has two EPS bearers, which are carrying video and email packets that require different qualities of service. These packets arrive at the PDN gateway using the normal transport mechanisms of the internet.

The PDN gateway now has to assign each incoming packet to the correct EPS bearer. To help it achieve this, each EPS bearer is associated with a *traffic flow template* (TFT). This comprises a set of *packet filters*, one for each of the packet flows that make up the bearer. In turn, each packet filter contains information such as the IP addresses of the source and destination devices, and the UDP or TCP port numbers of the source and destination applications. By inspecting every incoming packet and comparing it with all the packet filters that have been installed, the PDN gateway can assign every packet to the correct bearer.

The PDN gateway now looks up the corresponding GTP-U tunnel and adds a GTP-U header that contains the downlink TEID (17 for email packets in the example shown). It also looks up the mobile's serving gateway and adds an IP header that contains the serving gateway's IP address. It can then forward the packet to the serving gateway.

When the packet arrives, the serving gateway opens the GTP-U header and reads its TEID. It uses this information to identify the corresponding EPS bearer, and to look up the destination base station and the next TEID (5 for email in this example). It then forwards the packet to the base station in the manner described above and the base station transmits the packet to the mobile. A similar process happens in reverse on the uplink.

2.4.3 Tunnelling Using GRE and PMIP

The GRE protocol also uses tunnels, each of which is identified using a 32 bit key field in the GRE packet header. Unlike GTP-C, however, PMIP does not include any mechanism to specify the quality of service of a data stream.

System Architecture Evolution

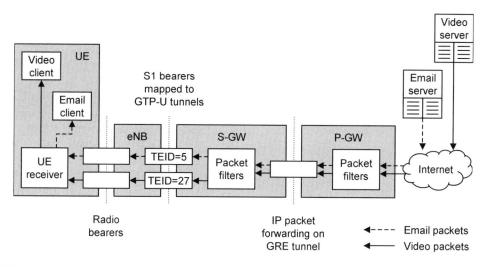

Figure 2.23 Implementation of tunnelling in the downlink, when using an S5/S8 interface based on PMIP. Reproduced by permission of ETSI.

If the network is using GRE and PMIP, then the EPS bearer only extends as far as the serving gateway [51]. On the S5/S8 interface, a mobile only has one GRE tunnel (Figure 2.23). This handles all the data packets that the mobile is transmitting or receiving, without any quality of service guarantees. The PDN gateway still contains a set of packet filters, which it uses to direct incoming packets to the correct GRE tunnel, and hence to the correct mobile. However, the serving gateway now contains packet filters as well, to handle the one-to-many mapping from GRE tunnels to EPS bearers.

2.4.4 Signalling Radio Bearers

LTE uses three special radio bearers, known as *signalling radio bearers* (SRBs), to carry signalling messages between the mobile and the base station [52]. The signalling radio bearers are listed in Table 2.2. Each of them is associated with a specific configuration of the air interface protocols, so that the mobile and base station can agree on how the signalling messages should be transmitted and received.

SRB0 is only used for a few RRC signalling messages, which the mobile and base station use to establish communications in a procedure known as *RRC connection establishment*. Its configuration is very simple and is defined in special RRC messages known

Table 2.2 Signalling radio bearers

Signalling radio bearer	Configured by	Used by
SRB 0	System information	RRC messages before establishment of SRB 1
SRB 1	RRC message on SRB 0	Subsequent RRC messages
		NAS messages before establishment of SRB 2
SRB 2	RRC message on SRB 1	Subsequent NAS messages

as *system information messages*, which the base station broadcasts across the whole of the cell to tell the mobiles about how the cell is configured.

SRB1 is configured using signalling messages that are exchanged on SRB0, at the time when a mobile establishes communications with the radio access network. It is used for all subsequent RRC messages, and also transports a few EMM and ESM messages that are exchanged prior to the establishment of SRB2. SRB2 is configured using signalling messages that are exchanged on SRB1, at the time when the mobile establishes communications with the evolved packet core. It is used to transport all the remaining EMM and ESM messages.

2.5 State Diagrams

2.5.1 EPS Mobility Management

A mobile's behaviour is defined using three state diagrams [53–55], which describe whether the mobile is registered with the EPC and whether it is active or idle. The first state diagram is the one for EPS mobility management (EMM). It is managed by the EMM protocol in the mobile and the MME, and is shown in Figure 2.24.

The mobile's EMM state depends on whether it is registered with the EPC. In the state EMM-REGISTERED, the mobile is switched on, and is registered with a serving MME and a serving gateway. The mobile has an IP address and a default EPS bearer, which gives it always-on connectivity with a default packet data network. In EMM-DEREGISTERED, the mobile is switched off or out of coverage and has none of these attributes.

2.5.2 EPS Connection Management

The second state diagram (Figure 2.25) is for *EPS connection management* (ECM). Once again, these states are managed by the EMM protocol. Each state has two names: TS 23.401 calls them ECM-CONNECTED and ECM-IDLE, while TS 24.301 calls them EMM-CONNECTED and EMM-IDLE. We will use the first of these.

The mobile's ECM state depends on whether it is active or on standby, from the viewpoint of the non access stratum protocols and the EPC. An active mobile is in ECM-CONNECTED state. In this state, all the data bearers and signalling radio bearers are in place. Using them, the mobile can freely exchange signalling messages with the MME through a logical connection that is known as a *signalling connection* and can freely exchange data with the serving gateway.

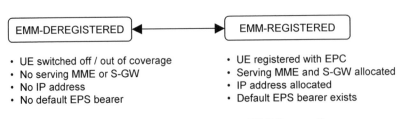

Figure 2.24 EPS mobility management (EMM) state diagram.

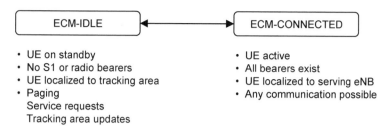

Figure 2.25 EPS connection management (ECM) state diagram.

When on standby, a mobile is in ECM-IDLE. In this state, it would be inappropriate to keep all the bearers in place, because the network would have to re-route them whenever the mobile moved from one cell to another, even though they would not be carrying any information. To avoid the resulting signalling overhead, the network tears down a mobile's S1 bearers and radio bearers whenever the mobile enters ECM-IDLE. The mobile can then freely move from one cell to another, without the need to re-route the bearers every time. However, the EPS bearers remain in place, so that a mobile retains its logical connections with the outside world. The S5/S8 bearers also remain in place, as the mobile changes its serving gateway only occasionally.

Furthermore, the MME does not know exactly where an idle mobile is located: instead, it just knows which tracking area the mobile is in. This allows the mobile to move from one cell to another without notifying the MME; instead, it only does so if it crosses a tracking area boundary. The MME can also register a mobile in more than one tracking area and can tell the mobile to send a notification only if it moves outside those tracking areas. This can be useful if the mobile is moving back and forth across a tracking area boundary.

Some limited communication is still possible, however. If the MME wishes to contact an idle mobile, then it can do so by sending an S1-AP *Paging* message to all the base stations in the mobile's tracking area(s). The base stations react by transmitting an RRC Paging message, in the manner described below. If the mobile wishes to contact the network or reply to a paging message, then it sends the MME an EMM message called a *Service Request* and the MME reacts by moving the mobile into ECM-CONNECTED. Finally, the mobile can send an EMM *Tracking Area Update Request* to the MME, if it notices that it has moved into a tracking area in which it is not currently registered.

2.5.3 Radio Resource Control

The final state diagram (Figure 2.26) is for radio resource control (RRC). As the name implies, these states are managed by the RRC protocol in the mobile and the serving eNB.

The mobile's RRC state depends on whether it is active or idle, from the viewpoint of the access stratum protocols and the E-UTRAN. An active mobile is in RRC_CONNECTED state. In this state, the mobile is assigned to a serving eNB, and can freely communicate with it using signalling messages on SRB 1.

When on standby, a mobile is in RRC_IDLE. In this state, SRB 1 is torn down, and there is no serving eNB assigned. As before, however, some limited communication is still possible. If the radio access network wishes to contact the mobile, typically because

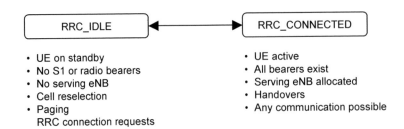

Figure 2.26 Radio resource control (RRC) state diagram.

it has received a paging request from the evolved packet core, then it can do so using an RRC *Paging* message. If the mobile wishes to contact the radio access network or reply to a paging message, then it can do so by initiating the RRC connection establishment procedure that we introduced above. In turn, the base station reacts by moving the mobile into RRC_CONNECTED.

The two RRC states handle moving devices in different ways. A mobile in RRC_CONNECTED state can be transmitting and receiving at a high data rate, so it is important for the radio access network to control which cell the mobile is communicating with. It does this using a procedure known as *handover*, in which the network switches the mobile's communication path from one cell to another. If the old and new cells are controlled by different base stations, then the network also re-routes the mobile's S1-U and S1-MME interfaces, so that they run directly between the new base station and the evolved packet core: the old base station drops out of the mobile's communication path altogether. In addition, the network will change the mobile's serving gateway and S5/S8 interface(s) if it moves into a new S-GW service area and will change the mobile's serving MME if it moves into a new MME pool area.

In RRC_IDLE state, the main motivation is to reduce signalling. To achieve this, the mobile decides which cell it will listen to, using a procedure known as *cell reselection*. The radio access network remains completely unaware of its location, while the EPC is only informed if a tracking area update is required. In turn, the tracking area update may lead to a change of serving gateway or serving MME in the manner described above.

Except in certain transient situations, the ECM and RRC state diagrams are always used together. An active mobile is always in ECM-CONNECTED and RRC_CONNECTED, while a mobile on standby is always in ECM-IDLE and RRC_IDLE.

2.6 Spectrum Allocation

The 3GPP specifications allow mobiles and base stations to use a large number of frequency bands [56, 57]. These are defined in response to decisions from the ITU and national regulators about the allocation of radio spectrum to mobile telecommunications. Table 2.3 lists the bands that support frequency division duplex (FDD) mode, while Table 2.4 lists the bands that support time division duplex (TDD).

The tables also show the first release in which each band was introduced. Note, however, that the bands are intended to be release independent: a mobile can support bands that were not introduced until Releases 9 and 10 of the 3GPP specifications, even if it otherwise

Table 2.3 FDD frequency bands. Reproduced by permission of ETSI

Band	Release	Uplink band (MHz)	Downlink band (MHz)	Main regions	Notes
1	R99	1920–1980	2110–2170	Europe, Asia, Africa	WCDMA
2	R99	1850–1910	1930–1990	Americas	GSM 1900, CDMA
3	R5	1710–1785	1805–1880	Europe, Asia, Africa	GSM 1800
4	R6	1710–1755	2110–2155	Americas	
5	R6	824–849	869–894	Americas	GSM 850, CDMA
6	–	–	–	–	Not used by LTE
7	R7	2500–2570	2620–2690	Europe	
8	R7	880–915	925–960	Europe, Asia, Africa	GSM 900
9	R7	1749.9–1784.9	1844.9–1879.9	Japan	
10	R7	1710–1770	2110–2170	Americas	
11	R8	1427.9–1447.9	1475.9–1495.9	Japan	
12	R8	699–716	729–746	USA	Digital dividend
13	R8	777–787	746–756	USA	Digital dividend
14	R8	788–798	758–768	USA	Digital dividend
15	–	–	–	–	Not used by 3GPP
16	–	–	–	–	Not used by 3GPP
17	R8	704–716	734–746	USA	Digital dividend
18	R9	815–830	860–875	Japan	
19	R9	830–845	875–890	Japan	
20	R9	832–862	791–821	Europe	Digital dividend
21	R9	1447.9–1462.9	1495.9–1510.9	Japan	
22	R10	3410–3490	3510–3590	Europe	
23	R10	2000–2020	2180–2200	North America	
24	R10	1626.5–1660.5	1525–1559	North America	
25	R10	1850–1915	1930–1995	Americas	

Table 2.4 TDD frequency bands. Reproduced by permission of ETSI

Band	Release	Frequency band (MHz)	Main regions
33	R99	1900–1920	Europe, Asia
34	R99	2010–2025	Europe, Asia
35	R99	1850–1910	Americas
36	R99	1930–1990	Americas
37	R99	1910–1930	Americas
38	R7	2570–2620	Europe
39	R8	1880–1920	China
40	R8	2300–2400	China
41	R10	2496–2690	USA
42	R10	3400–3600	Europe
43	R10	3600–3800	Europe

conforms to Release 8. In addition, most of the bands are also supported by other systems such as UMTS and GSM.

Some of these frequency bands are being newly released for use by mobile telecommunications. In 2008, for example, the US *Federal Communications Commission* (FCC) auctioned frequencies around 700 MHz (FDD bands 12, 13, 14 and 17) that had previously been used for analogue television broadcasting. In Europe, similar auctions have been taking place for frequencies around 800 and 2600 MHz (FDD bands 7 and 20, and TDD band 38). Network operators can also re-allocate frequencies that they have previously used for other mobile communication systems, as their users migrate to LTE. Likely examples include FDD bands 1, 3 and 8 in Europe (originally used by WCDMA, GSM 1800 and GSM 900 respectively) and FDD bands 2, 4 and 5 in the USA.

The result is that LTE is likely to be deployed in a large number of frequency bands, with different bands used by different regions, countries and network operators. Research by Informa Telecoms & Media in 2011 [58] suggested that eight LTE frequencies would be particularly important, namely 700, 800, 900, 1800, 2100 and 2600 MHz in FDD mode, and 2300 and 2600 MHz in TDD mode. The situation is rapidly evolving, but there are regularly updated lists of LTE deployments available online [59].

In turn, this situation means that an LTE device will have to support a large number of carrier frequencies if it is to be truly usable worldwide. This proliferation of carrier frequencies may yet delay the large-scale adoption of LTE and will certainly cause several headaches for the designers of mobile communication devices.

References

1. 3GPP TS 23.401 (September 2011) *General Packet Radio Service (GPRS) Enhancements for Evolved Universal Terrestrial Radio Access Network (E-UTRAN) Access*, Release 10.
2. 3GPP TS 36.300 (October 2011) *Evolved Universal Terrestrial Radio Access (E-UTRA) and Evolved Universal Terrestrial Radio Access Network (E-UTRAN); Overall Description; Stage* 2, Release 10.
3. 3GPP TS 23.002 (September 2011) *Network Architecture*, Release 10.
4. 3GPP TS 36.401 (September 2011) *Evolved Universal Terrestrial Radio Access Network (E-UTRAN); Architecture Description*, Release 10.
5. 3GPP TS 27.001 (September 2011) *General on Terminal Adaptation Functions (TAF) for Mobile Stations (MS)*, Release 10, section 4.
6. 3GPP TS 31.102 (October 2011) *Characteristics of the Universal Subscriber Identity Module (USIM) Application*, Release 10.
7. 3GPP TS 36.306 (October 2011) *Evolved Universal Terrestrial Radio Access (E-UTRA); User Equipment (UE) Radio Access Capabilities*, Release 10.
8. 3GPP TS 36.401 (September 2011) *Evolved Universal Terrestrial Radio Access Network (E-UTRAN); Architecture Description*, Release 10, section 6.
9. 3GPP TS 36.300 (October 2011) *Evolved Universal Terrestrial Radio Access (E-UTRA) and Evolved Universal Terrestrial Radio Access Network (E-UTRAN); Overall Description; Stage 2*, Release 10, section 4.6.
10. 3GPP TS 23.401 (September 2011) *General Packet Radio Service (GPRS) Enhancements for Evolved Universal Terrestrial Radio Access Network (E-UTRAN) Access*, Release 10, sections 4.2.1, 4.4.
11. 3GPP TS 23.002 (September 2011) *Network Architecture*, Release 10, section 4.1.4.
12. 3GPP TS 23.003 (September 2011) *Numbering, Addressing and Identification*, Release 10, section 9.
13. 3GPP TS 22.168 (September 2011) *Earthquake and Tsunami Warning System (ETWS) Requirements; Stage 1*, Release 8.
14. 3GPP TS 23.401 (September 2011) *General Packet Radio Service (GPRS) Enhancements for Evolved Universal Terrestrial Radio Access Network (E-UTRAN) Access*, Release 10, section 4.2.2.
15. 3GPP TS 23.401 (September 2011) *General Packet Radio Service (GPRS) Enhancements for Evolved Universal Terrestrial Radio Access Network (E-UTRAN) Access*, Release 10, section 4.3.8.1.

16. 3GPP TS 23.401 (September 2011) *General Packet Radio Service (GPRS) Enhancements for Evolved Universal Terrestrial Radio Access Network (E-UTRAN) Access*, Release 10, section 3.1.
17. 3GPP TS 23.003 (September 2011) *Numbering, Addressing and Identification*, Release 10, sections 2, 6, 19.
18. 3GPP TS 36.401 (September 2011) *Evolved Universal Terrestrial Radio Access Network (E-UTRAN); Architecture Description*, Release 10, section 5.
19. 3GPP TS 36.321 (October 2011) *Evolved Universal Terrestrial Radio Access (E-UTRA); Medium Access Control (MAC) Protocol Specification*, Release 10.
20. 3GPP TS 36.322 (December 2010) *Evolved Universal Terrestrial Radio Access (E-UTRA); Radio Link Control (RLC) Protocol Specification*, Release 10.
21. 3GPP TS 36.323 (March 2011) *Evolved Universal Terrestrial Radio Access (E-UTRA); Packet Data Convergence Protocol (PDCP) Specification*, Release 10.
22. IETF RFC 791 (September 1981) *Internet Protocol*.
23. IETF RFC 2460 (December 1998), *Internet Protocol, Version 6 (IPv6) Specification*.
24. 3GPP TS 29.281 (September 2011) *General Packet Radio System (GPRS) Tunnelling Protocol User Plane (GTPv1-U)*, Release 10, section 4.4.1.
25. 3GPP TS 36.414 (June 2011) *Evolved Universal Terrestrial Radio Access Network (E-UTRAN); S1 Data Transport*, Release 10, section 5.3.
26. 3GPP TS 36.424 (June 2011) *Evolved Universal Terrestrial Radio Access Network (E-UTRAN); X2 Data Transport*, Release 10, section 5.3.
27. IETF RFC 768 (August 1980) *User Datagram Protocol*.
28. IETF RFC 793 (September 1981) *Transmission Control Protocol*.
29. IETF RFC 4960 (September 2007) *Stream Control Transmission Protocol*.
30. 3GPP TS 29.281 (September 2011) *General Packet Radio System (GPRS) Tunnelling Protocol User Plane (GTPv1-U)*, Release 10.
31. IETF RFC 2784 (March 2000) *Generic Routing Encapsulation (GRE)*.
32. 3GPP TS 36.331 (October 2011) *Evolved Universal Terrestrial Radio Access (E-UTRA); Radio Resource Control (RRC); Protocol Specification*, Release 10.
33. 3GPP TS 36.413 (September 2011) *Evolved Universal Terrestrial Radio Access Network (E-UTRAN); S1 Application Protocol (S1AP)*, Release 10.
34. 3GPP TS 36.423 (September 2011) *Evolved Universal Terrestrial Radio Access Network (E-UTRAN); X2 Application Protocol (X2AP)*, Release 10.
35. 3GPP TS 24.301 (September 2011) *Non-Access-Stratum (NAS) Protocol for Evolved Packet System (EPS); Stage 3*, Release 10.
36. IETF RFC 3588 (September 2003) *Diameter Base Protocol*.
37. IETF RFC 2865 (June 2000) *Remote Authentication Dial-In User Service (RADIUS)*.
38. IETF RFC 2866 (June 2000) *RADIUS Accounting*.
39. 3GPP TS 29.272 (September 2011) *Evolved Packet System (EPS); Mobility Management Entity (MME) and Serving GPRS Support Node (SGSN) Related Interfaces Based on Diameter Protocol*, Release 10.
40. 3GPP TS 29.274 (September 2011) *3GPP Evolved Packet System (EPS); Evolved General Packet Radio Service (GPRS) Tunnelling Protocol for Control Plane (GTPv2-C); Stage 3*, Release 10.
41. 3GPP TS 29.275 (September 2011) *Proxy Mobile IPv6 (PMIPv6) Based Mobility and Tunnelling Protocols; Stage 3*, Release 10.
42. IETF RFC 5213 (August 2008) *Proxy Mobile IPv6*.
43. 3GPP TS 23.402 (September 2011). *Architecture Enhancements for Non-3GPP Accesses*, Release 10.
44. 3GPP TS 36.331 (October 2011) *Radio Resource Control (RRC); Protocol Specification*, Release 10, section 5.6.3.
45. 3GPP TS 24.301 (September 2011) *Non-Access-Stratum (NAS) Protocol for Evolved Packet System (EPS); Stage 3*, Release 10, section 5.4.1.
46. 3GPP TS 36.413 (September 2011) *Evolved Universal Terrestrial Radio Access Network (E-UTRAN); S1 Application Protocol (S1AP)*', Release 10, section 8.6.2.
47. 3GPP TS 36.331 (October 2011) *Radio Resource Control (RRC); Protocol Specification*, Release 10, sections 5.6.1, 5.6.2.
48. 3GPP TS 36.300 (October 2011) *Evolved Universal Terrestrial Radio Access (E-UTRA) and Evolved Universal Terrestrial Radio Access Network (E-UTRAN); Overall Description; Stage 2*, Release 10, section 13.

49. 3GPP TS 23.401 (September 2011) *General Packet Radio Service (GPRS) Enhancements for Evolved Universal Terrestrial Radio Access Network (E-UTRAN) Access*, Release 10, section 4.7.
50. 3GPP TS 24.007 (March 2011) *Mobile Radio Interface Signalling Layer 3; General Aspects*, Release 10, section 11.2.3.1.5.
51. 3GPP TS 23.402 (September 2011) *Architecture Enhancements for Non-3GPP Accesses*, Release 10, section 4.10.
52. 3GPP TS 36.331 (October 2011) *Radio Resource Control (RRC); Protocol Specification*, Release 10, section 4.2.2.
53. 3GPP TS 23.401 (September 2011) *General Packet Radio Service (GPRS) Enhancements for Evolved Universal Terrestrial Radio Access Network (E-UTRAN) Access*, Release 10, section 4.6.
54. 3GPP TS 24.301 (September 2011) *Non-Access-Stratum (NAS) Protocol for Evolved Packet System (EPS); Stage 3*, Release 10, section 3.1.
55. 3GPP TS 36.331 (October 2011) *Radio Resource Control (RRC); Protocol Specification*, Release 10, section 4.2.8.
56. 3GPP TS 36.101 (October 2011) *Evolved Universal Terrestrial Radio Access (E-UTRA); User Equipment (UE) Radio Transmission and Reception*, Release 10, section 5.5.
57. 3GPP TS 36.104 (October 2011) *Evolved Universal Terrestrial Radio Access (E-UTRA); Base Station (BS) Radio Transmission and Reception*, Release 10, section 5.5.
58. Informa Telecoms & Media (2011) *Latest Research from Informa Telecoms & Media Identifies Eight Core LTE Bands*. Available at: http://blogs.informatandm.com/2094/press-release-latest-research-from-informa-telecoms-media-identifies-eight-core-lte-bands/ (accessed 12 December, 2011).
59. LteMaps (2011) *Mapping LTE Deployments*. Available at: http://ltemaps.org (accessed 12 December, 2011).

3

Digital Wireless Communications

The next three chapters describe the principles of radio transmission and reception in LTE. Here, we begin by reviewing the radio transmission techniques that LTE has inherited from 2G and 3G communication systems. The chapter covers the principles of modulation, signal reception and channel estimation, and shows how the received signal can be degraded in a multipath environment by fading and inter-symbol interference. It then discusses the techniques that are used to minimize the number of errors in the received signal, notably forward error correction, re-transmissions and hybrid automatic repeat request. For some detailed accounts of the material covered in this chapter, see for example, References [1–6].

3.1 Radio Transmission and Reception

3.1.1 Signal Transmission

Figure 3.1 shows the most important components of a wireless transmission system. In the figure, the transmitter accepts a stream of bits from the application software. It then encodes these bits onto a radio wave, known as a *carrier*, by adjusting parameters of the wave such as its amplitude or phase.

As shown in the figure, the transmitter usually processes the information in two stages. In the first stage, a *modulator* accepts the incoming bits, and computes *symbols* that represent the amplitude and phase of the outgoing wave. It then passes these to the analogue transmitter, which generates the radio wave itself.

The modulation scheme used in Figure 3.1 is known as *quadrature phase shift keying* (QPSK). A QPSK modulator takes the incoming bits two at a time and transmits them using a radio wave that can have four different states. These have phases of 45°, 135°, 225° and 315° (Figure 3.2a), which correspond to bit combinations of 00, 10, 11 and 01 respectively. We can represent the four states of QPSK using the *constellation diagram* shown in Figure 3.2(b). In this diagram, the distance of each state from the origin represents the amplitude of the transmitted wave, while the angle (measured anti-clockwise from the x-axis) represents its phase.

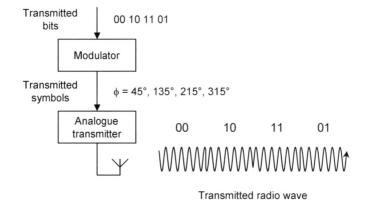

Figure 3.1 Architecture of a wireless communication transmitter.

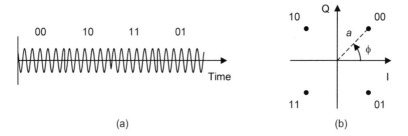

Figure 3.2 Quadrature phase shift keying. (a) Example QPSK waveform. (b) QPSK constellation diagram.

Usually, it is more convenient to represent each symbol using two other numbers, which are known as the *in-phase* (I) and *quadrature* (Q) components. These are computed as follows:

$$I = a \cos \phi$$
$$Q = a \sin \phi \qquad (3.1)$$

where a is the amplitude of the transmitted wave and ϕ is its phase. Mathematicians will recognize the in-phase and quadrature components as the real and imaginary parts of a complex number.

As shown in Figure 3.3, LTE uses four modulation schemes altogether. *Binary phase shift keying* (BPSK) sends bits one at a time, using two states that can be interpreted as starting phases of $0°$ and $180°$, or as signal amplitudes of $+1$ and -1. LTE uses this scheme for a limited number of control streams, but does not use it for normal data transmissions. *16 quadrature amplitude modulation* (16-QAM) sends bits four at a time, using 16 states that have different amplitudes and phases. Similarly, 64-QAM sends bits six at a time using 64 different states, so it has a data rate six times greater than that of BPSK.

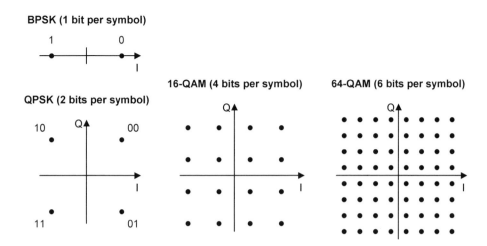

Figure 3.3 Modulation schemes used by LTE.

3.1.2 Signal Reception

In a wireless communication system, the signal spreads out as it travels from the transmitter to the receiver, so the received power P_R is less than the transmitted power P_T. The *propagation loss* or *path loss*, PL, is the ratio of the two:

$$\text{PL} = \frac{P_T}{P_R} \qquad (3.2)$$

If the signal is travelling through empty space, then at a distance r from the transmitter, it occupies a spherical surface with an area of $4\pi r^2$. The propagation loss is therefore proportional to r^2. In a cellular network, the signal can also be absorbed and reflected by obstacles such as buildings and the ground, which in turn affects the propagation loss. Experimentally, we find that the propagation loss in a cellular network is roughly proportional to r^m, where m typically lies between 3.5 and 4.

By itself, propagation loss would not be a problem. As shown in Figure 3.4, however, the received signal is also distorted by thermal noise and by interference from other transmitters. These effects mean that the receiver cannot make a completely accurate estimate of the transmitted amplitude and phase.

The receiver deals with this issue as follows. The *symbol estimation* stage extracts the amplitude and phase of the incoming signal in the form of continuously varying real numbers. The *demodulator* then uses this information to estimate the received bits, which can take the form of *hard decisions* in which the bits are either 1 or 0, or *soft decisions* that include some measure of confidence. As shown in the figure, a typical demodulator will first estimate the soft decisions and then convert these to hard decisions later on.

If the noise and interference are large enough, then a bit of 1 can be misinterpreted as a bit of 0 and vice versa, leading to bit errors in the receiver. The error rate depends on the signal to interference plus noise ratio (SINR) at the receiver. In a fast modulation scheme such as 64-QAM, the signal can be transmitted in many different ways, using states in the constellation diagram that are packed closely together. As a result, 64-QAM

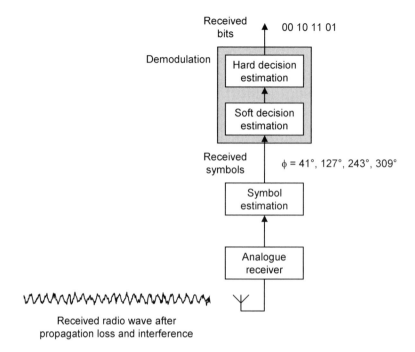

Figure 3.4 Basic architecture of a wireless communication receiver.

is vulnerable to errors and can only be used if the SINR is high. In contrast, QPSK only has a few states, so is less vulnerable to errors and can be successfully used at a lower SINR. LTE exploits this by switching dynamically between different modulation schemes: it uses 64-QAM at high SINR to give a high data rate, but falls back to 16-QAM or QPSK at lower SINR to reduce the number of errors.

3.1.3 Channel Estimation

There is one remaining problem with the receiver from Figure 3.4. The phase of the received signal depends not only on the phase of the transmitted signal, but also on the receiver's exact position. If the receiver moves through a half a wavelength of the carrier signal (a distance of 10 cm at a carrier frequency of 1500 MHz, for example), then the phase of the received signal changes by 180°. When using QPSK, this phase change turns bit pairs of 00 into 11 and vice versa, and completely destroys the received information.

To deal with this problem, the transmitter inserts occasional *reference symbols* into the data stream, which have a pre-defined amplitude and phase. In the receiver, a *channel estimation* function measures the reference symbols, compares them with the ones transmitted and estimates the phase shift that the air interface introduced. It can then remove this phase shift from the information symbols, and can recover the information bits.

The resulting receiver architecture is shown in Figure 3.5. The incoming signal arrives with a different phase angle from the one shown earlier. However, the channel estimator detects this phase shift, which allows the receiver to reconstruct the transmitted bits in

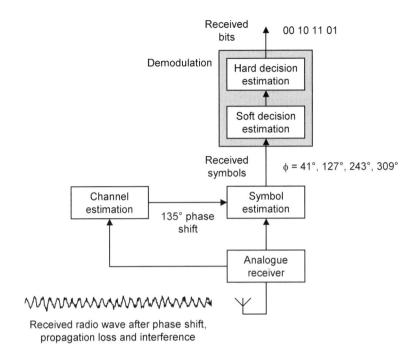

Figure 3.5 Architecture of a wireless communication receiver, including the use of channel estimation.

the same way as before. The phase shift does not change much from one symbol to the next, so the reference symbols only need to take up a small part of the transmitted data stream. The resulting overhead in LTE is about 10%.

3.1.4 Multiple Access Techniques

The techniques described so far work well for one-to-one communications. In a cellular network, however, a base station has to transmit to many different mobiles at once. It does this by sharing the resources of the air interface, in a technique known as *multiple access*.

Mobile communication systems use a few different multiple access techniques, two of which are shown in Figure 3.6. *Frequency division multiple access* (FDMA) was used by the first generation analogue systems. In this technique, each mobile receives on its own carrier frequency, which it distinguishes from the others by the use of analogue filters. The carriers are separated by unused *guard bands*, which minimizes the interference between them. In *time division multiple access* (TDMA), mobiles receive information on the same carrier frequency but at different times.

GSM uses a mix of frequency and time division multiple access, in which every cell has several carrier frequencies that are each shared amongst eight different mobiles. LTE uses another mixed technique known as *orthogonal frequency division multiple access* (OFDMA), which we will cover in Chapter 4.

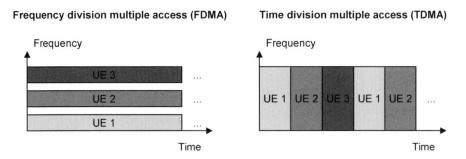

Figure 3.6 Example multiple access techniques.

Third generation communication systems used a different technique altogether, known as *code division multiple access* (CDMA). In this technique, mobiles receive on the same carrier frequency and at the same time, but the signals are labelled by the use of codes, which allow a mobile to separate its own signal from those of the others. LTE uses a few of the concepts from CDMA for some of its control signals, but does not implement the technique otherwise.

Multiple access is actually a generalization of a simpler technique known as *multiplexing*. The difference between the two is that a multiple access system can dynamically change the allocation of resources to different mobiles, while in a multiplexing system the resource allocation is fixed.

3.1.5 FDD and TDD Modes

By using the multiple access techniques described above, a base station can distinguish the transmissions to and from the individual mobiles in the cell. However, we still need a way to distinguish the mobiles' transmissions from those of the base stations themselves.

To do this, a mobile communication system can operate in the transmission modes that we introduced in Chapter 1 (Figure 3.7). When using frequency division duplex (FDD), the base station and mobile transmit and receive at the same time, but using different carrier frequencies. Using time division duplex (TDD), they transmit and receive on the same carrier frequency but at different times.

FDD and TDD modes have different advantages and disadvantages. In FDD mode, the bandwidths of the uplink and downlink are fixed and are usually the same. This makes it suitable for voice communications, in which the uplink and downlink data rates are very similar. In TDD mode, the system can adjust how much time is allocated to the uplink and downlink. This makes it suitable for applications such as web browsing, in which the downlink data rate can be much greater than the rate on the uplink.

TDD mode can be badly affected by interference if, for example, one base station is transmitting while a nearby base station is receiving. To avoid this, nearby base stations must be carefully time synchronized and must use the same allocations for the uplink and downlink, so that they all transmit and receive at the same time. This makes TDD suitable for networks that are made from isolated hotspots, because each hotspot can have a different timing and resource allocation. In contrast, FDD is often preferred for wide-area networks that have no isolated regions.

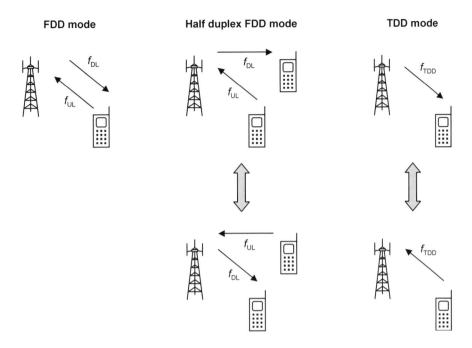

Figure 3.7 Operation of FDD and TDD modes.

When operating in FDD mode, the mobile usually has to contain a high attenuation duplex filter that isolates the uplink transmitter from the downlink receiver. In a variation known as *half duplex FDD mode*, a base station can still transmit and receive at the same time, but a mobile can only do one or the other. This means that the mobile does not have to isolate the transmitter and receiver to the same extent, which eases the design of its radio hardware.

LTE supports each of the modes described above. A cell can use either FDD or TDD mode. A mobile can support any combination of full duplex FDD, half duplex FDD and TDD, although it will only use one of these at a time.

3.2 Multipath, Fading and Inter-Symbol Interference

3.2.1 Multipath and Fading

Propagation loss and noise are not the only problem. As a result of reflections, rays can take several different paths from the transmitter to the receiver. This phenomenon is known as *multipath*.

At the receiver, the incoming rays can add together in different ways, which are shown in Figure 3.8. If the peaks of the incoming rays coincide then they reinforce each other, a situation known as *constructive interference*. If, however, the peaks of one ray coincide with the troughs of another, then the result is *destructive interference*, in which the rays cancel. Destructive interference can make the received signal power drop to a very low level, a situation known as *fading*. The resulting increase in the error rate makes fading a serious problem for any mobile communication system.

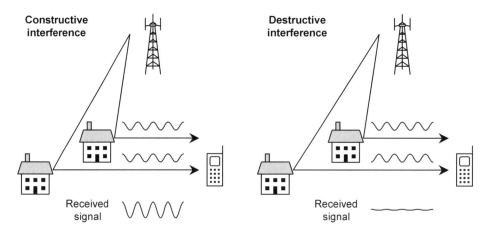

Figure 3.8 Generation of constructive interference, destructive interference and fading in a multipath environment.

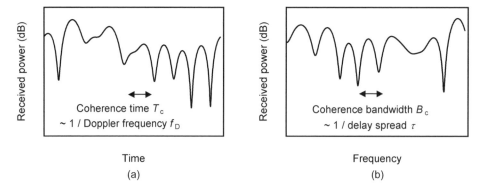

Figure 3.9 Fading as a function of (a) time and (b) frequency.

If the mobile moves from one place to another, then the ray geometry changes, so the interference pattern changes between constructive and destructive. Fading is therefore a function of time, as shown in Figure 3.9(a). The amplitude and phase of the received signal vary over a timescale called the *coherence time*, T_c, which can be estimated as follows:

$$T_c \approx \frac{1}{f_D} \qquad (3.3)$$

Here f_D is the mobile's Doppler frequency:

$$f_D = \frac{v}{c} f_C \qquad (3.4)$$

where f_C is the carrier frequency, v is the speed of the mobile and c is the speed of light (3×10^8 ms^{-1}). For example, a pedestrian might walk with a speed of 1 ms^{-1}

(3.6 km hr^{-1}). At a carrier frequency of 1500 MHz, the resulting Doppler shift is 5 Hz, giving a coherence time of about 200 milliseconds. Faster mobiles move through the interference pattern more quickly, so their coherence time is correspondingly less.

If the carrier frequency changes, then the wavelength of the radio signal changes. This also makes the interference pattern change between constructive and destructive, so fading is a function of frequency as well (Figure 3.9b). The amplitude and phase of the received signal vary over a frequency scale called the *coherence bandwidth*, B_c, which can be estimated as follows:

$$B_c \approx \frac{1}{\tau} \qquad (3.5)$$

Here, τ is the *delay spread* of the radio channel, which is the difference between the arrival times of the earliest and latest rays. It can be calculated as follows:

$$\tau = \frac{\Delta L}{c} \qquad (3.6)$$

where ΔL is the difference between the path lengths of the longest and shortest rays. In a macrocell, a typical path difference might be around 300 metres, giving a delay spread of 1 μs and a coherence bandwidth of around 1 MHz. Smaller cells have a smaller delay spread, so have a larger coherence bandwidth.

3.2.2 Inter-Symbol Interference

If the path lengths of the longest and shortest rays are different, then symbols travelling on those rays will reach the receiver at different times. In particular, the receiver can start to receive one symbol on a short direct ray, while it is still receiving the previous symbol on a longer reflected ray. The two symbols therefore overlap at the receiver (Figure 3.10), causing another problem known as *inter-symbol interference* (ISI).

Let us continue the previous example, in which the delay spread τ was 1 μs. If the symbol rate is 400 ksps, then the symbol duration is 2.5 μs, so the symbols on the longest

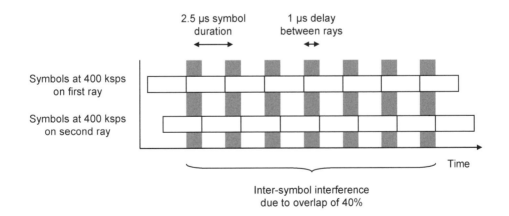

Figure 3.10 Generation of inter-symbol interference in a multipath environment.

and shortest rays overlap by 40%. This causes a large amount of inter-symbol interference, which will greatly increase the error rate in the receiver. As the data rate increases, so the symbol duration falls and the problem becomes progressively worse. This makes inter-symbol interference a problem for any high data rate communication system.

In these discussions, we have seen that frequency-dependent fading and inter-symbol interference are both important if the delay spread is large. In fact they are different ways of looking at the same underlying phenomenon: a large delay spread causes both frequency-dependent fading and ISI. 2G and 3G communication systems often combat the two effects using a device known as an *equalizer*, which passes the received signal through a filter that tries to model the time delays and undo their effect. Unfortunately equalizers are complex devices and are far from perfect. In Chapter 4, we will see how the multiple access technique of OFDMA can deal with these issues in a far more direct way.

3.3 Error Management

3.3.1 Forward Error Correction

In the earlier sections, we saw that noise and interference lead to errors in a wireless communication receiver. These are bad enough during voice calls, but are even more damaging to important information such as web pages and emails. Fortunately there are several ways to solve the problem.

The most important technique is *forward error correction*. In this technique, the transmitted information is represented using a *codeword* that is typically two or three times as long. The extra bits supply additional, redundant data that allow the receiver to recover the original information sequence. For example, a transmitter might represent the information sequence 101 using the codeword 110010111. After an error in the second bit, the receiver might recover the codeword 100010111. If the coding scheme has been well designed, then the receiver can conclude that this is not a valid codeword, and that the most likely transmitted codeword was 110010111. The receiver has therefore corrected the bit error and can recover the original information. The effect is very like written English, which contains redundant letters that allow the reader to understand the underlying information, even in the presence of spelling mistakes.

The *coding rate* is the number of information bits divided by the number of transmitted bits ($1/3$ in the example above). Usually, forward error correction algorithms operate with a fixed coding rate. Despite this, a wireless transmitter can still adjust the coding rate using the two-stage process shown in Figure 3.11. In the first stage, the information bits are passed through a fixed-rate coder. The main algorithm used by LTE is known as *turbo coding* and has a fixed coding rate of $1/3$. In the second stage, called *rate matching*, some of the coded bits are selected for transmission, while the others are discarded in a process known as *puncturing*. The receiver has a copy of the puncturing algorithm, so it can insert dummy bits at the points where information was discarded. It can then pass the result through a turbo decoder for error correction.

Changes in the coding rate have a similar effect to changes in the modulation scheme. If the coding rate is low, then the transmitted data contain many redundant bits. This allows the receiver to correct a large number of errors and to operate successfully at a low SINR, but at the expense of a low information rate. If the coding rate is close to 1, then the

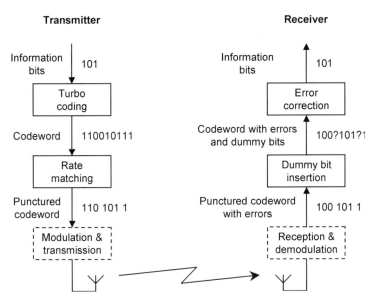

Figure 3.11 Block diagram of a transmitter and receiver using forward error correction and rate matching.

information rate is higher but the system is more vulnerable to errors. LTE exploits this with a similar trade-off to the one we saw earlier, by transmitting with a high coding rate if the received SINR is high and vice versa.

3.3.2 Automatic Repeat Request

Automatic repeat request (ARQ) is another error management technique, which is shown in Figure 3.12. Here, the transmitter takes a block of information bits and uses them to

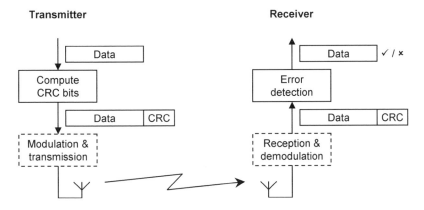

Figure 3.12 Block diagram of a transmitter and receiver using automatic repeat request.

compute some extra bits that are known as a *cyclic redundancy check* (CRC). It appends these to the information block and then transmits the two sets of data in the usual way.

The receiver separates the two fields and uses the information bits to compute the expected CRC bits. If the observed and expected CRC bits are the same, then it concludes that the information has been received correctly and sends a positive acknowledgement back to the transmitter. If the CRC bits are different, it concludes that an error has occurred and sends a negative acknowledgement to request a re-transmission. Positive and negative acknowledgements are often abbreviated to ACK and NACK respectively.

A wireless communication system often combines the two error management techniques that we have been describing. Such a system corrects most of the bit errors by the use of forward error correction and then uses automatic repeat requests to handle the remaining errors that leak through.

Normally, ARQ uses a technique called *selective re-transmission* (Figure 3.13), in which the receiver waits for several blocks of data to arrive before acknowledging them all. This allows the transmitter to continue sending data without waiting for an acknowledgement, but it means that any re-transmitted data can take a long time to arrive. Consequently, this technique is only suitable for non real-time streams such as web pages and emails.

3.3.3 Hybrid ARQ

The ARQ technique from the previous section works well, but has one shortcoming: if a block of data fails the cyclic redundancy check, then the receiver throws it away, despite the fact that it contains some useful signal energy. If we could find a way to use that signal energy, then we might be able to design a more powerful receiver.

This idea is implemented in a technique known as *hybrid ARQ* (HARQ) which is shown in Figure 3.14. Here, the transmitter sends the data as before. The receiver demodulates the incoming data, but this time it passes the soft decisions up to the next stage, instead of the hard decisions. It inserts zero soft decisions to account for any bits that the transmitter removed and stores the resulting codeword in a buffer. It then passes the codeword through the stages of error correction and error detection, and sends an acknowledgement back to the transmitter.

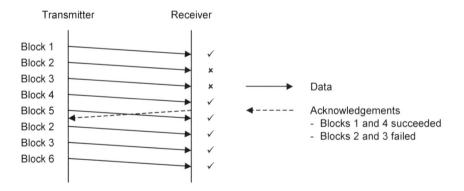

Figure 3.13 Operation of a selective re-transmission ARQ scheme.

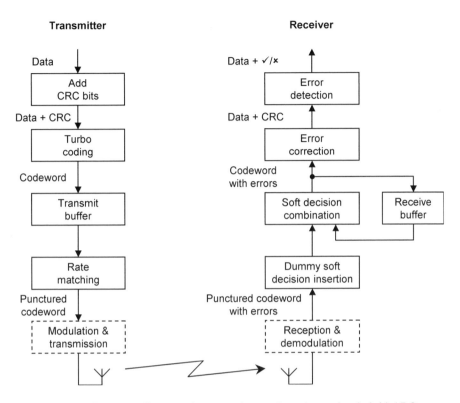

Figure 3.14 Block diagram of a transmitter and receiver using hybrid ARQ.

If the cyclic redundancy check fails, then the transmitter sends the data again. This time, however, the receiver combines the data from the first transmission and the re-transmission, by adding the soft decisions. This increases the signal energy at the receiver, so it increases the likelihood of a CRC pass. As a result, this scheme performs better than the basic ARQ technique, in which the first transmission was discarded.

Normally, hybrid ARQ uses a re-transmission technique called *stop-and-wait*, in which the transmitter waits for an acknowledgment before sending new data or a re-transmission. This simplifies the design and reduces the time delays in the system, which can make hybrid ARQ acceptable even for real-time streams such as voice. However, it also means that the transmitter has to pause while waiting for the acknowledgement to arrive. To prevent the throughput from falling, the system shares the data amongst several *hybrid ARQ processes*, which are multiple copies of Figure 3.14. One process can then transmit while the others are waiting for acknowledgements, in the manner shown in Figure 3.15.

The use of multiple hybrid ARQ processes means that the receiver decodes data blocks in a different order from the one in which they were transmitted. In Figure 3.15, for example, block 3 is transmitted four times and is only decoded some time after block 4. To deal with that problem, the receiver includes a re-ordering function that accepts the decoded blocks and returns them to their initial order.

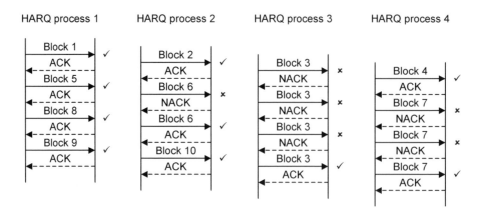

Figure 3.15 Operation of multiple hybrid ARQ processes in conjunction with a stop-and-wait re-transmission scheme.

There is one final problem. If an initial transmission is badly corrupted by interference, then it may take several re-transmissions before the interference is overcome. To limit the resulting time delays, a hybrid ARQ process is usually configured so that it gives up after a few unsuccessful attempts to transfer a block of data. At a higher level, a basic ARQ receiver can detect the problem, and can instruct the transmitter to send the block again from the beginning. LTE implements this technique, by the use of hybrid ARQ in the physical layer, backed up by a basic ARQ scheme in the radio link control protocol.

References

1. Goldsmith, A. (2005) *Wireless Communications*, Cambridge University Press.
2. Molisch, A. F. (2010) *Wireless Communications*, 2nd edn, John Wiley & Sons, Ltd, Chichester.
3. Rappaport, T. S. (2001) *Wireless Communications: Principles and Practice*, 2nd edn, Prentice Hall.
4. Tse, D. and Viswanath, P. (2005) *Fundamentals of Wireless Communication*, Cambridge University Press.
5. Parsons, J. D. (2000) *The Mobile Radio Propagation Channel*, 2nd edn, John Wiley & Sons, Ltd, Chichester.
6. Saunders, S. and Aragón-Zavala, A. (2007) *Antennas and Propagation for Wireless Communication Systems*, 2nd edn, John Wiley & Sons, Ltd, Chichester.

4

Orthogonal Frequency Division Multiple Access

The technique used for radio transmission and reception in LTE is known as orthogonal frequency division multiple access (OFDMA). OFDMA carries out the same functions as any other multiple access technique, by allowing the base station to communicate with several different mobiles at the same time. However, it is also a powerful way to minimize the problems of fading and inter-symbol interference that we introduced in Chapter 3. In this chapter, we will describe the basic principles of OFDMA, and show how it is applied to a mobile cellular network. We will also cover a modified radio transmission technique, known as single carrier frequency division multiple access (SC-FDMA), which is used for the LTE uplink.

OFDMA is also used by several other radio communication systems, such as wireless local area networks (IEEE 802.11 versions a, g and n) and WiMAX (IEEE 802.16), as well as in digital television and radio broadcasting. However, LTE is the first system to have made use of SC-FDMA.

4.1 Orthogonal Frequency Division Multiplexing

4.1.1 Reduction of Inter-Symbol Interference using OFDM

In the last chapter, we saw how high data rate transmission in a multipath environment leads to inter symbol interference (ISI). In Figure 3.10, for example, the delay spread was 1 μs and the data rate was 400 ksps, so the symbols overlapped at the receiver by 40%. That led to interference and bit errors at the receiver.

Orthogonal frequency division multiplexing (OFDM) is a powerful way to solve the problem. Instead of sending the information as a single stream, an OFDM transmitter divides the information into several parallel sub-streams, and sends each sub-stream on a different frequency known as a *sub-carrier*. If the total data rate stays the same, then the data rate on each sub-carrier is less than before, so the symbol duration is longer. This reduces the amount of ISI, and reduces the error rate.

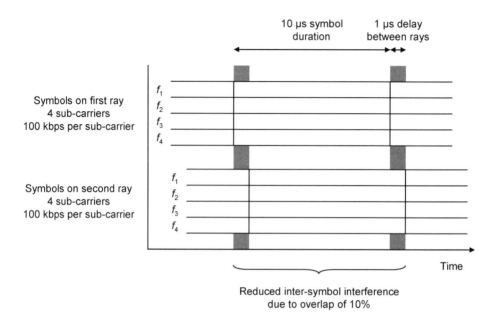

Figure 4.1 Reduction of inter-symbol interference by transmission on multiple sub-carriers.

Figure 4.1 shows a simple example. Here, we have divided the original data stream amongst four sub-carriers with frequencies f_1 to f_4. The data rate on each sub-carrier is now 100 ksps, so the symbol duration has increased to 10 μs. If the delay spread remains at 1 μs, then the symbols only overlap by 10%. This reduces the amount of ISI to one quarter of what it was before and reduces the number of errors in the receiver. In practice, LTE can use a very large number of sub-carriers, up to a maximum of 1200 in Release 8, which reduces the amount of ISI to negligible levels.

4.1.2 The OFDM Transmitter

Figure 4.2 is a block diagram of an analogue OFDM transmitter. The diagram contains some simplifications that we will deal with shortly, but it serves to illustrate the basic principles of the technique.

The transmitter accepts a stream of bits from higher layer protocols and converts them to symbols using the chosen modulation scheme, for example quadrature phase shift keying (QPSK). The serial-to-parallel converter then takes a block of symbols, four in this example, and mixes each symbol with one of the sub-carriers by adjusting its amplitude and phase.

LTE uses a fixed sub-carrier spacing of 15 kHz, so the sub-carriers in Figure 4.2 have frequencies of 0, 15, 30 and 45 kHz. (We will mix the signals up to radio frequency at the end.) The symbol duration is the reciprocal of the sub-carrier spacing, so is about 66.7 μs. For the moment this is just an arbitrary choice: the reasons will become clear in due course. However, it means that the 15 kHz sub-carrier goes through one cycle during the 66.7 μs symbol duration, while the sub-carriers at 30 and 45 kHz go through two and three cycles respectively.

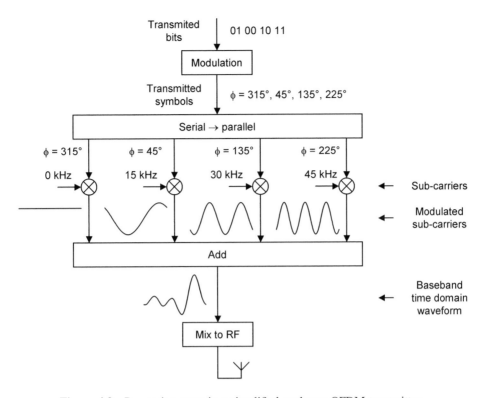

Figure 4.2 Processing steps in a simplified analogue OFDM transmitter.

We now have four sine waves, at frequencies of 0, 15, 30 and 45 kHz, whose amplitudes and phases represent the eight transmitted bits. By adding these sine waves together, we can generate a single time-domain waveform, which is a low frequency representation of the signal that we need to send. The only remaining task is to mix the waveform up to radio frequency (RF) for transmission.

Figure 4.3 includes three extensions. Firstly, we have added four more sub-carriers, at frequencies of -15, -30, -45 and -60 kHz. The distinction between positive and negative frequencies is that the latter are eventually transmitted below the carrier frequency, not above it. At a carrier frequency of 800 MHz, for example, the 15 kHz sub-carrier ends up at 800.015 MHz, while the -15 kHz sub-carrier ends up at 799.985 MHz.

Secondly, we distinguish the positive and negative frequencies by retaining the in-phase and quadrature components of each sub-carrier through most of the transmission process. In Figure 4.3, for example, the in-phase components of the 15 kHz and -15 kHz signals are exactly the same, but we can distinguish them because their quadrature components are different. After mixing the information up to radio frequency, all the frequencies are positive and the quadrature components can be discarded.

Thirdly, it is highly desirable to do the processing digitally, rather than in analogue form. In Figure 4.3, we sample the in-phase and quadrature components eight times per symbol, which allows us to sample the -60 kHz sub-carrier twice in every cycle. More generally, the minimum number of samples per symbol equals the number of sub-carriers. We can then do the mixing and addition operations digitally, which results in a digital

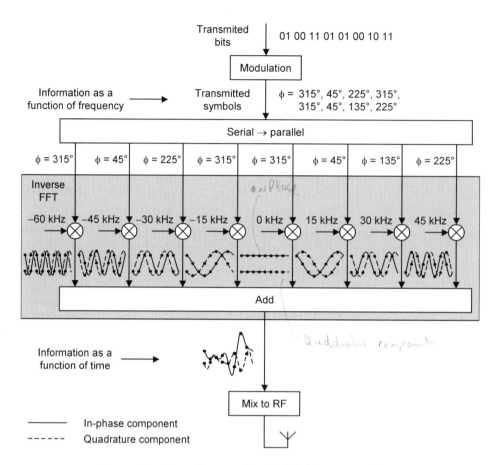

Figure 4.3 Processing steps in a digital OFDM transmitter.

time-domain waveform that contains all the information we require. We can then convert the waveform from digital to analogue form, filter it and mix it up to radio frequency for transmission.

Now let us look at two important points in the processing chain. At the serial-to-parallel conversion stage, the data represent the amplitude and phase of each sub-carrier, as a function of frequency. After the addition stage towards the end, the data represent the in-phase and quadrature components of the transmitted signal, as a function of time. We can see that the mixing and addition steps have simply converted the data from a function of frequency to a function of time.

This conversion is actually a well-known computational technique called the *inverse discrete Fourier transform* (DFT). (The Fourier transform converts data from the time domain to the frequency domain, so the transmitter requires an inverse transform, which carries out the reverse process.) By using this technique, we can hide the explicit mixing steps in Figures 4.2 and 4.3: instead, we just pass the symbols into an inverse Fourier transform and pick up the time-domain signal from the output.

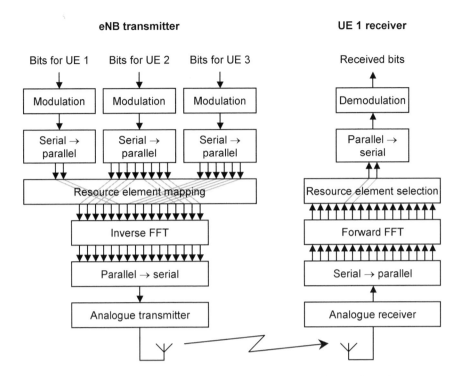

Figure 4.4 Initial block diagram of an OFDM transmitter and receiver.

In turn, the discrete Fourier transform can be implemented extremely quickly using an algorithm known as the *fast Fourier transform* (FFT). This limits the computational load on the transmitter and receiver, and allows the two devices to be implemented in a computationally efficient way. However, there is one important restriction: for the FFT to work efficiently, the number of data points should be either an exact power of two or a product of small prime numbers alone. There are several books with more details about the Fourier transform, for example [1] and [2].

4.1.3 Initial Block Diagram

Figure 4.4 is a block diagram of an OFDM transmitter and receiver, using the principles that we have discussed so far. We assume that the system is operating on the downlink, so that the transmitter is in the base station and the receiver is in the mobile. The diagram still contains a few simplifications, but we will deal with these shortly.

In the diagram, the base station is sending streams of bits to three different mobiles. It modulates each bit stream independently, possibly using a different modulation scheme for each one. It then passes each symbol stream through a serial-to-parallel converter, to divide it into sub-streams. The number of sub-streams per mobile depends on the data rate: for example, a voice application might only use a few sub-streams, while a video application might use many more.

The *resource element mapper* takes the individual sub-streams and chooses the sub-carriers on which to transmit them. A mobile's sub-carriers may lie in one contiguous block (as in the case of mobiles 1 and 3), or they may be divided (as for mobile 2). The resulting information is the amplitude and phase of each sub-carrier as a function of frequency. By passing it through an inverse FFT, we can compute the in-phase and quadrature components of the corresponding time-domain waveform. This can then be digitized, filtered and mixed up to radio frequency for transmission.

The mobile reverses the process. It starts by sampling the incoming signal, filtering it, and converting it down to baseband. It then passes the data through a forward FFT, to recover the amplitude and phase of each sub-carrier. We now assume that the base station has already told the mobile which sub-carriers to use, through scheduling techniques that we will cover in Chapter 8. Using this knowledge, the mobile selects the required sub-carriers and recovers the transmitted information, while discarding the remainder.

4.2 OFDMA in a Mobile Cellular Network

4.2.1 Multiple Access

In Figure 4.4, the base station transmitted to three mobiles at the same time using orthogonal frequency division multiplexing. We can take this idea a step further, by sharing the resources dynamically amongst all the mobiles in the cell. The resulting technique is known as *orthogonal frequency division multiple access* (OFDMA), and is illustrated in Figure 4.5.

In OFDMA, the base station shares its resources by transmitting to the mobiles at different times and frequencies, so as to meet the requirements of the individual applications. For example, mobile 1 is receiving a voice over IP stream, so the data rate, and hence the number of sub-carriers, is low but constant. On the other hand, mobile 2 is receiving a stream of non real time packet data. The average data rate is higher, but the data come in bursts, so the number of sub-carriers can vary.

The base station can also respond to frequency dependent fading, by allocating sub-carriers on which the mobile is receiving a strong signal. In the figure, mobile 3 is receiving a VoIP stream, but it is also affected by frequency dependent fading. In response, the base

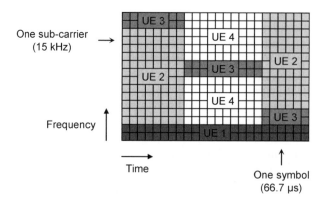

Figure 4.5 Implementation of time and frequency division multiple access when using OFDMA.

station allocates sub-carriers on which the mobile is receiving a strong signal, and changes this allocation as the fading pattern changes. In a similar way, it can transmit to mobile 4 using two separate blocks of sub-carriers, which are separated by a fade.

By allocating sub-carriers in response to changes in the fading patterns, an OFDMA transmitter can greatly reduce the impact of time- and frequency-dependent fading. The process requires feedback from the mobile, which we will cover as part of Chapter 8.

4.2.2 Fractional Frequency Re-Use

Using the techniques described above, one base station can send information to a large number of mobiles. However, a mobile communication system also has a large number of base stations, so every mobile has to receive a signal from one base station in the presence of interference from the others. We need a way to minimize the interference, so that the mobile can receive the information successfully.

Previous systems have used two different techniques. In GSM, nearby cells transmit using different carrier frequencies. Typically, each cell might use a quarter of the total bandwidth, with a *re-use factor* of 25%. This technique reduces the interference between nearby cells, but it means that the frequency band is used inefficiently. In UMTS, each cell has the same carrier frequency, with a re-use factor of 100%. This technique uses the frequency band more efficiently than before, at the expense of increasing the interference in the system.

In an LTE network, every base station can transmit in the same frequency band. However, it can allocate the sub-carriers within that band in a flexible way, using a technique known as *fractional frequency re-use* that gives the best of both worlds.

Figure 4.6 shows a simple example, in which every base station is controlling one cell and every cell is sharing the same frequency band. Within that band, each cell transmits to nearby mobiles using the same set of sub-carriers, denoted f_0. This works well, because the mobiles are close to their respective base stations, so the received signals are strong enough to overwhelm any interference. Distant mobiles receive much weaker signals, which are easily damaged by interference. To avoid this, neighbouring cells can transmit to those mobiles using different sets of sub-carriers. In the example shown, half the

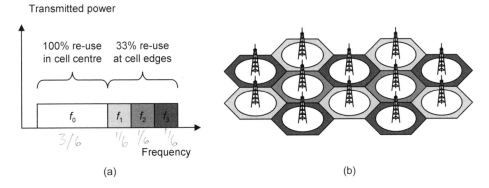

Figure 4.6 Example implementation of fractional frequency re-use when using OFDMA. (a) Use of the frequency domain. (b) Resulting network plan.

frequency band is reserved for nearby mobiles, while the remainder is divided into three sets, denoted f_1, f_2 and f_3, for use by distant mobiles. The resulting re-use factor is 67%.

More flexible implementations are possible: for example, one cell might use a set of sub-carriers for distant mobiles, while its neighbours use the same set for nearby mobiles. To support this flexibility, base stations can exchange signalling messages across the X2 interface, in which they tell each other about how they are using the frequency band. We will discuss these messages further in Chapter 19.

4.2.3 Channel Estimation

Figure 4.7 is a detailed block diagram of OFDMA. It is very like the block diagram shown earlier, but with two extra processes. Firstly, the receiver contains the extra steps of channel estimation and equalization. Secondly, the transmitter inserts a cyclic prefix into the data stream, which is then removed in the receiver.

First consider channel estimation. As we noted in Chapter 3, each sub-carrier can reach the receiver with a completely arbitrary amplitude and phase. To deal with this,

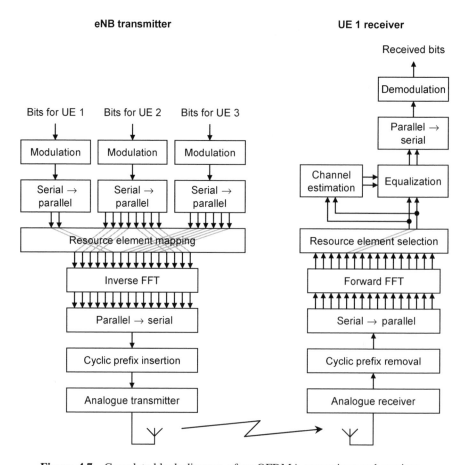

Figure 4.7 Complete block diagram of an OFDMA transmitter and receiver.

the OFDMA transmitter injects reference symbols into the transmitted data stream. The receiver measures the incoming reference symbols, compares them with the ones transmitted, and uses the result to remove the amplitude changes and phase shifts from the incoming signal.

In the presence of frequency-dependent fading, the amplitude changes and phase shifts are functions of frequency as well as time and affect the different sub-carriers in different ways. To ensure that the receiver can measure all the information it requires, the LTE reference symbols are scattered across the time and frequency domains, in the manner that will be described in Chapter 7. The reference symbols take up about 10% of the transmitted data stream, so do not cause a significant overhead.

4.2.4 Cyclic Prefix Insertion

Earlier, we saw how OFDMA reduces the amount of inter symbol interference by transmitting data on multiple sub-carriers. A final technique allows us to get rid of ISI altogether.

The basic idea is to insert a *guard period* (GP) before each symbol, in which nothing is transmitted. If the guard period is longer than the delay spread, then the receiver can be confident of reading information from just one symbol at a time, without any overlap with the symbols that precede or follow. Naturally the symbol reaches the receiver at different times on different rays and some extra processing is required to tidy up the confusion. The extra processing is relatively straightforward, however.

LTE uses a slightly more complex technique, known as *cyclic prefix* (CP) *insertion* (Figure 4.8). Here, the transmitter starts by inserting a guard period before each symbol, as before. However, it then copies data from the end of the symbol following, so as to fill up the guard period. If the cyclic prefix is longer than the delay spread, then the receiver can still be confident of reading information from just one symbol at a time.

We can see how cyclic prefix insertion works by looking at one sub-carrier (Figure 4.9). The transmitted signal is a sine wave, whose amplitude and phase change from one symbol

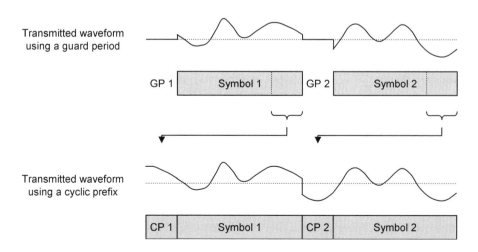

Figure 4.8 Operation of cyclic prefix insertion.

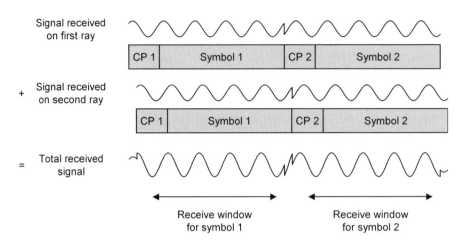

Figure 4.9 Operation of the cyclic prefix on a single sub-carrier.

to the next. As noted earlier, each symbol contains an exact number of cycles of the sine wave, so the amplitude and phase at the start of each symbol equal the amplitude and phase at the end. Because of this, the transmitted signal changes smoothly as we move from each cyclic prefix to the symbol following.

In a multipath environment, the receiver picks up multiple copies of the transmitted signal with multiple arrival times. These add together at the receive antenna, giving a sine wave with the same frequency but a different amplitude and phase. The received signal still changes smoothly at the transition from a cyclic prefix to the symbol that follows. There are a few glitches, but these are only at the start of the cyclic prefix and the end of the symbol, where the preceding and following symbols start to interfere.

The receiver processes the received signal within a window whose length equals the symbol duration, and discards the remainder. If the window is correctly placed, then the received signal is exactly what was transmitted, without any glitches, and subject only to an amplitude change and a phase shift. But the receiver can compensate for these using the channel estimation and equalization techniques described above. It can therefore handle the cyclic prefix without any extra processing at all.

Admittedly the system uses multiple sub-carriers, not just one. However we have already seen that the sub-carriers do not interfere with each other and can be treated independently, so the existence of multiple sub-carriers does not affect this argument at all.

Normally, LTE uses a cyclic prefix of about 4.7 μs. This corresponds to a maximum path difference of about 1.4 km between the lengths of the longest and shortest rays, which is enough for all but the very largest and most cluttered cells. The cyclic prefix reduces the data rate by about 7%, but this is a small price to pay for the removal of ISI.

4.2.5 Use of the Frequency Domain

Let us now look in more detail at the way in which a mobile communication system uses the frequency domain. In traditional analogue FDMA, a mobile has to measure the

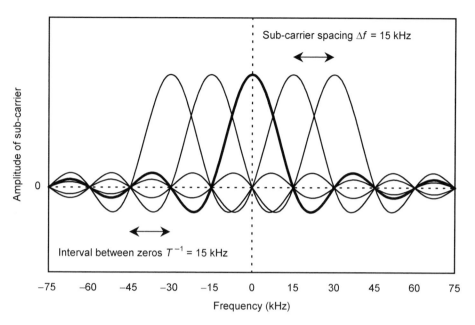

Figure 4.10 Amplitudes of the signals transmitted on neighbouring sub-carriers, as a function of frequency.

signal on one sub-carrier in the presence of interference from all the others. To minimize the amount of interference, the sub-carriers have to be separated by wide guard bands. The need for these guard bands implies that the system uses the frequency domain in an inefficient way.

Now consider the situation with OFDMA. In the time domain, each sub-carrier starts life as a sine wave, but the modulation process makes its amplitude and phase change at intervals of the symbol duration T, which equals 66.7 μs. This broadens the signal in the frequency domain, to a bandwidth of about T^{-1}. Figure 4.10 shows the details. In the frequency domain, the amplitude of each sub-carrier oscillates either side of zero and crosses through zero at regular intervals of T^{-1}. (Mathematicians will recognize this response as a sinc function ($x^{-1} \sin x$).)

Now, the interval between adjacent sub-carriers is the sub-carrier spacing Δf. If $\Delta f = T^{-1}$, then the sub-carriers overlap in the frequency domain, but the peak response of one sub-carrier coincides with zeros of all the others. As a result, the mobile can sample one sub-carrier and can measure its amplitude and phase without any interference from the others, despite the fact that they are closely packed together. Sub-carriers with this property are said to be *orthogonal*.

This property means that OFDMA uses the frequency domain in a very efficient way and is one of the reasons why the spectral efficiency of LTE is so much greater than that of previous mobile telecommunication systems. It also justifies the decision made in Section 4.1.2, when we set the symbol duration T equal to the reciprocal of the sub-carrier spacing Δf.

4.2.6 Choice of Sub-Carrier Spacing

The argument in the previous section works fine if the mobile is stationary. If the mobile is moving, then any incoming rays are Doppler shifted to higher or lower frequencies. The same applies to each of the OFDMA sub-carriers.

In a multipath environment, a mobile can be moving towards some rays, which are shifted to higher frequencies, but away from others, whose frequencies move lower. As a result, the sub-carriers are not simply shifted: instead, they are blurred across a range of frequencies. If a mobile tries to measure the peak response of one sub-carrier, then it will now receive interference from all the others. We have therefore lost the orthogonality property from the previous section.

The amount of interference will still be acceptable, however, if the Doppler shift is much less than the sub-carrier spacing. We therefore need to choose the sub-carrier spacing Δf as follows:

$$\Delta f \gg f_D \qquad (4.1)$$

where f_D is the Doppler shift from Equation (3.4). LTE is designed to operate with a maximum mobile speed of 350 km hr^{-1} and a maximum carrier frequency of about 3.5 GHz, which gives a maximum Doppler shift of about 1.1 kHz. This is 7% of the sub-carrier spacing, so it satisfies the constraint above.

There is another constraint on the parameters used by LTE. To minimize the impact of inter-symbol interference, we need to choose the symbol duration T as follows:

$$T \gg \tau \qquad (4.2)$$

where τ is the delay spread from Equation (3.6). As we noted earlier, LTE normally works with a maximum delay spread of about 4.7 μs. This is 7% of the 66.7 μs symbol duration, so it satisfies this second constraint.

We can draw the following conclusions. If the sub-carrier spacing were much less than 15 kHz, then the system would be prone to interference between the sub-carriers at high mobile speeds. If it were much greater, then the system would be prone to inter symbol interference in large, cluttered cells. The chosen sub-carrier spacing is the result of a trade-off between these two extremes.

4.3 Single Carrier Frequency Division Multiple Access

4.3.1 Power Variations from OFDMA

OFDMA works well on the LTE downlink. However, it has one disadvantage: the power of the transmitted signal is subject to rather large variations. To illustrate this, Figure 4.11(a) shows a set of sub-carriers that have been modulated using QPSK, and which therefore have constant power. The amplitude of the resulting signal (Figure 4.11b) varies widely, with maxima where the peaks of the sub-carriers coincide and zeros where they cancel. In turn, these variations are reflected in the power of the transmitted signal (Figure 4.11c).

These power variations can cause problems for the transmitter's power amplifier. If the amplifier is linear, then the output power is proportional to the input, so the output waveform is exactly the shape that we require. If the amplifier is non-linear, then the output power is no longer proportional to the input, so the output waveform is distorted.

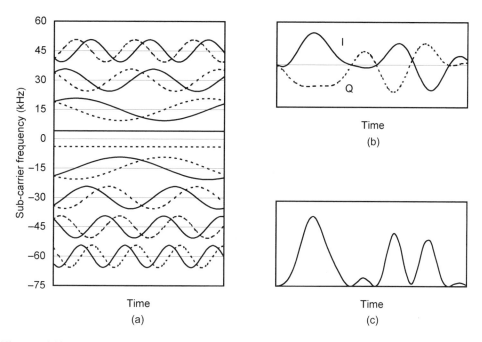

Figure 4.11 Example OFDMA waveform. (a) Amplitudes of the individual sub-carriers. (b) Amplitude of the resulting OFDMA waveform. (c) Power of the OFDMA waveform.

Any distortion of the time-domain waveform will distort the frequency-domain power spectrum as well, so the signal will leak into adjacent frequency bands and will cause interference to other receivers.

In the downlink, the base station transmitters are large, expensive devices, so they can avoid the problem by using expensive power amplifiers that are very close to linear. In the uplink, a mobile transmitter has to be cheap, so does not have this option. This makes OFDMA unsuitable for the LTE uplink.

4.3.2 Block Diagram of SC-FDMA

The power variations described above arise because there is a one-to-one mapping between symbols and sub-carriers. If we mixed the symbols together before placing them on the sub-carriers, then we might be able to adjust the transmitted signal and reduce its power variations. For example, when transmitting two symbols x_1 and x_2 on two sub-carriers, we might send their sum $x_1 + x_2$ on one sub-carrier, and their difference $x_1 - x_2$ on the other. We can use any mixing operation at all, as the receiver can reverse it: we just need to find one that minimizes the power variations in the transmitted signal.

It turns out that a suitable mixing operation is another FFT, this time a forward FFT. By including this operation, we arrive at a technique known as *single carrier frequency division multiple access* (SC-FDMA), which is illustrated in Figure 4.12.

In this diagram, there are three differences from OFDMA. The main difference is that the SC-FDMA transmitter includes an extra forward FFT, between the steps of serial-to-parallel conversion and resource element mapping. This mixes the symbols together in

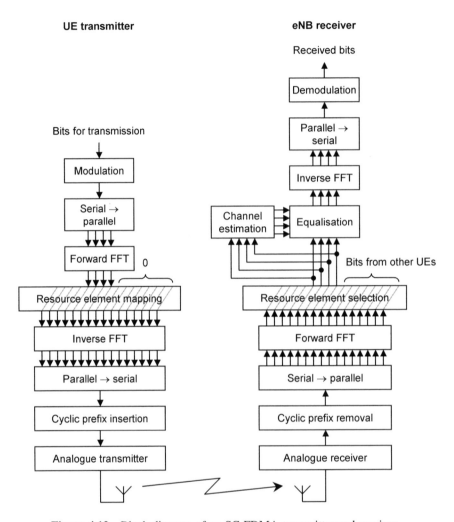

Figure 4.12 Block diagram of an SC-FDMA transmitter and receiver.

the manner required to minimize the power variations and is reversed by an inverse FFT in the receiver.

The second difference arises because the technique is used on the uplink. Because of this, the mobile transmitter only uses some of the sub-carriers: the others are set to zero, and are available for the other mobiles in the cell. Finally, each mobile transmits using a single, contiguous block of sub-carriers, without any internal gaps. This is implied by the name SC-FDMA and is necessary to keep the power variations to the lowest possible level.

We can understand how SC-FDMA works by looking at three key transmission steps: the forward FFT, the resource element mapper and the inverse FFT. The input to the forward FFT is a sequence of symbols in the time domain. The forward FFT converts these symbols to the frequency domain, the resource element mapper shifts them to the desired centre frequency and the inverse FFT converts them back to the time domain. Looking at

OFDMA

these steps as a whole, we can see that the transmitted signal should be much the same as the original modulated waveform, except for a shift to another centre frequency. But the power of a QPSK signal is constant (at least in the absence of additional filtering), and it hardly varies at all in the cases of 16-QAM and 64-QAM. We have therefore achieved the result we require, of transmitting a signal with a roughly constant power.

Figure 4.13 shows the resulting waveforms, from an example in which the mobile is using four sub-carriers from a total of 256. The input (Figure 4.13a) is a sequence of four

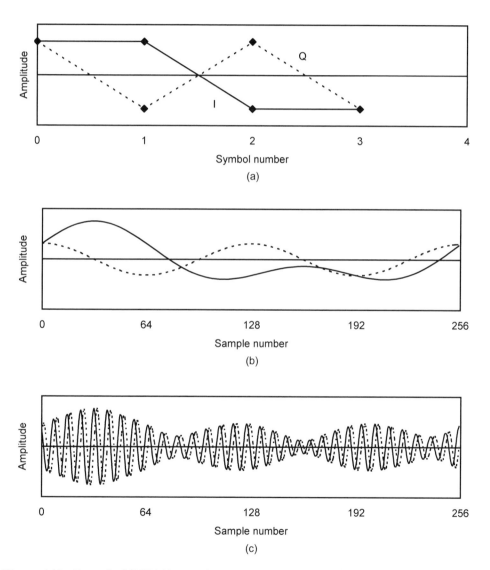

Figure 4.13 Example SC-FDMA waveform. (a) Transmitted symbols. (b) Resulting SC-FDMA waveform, if the data are transmitted on the central 4 sub-carriers out of 256. (c) SC-FDMA waveform, if the data are shifted by 32 sub-carriers.

QPSK symbols, with [I, Q] values of [1, 1], [1, −1], [−1, 1] and [−1, −1]. If the data are transmitted on the central four sub-carriers, then the result (Figure 4.13b) looks very like the original QPSK waveform. The only difference is a smooth interpolation between the 256 samples in the time domain, which wraps round the ends of the data sequence due to the cyclic nature of the FFT. If we instead shift the data by 32 sub-carriers, then the only change (Figure 4.13c) is the introduction of some extra phase rotation into the resulting waveform.

We don't use SC-FDMA in the downlink, because the base station has to transmit to several mobiles, not just one. We could add one forward FFT per mobile to Figure 4.7, but that would destroy the single carrier nature of the transmission, and would allow the high power variations to return. Alternatively, we could add a single forward FFT across the whole of the downlink band. Unfortunately that would spread every mobile's data across the whole of the frequency domain, and would remove our ability to carry out frequency-dependent scheduling. Either way, SC-FDMA is unsuitable for the LTE downlink.

References

1. Smith, S. W. (1998) *The Scientist and Engineer's Guide to Digital Signal Processing*, California Technical Publishing.
2. Lyons, R. G. (2010) *Understanding Digital Signal Processing*, 3rd edn, Prentice Hall.

5

Multiple Antenna Techniques

From the beginning, LTE was designed so that the base station and mobile could both use multiple antennas for radio transmission and reception. This chapter covers the three main multiple antenna techniques, which have different objectives and which are implemented in different ways.

The most familiar is diversity processing, which increases the received signal power and reduces the amount of fading by using multiple antennas at the transmitter, the receiver or both. Diversity processing has been used since the early days of mobile communications, so we will only review it briefly.

In spatial multiplexing, the transmitter and receiver both use multiple antennas so as to increase the data rate. Spatial multiplexing is a relatively new technique that has only recently been introduced into mobile communications, so we will cover it in more detail than the others. It also relies rather heavily on the underlying maths, so our treatment of spatial multiplexing will, from necessity, be more mathematical than that of the other topics in this book. Finally, beamforming uses multiple antennas at the base station in order to increase the coverage of the cell.

Spatial multiplexing is often described as the use of *multiple input multiple output* (MIMO) antennas. This name is derived from the inputs and outputs to the air interface, so that 'multiple input' refers to the transmitter and 'multiple output' to the receiver. Unfortunately the name is a little ambiguous, as it can either refer to spatial multiplexing alone, or include the use of transmit and receive diversity as well. For this reason, we will generally use the term 'spatial multiplexing' instead. For some reviews of multiple antenna techniques and their use in LTE, see References [1–4].

5.1 Diversity Processing

5.1.1 Receive Diversity

Receive diversity is most often used in the uplink, in the manner shown in Figure 5.1. Here, the base station uses two antennas to pick up two copies of the received signal. The signals reach the receive antennas with different phase shifts, but these can be removed

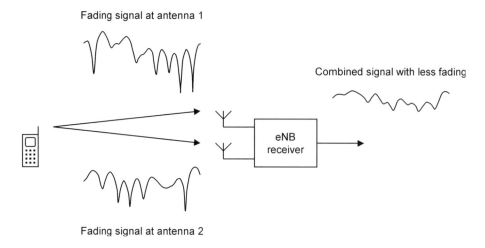

Figure 5.1 Reduction in fading by the use of a diversity receiver.

by antenna-specific channel estimation. The base station can then add the signals together in phase, without any risk of destructive interference between them.

The signals are both made up from several smaller rays, so they are both subject to fading. If the two individual signals undergo fades at the same time, then the power of the combined signal will be low. But if the antennas are far enough apart (a few wavelengths of the carrier frequency), then the two sets of fading geometries will be very different, so the signals will be far more likely to undergo fades at completely different times. We have therefore reduced the amount of fading in the combined signal, which in turn reduces the error rate.

Base stations usually have more than one receive antenna. In LTE, the mobile's test specifications assume that the mobile is using two receive antennas [5], so LTE systems are expected to use receive diversity on the downlink as well as the uplink. A mobile's antennas are closer together than a base station's, which reduces the benefit of receive diversity, but the situation can often be improved using antennas that measure two independent polarizations of the incoming signal.

5.1.2 Closed Loop Transmit Diversity

Transmit diversity reduces the amount of fading by using two or more antennas at the transmitter. It is superficially similar to receive diversity, but with a crucial problem: the signals add together at the single receive antenna, which brings a risk of destructive interference. There are two ways to solve the problem, the first of which is *closed loop transmit diversity* (Figure 5.2).

Here, the transmitter sends two copies of the signal in the expected way, but it also applies a phase shift to one or both signals before transmission. By doing this, it can ensure that the two signals reach the receiver in phase, without any risk of destructive interference. The phase shift is determined by a *precoding matrix indicator* (PMI), which is calculated by the receiver and fed back to the transmitter. A simple PMI might indicate two options: either transmit both signals without any phase shifts, or transmit the second

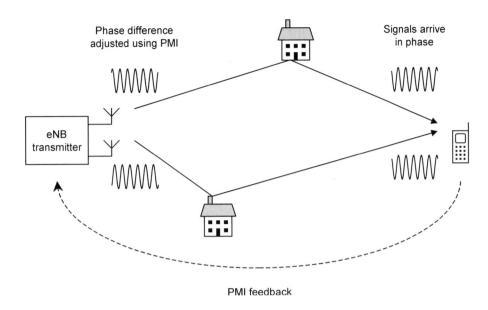

Figure 5.2 Operation of closed loop transmit diversity.

with a phase shift of 180°. If the first option leads to destructive interference, then the second will automatically work. Once again, the amplitude of the combined signal is only low in the unlikely event that the two received signals undergo fades at the same time.

The phase shifts introduced by the radio channel depend on the wavelength of the carrier signal and hence on its frequency. This implies that the best choice of PMI is a function of frequency as well. However, this is easily handled in an OFDMA system, as the receiver can feed back different PMI values for different sets of sub-carriers. The best choice of PMI also depends on the position of the mobile, so a fast moving mobile will have a PMI that frequently changes. Unfortunately the feedback loop introduces time delays into the system, so in the case of fast moving mobiles, the PMI may be out of date by the time it is used. For this reason, closed loop transmit diversity is only suitable for mobiles that are moving sufficiently slowly. For fast moving mobiles, it is better to use the open loop technique described in the next section.

5.1.3 Open Loop Transmit Diversity

Figure 5.3 shows an implementation of *open loop transmit diversity* that is known as *Alamouti's technique* [6]. Here, the transmitter uses two antennas to send two symbols, denoted s_1 and s_2, in two successive time steps. In the first step, the transmitter sends s_1 from the first antenna and s_2 from the second, while in the second step, it sends $-s_2^*$ from the first antenna and s_1^* from the second. (The symbol * indicates that the transmitter should change the sign of the quadrature component, in a process known as complex conjugation.)

The receiver can now make two successive measurements of the received signal, which correspond to two different combinations of s_1 and s_2. It can then solve the resulting

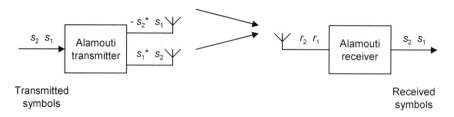

Figure 5.3 Operation of Alamouti's technique for open loop transmit diversity.

equations, so as to recover the two transmitted symbols. There are only two requirements: the fading patterns must stay roughly the same between the first time step and the second, and the two signals must not undergo fades at the same time. Both requirements are usually met.

There is no equivalent to Alamouti's technique for systems with more than two antennas. Despite this, some extra diversity gain can still be achieved in four antenna systems, by swapping back and forth between the two constituent antenna pairs. This technique is used for four antenna open loop diversity in LTE.

We can combine open and closed loop transmit diversity with the receive diversity techniques from earlier, giving a system that carries out diversity processing using multiple antennas at both the transmitter and the receiver. The technique is different from the spatial multiplexing techniques that we will describe next, although, as we will see, a spatial multiplexing system can fall back to diversity transmission and reception if the conditions require.

5.2 Spatial Multiplexing

5.2.1 Principles of Operation

Spatial multiplexing has a different purpose from diversity processing. If the transmitter and receiver both have multiple antennas, then we can set up multiple parallel data streams between them, so as to increase the data rate. In a system with N_T transmit and N_R receive antennas, often known as an $N_T \times N_R$ spatial multiplexing system, the peak data rate is proportional to $\min(N_T, N_R)$.

Figure 5.4 shows a basic spatial multiplexing system, in which the transmitter and receiver both have two antennas. In the transmitter, the antenna mapper takes symbols from the modulator two at a time, and sends one symbol to each antenna. The antennas transmit the two symbols simultaneously, so as to double the transmitted data rate.

The symbols travel to the receive antennas by way of four separate radio paths, so the received signals can be written as follows:

$$y_1 = H_{11}x_1 + H_{12}x_2 + n_1 \\ y_2 = H_{21}x_1 + H_{22}x_2 + n_2 \quad (5.1)$$

Here, x_1 and x_2 are the signals sent from the two transmit antennas, y_1 and y_2 are the signals that arrive at the two receive antennas, and n_1 and n_2 represent the received noise and interference. H_{ij} expresses the way in which the transmitted symbols are attenuated

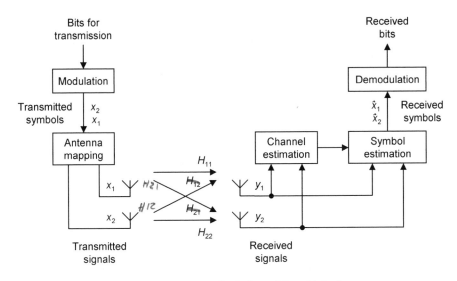

Figure 5.4 Basic principles of a 2x2 spatial multiplexing system.

and phase-shifted, as they travel to receive antenna i from transmit antenna j. (The subscripts i and j may look the wrong way round, but this is for consistency with the usual mathematical notation for matrices.)

In general, all the terms in the equation above are complex. In the transmitted and received symbols x_j and y_i and the noise terms n_i, the real and imaginary parts are the amplitudes of the in-phase and quadrature components. Similarly, in each of the channel elements H_{ij}, the magnitude represents the attenuation of the radio signal, while the phase represents the phase shift. However, the use of complex numbers would make the examples unnecessarily complicated without adding much extra information, so we will simplify the examples by using real numbers alone. To do this, we will assume that the transmitter is modulating the bits using binary phase shift keying, so that the in-phase components are $+1$ and -1, and the quadrature components are zero. We will also assume that the radio channel can attenuate or invert the signal, but does not introduce any other phase shifts.

Consistent with these assumptions, let us consider the following example:

$$\begin{aligned} H_{11} &= 0.8 & H_{12} &= 0.6 & x_1 &= +1 & n_1 &= +0.02 \\ H_{21} &= 0.2 & H_{22} &= 0.4 & x_2 &= -1 & n_2 &= -0.02 \end{aligned} \qquad (5.2)$$

Substituting these numbers into Equation (5.1) shows that the received signals are as follows:

$$\begin{aligned} y_1 &= +0.22 \\ y_2 &= -0.22 \end{aligned} \qquad (5.3)$$

The receiver's first task is to estimate the four channel elements H_{ij}. To help it do this, the transmitter broadcasts reference symbols that follow the basic technique described in Chapter 3, but with one extra feature: when one antenna transmits a reference symbol, the other antenna keeps quiet and sends nothing at all. The receiver can then estimate the channel elements H_{11} and H_{21}, by measuring the two received signals at the times when transmit antenna 1 is sending a reference symbol. It can then wait until transmit antenna 2 sends a reference symbol, before estimating the channel elements H_{12} and H_{22}.

The receiver now has enough information to estimate the transmitted symbols x_1 and x_2. There are several ways for it to do this, but the simplest is a *zero-forcing detector*, which operates as follows. If we ignore the noise and interference, then Equation (5.1) is a pair of simultaneous equations for two unknown quantities, x_1 and x_2. These equations can be inverted as follows:

$$\hat{x}_1 = \frac{\hat{H}_{22} y_1 - \hat{H}_{12} y_2}{\hat{H}_{11} \hat{H}_{22} - \hat{H}_{21} \hat{H}_{12}}$$
$$\hat{x}_2 = \frac{\hat{H}_{11} y_2 - \hat{H}_{21} y_1}{\hat{H}_{11} \hat{H}_{22} - \hat{H}_{21} \hat{H}_{12}}$$
(5.4)

Here, $\hat{H}_{ij}$ is the receiver's estimate of the channel element H_{ij}. (This quantity may be different from H_{ij}, because of noise and other errors in the channel estimation process.) Similarly, $\hat{x}_1$ and $\hat{x}_2$ are the receiver's estimates of the transmitted symbols x_1 and x_2. Substituting the numbers from Equations (5.2) and (5.3) gives the following result:

$$\hat{x}_1 = +1.1$$
$$\hat{x}_2 = -1.1$$
(5.5)

This is consistent with transmitted symbols of +1 and −1. We have therefore transferred two symbols at the same time using the same sub-carriers, and have doubled the data rate.

5.2.2 Open Loop Spatial Multiplexing

There is a problem with the technique described above. To illustrate this, let us change one of the channel elements, H_{11}, to give the following example:

$$H_{11} = 0.3 \quad H_{12} = 0.6$$
$$H_{21} = 0.2 \quad H_{22} = 0.4$$
(5.6)

If we try to estimate the transmitted symbols using Equation (5.4), we find that $H_{11}H_{22} - H_{21}H_{12}$ is zero. We therefore end up dividing by zero, which is nonsense. So, for this choice of channel elements, the technique has failed. We can see what has gone wrong by substituting the channel elements into Equation (5.1), and writing the received signals as follows:

$$y_1 = 0.3(x_1 + 2x_2) + n_1$$
$$y_2 = 0.2(x_1 + 2x_2) + n_2$$
(5.7)

By measuring the received signals y_1 and y_2, we were expecting to measure two different pieces of information, from which we could recover the transmitted data. This time, however, we have measured the same piece of information, namely $x_1 + 2x_2$, twice. As a result, we do not have enough information to recover x_1 and x_2 independently. Furthermore, this is not just an isolated special case. If $H_{11}H_{22} - H_{21}H_{12}$ is small but non-zero, then our estimates of x_1 and x_2 turn out to be badly corrupted by noise and are completely unusable.

Multiple Antenna Techniques

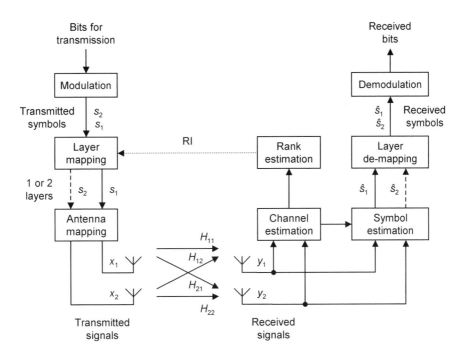

Figure 5.5 Operation of a 2x2 open loop spatial multiplexing system.

The solution comes from the knowledge that we can still send one symbol at a time, by the use of diversity processing. We therefore require an adaptive system, which can use spatial multiplexing to send two symbols at a time if the channel elements are well behaved and can fall back to diversity processing otherwise. Such a system is shown in Figure 5.5. Here, the receiver measures the channel elements and works out a *rank indication* (RI), which indicates the number of symbols that it can successfully receive. It then feeds the rank indication back to the transmitter.

If the rank indication is two, then the system operates in the same way that we described earlier. The transmitter's *layer mapper* grabs two symbols, s_1 and s_2, from the transmit buffer, so as to create two independent data streams that are known as *layers*. The *antenna mapper* then sends one symbol to each antenna, by a straightforward mapping operation:

$$x_1 = s_1 \\ x_2 = s_2 \quad (5.8)$$

The receiver measures the incoming signals and recovers the transmitted symbols as before.

If the rank indication is one, then the layer mapper only grabs one symbol, s_1, which the antenna mapper sends to both transmit antennas as follows:

$$x_1 = s_1 \\ x_2 = s_1 \quad (5.9)$$

Under these assumptions, Equation (5.7) becomes the following:

$$y_1 = 0.9s_1 + n_1$$
$$y_2 = 0.6s_1 + n_2 \qquad (5.10)$$

The receiver now has two measurements of the transmitted symbol s_1, and can combine these in a diversity receiver so as to recover the transmitted data.

The effect is as follows. If the channel elements are well behaved, then the transmitter sends two symbols at a time and the receiver recovers them using a spatial multiplexing receiver. Sometimes this is not possible, in which case the transmitter falls back to sending one symbol at a time and the receiver falls back to diversity reception. This technique is implemented in LTE and, for reasons that will become clear in the next section, is known as *open loop spatial multiplexing*.

5.2.3 Closed Loop Spatial Multiplexing

There is one remaining problem. To illustrate this, let us change two more of the channel elements, so that:

$$H_{11} = 0.3 \quad H_{12} = -0.3$$
$$H_{21} = 0.2 \quad H_{22} = -0.2 \qquad (5.11)$$

These channel elements are badly behaved, in that $H_{11}H_{22} - H_{21}H_{12}$ is zero. But if we try to handle the situation in the manner described above, by sending the same symbol from both transmit antennas, then the received signals are as follows:

$$y_1 = 0.3s_1 - 0.3s_1 + n_1$$
$$y_2 = 0.2s_1 - 0.2s_1 + n_2 \qquad (5.12)$$

So the transmitted signals cancel out at both receive antennas and we are left with measurements of the incoming noise and interference. We therefore have insufficient information even to recover s_1.

To see the way out, consider what happens if we send one symbol at a time as before, but invert the signal that is sent from the second antenna:

$$x_1 = s_1$$
$$x_2 = -s_1 \qquad (5.13)$$

The received signal can now be written as follows:

$$y_1 = 0.3s_1 + 0.3s_1 + n_1$$
$$y_2 = 0.2s_1 + 0.2s_1 + n_2 \qquad (5.14)$$

This time, we can recover the transmitted symbol s_1.

So we now require two levels of adaptation. If the rank indication is two, then the transmitter sends two symbols at a time using the antenna mapping of Equation (5.8). If the rank indication is one, then the transmitter falls back to diversity processing and sends one symbol at a time. In doing so, it chooses an antenna mapping such as Equation (5.9) or (5.13), which depends on the exact nature of the channel elements and which guarantees a strong signal at the receiver.

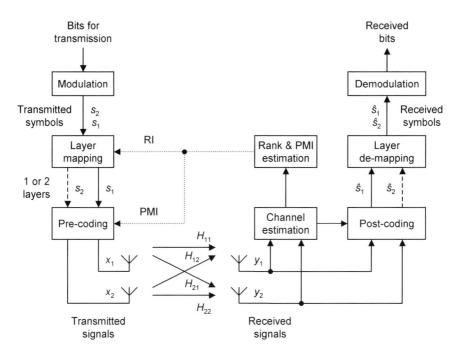

Figure 5.6 Operation of a 2x2 closed loop spatial multiplexing system.

Such a system is shown in Figure 5.6. Here, the receiver measures the channel elements as before and uses them to feed back two quantities, namely the rank indication and a precoding matrix indicator (PMI). The PMI controls a *precoding* step in the transmitter, which implements an adaptive antenna mapping using (for example) Equations (5.8), (5.9) and (5.13), to ensure that the signals reach the receiver without cancellation. (In fact the PMI has exactly the same role that we saw earlier when discussing closed loop transmit diversity, which is why its name is the same.) In the receiver, the *post-coding* step reverses the effect of precoding and also includes the soft decision estimation step from earlier.

This technique is also implemented in LTE, and is known as *closed loop spatial multiplexing*. In this expression, the term 'closed loop' refers specifically to the loop that is created by feeding back the PMI. The technique from Section 5.2.2 is known as 'open loop spatial multiplexing', even though the receiver is still feeding back a rank indication.

5.2.4 *Matrix Representation*

We have now covered the basic principles of spatial multiplexing. To go further, we need a more mathematical description in terms of matrices. Readers who are unfamiliar with matrices may prefer to skip this section and to resume the discussion in Section 5.2.5 below.

In matrix notation, we can write the received signal (Equation 5.1) as follows:

$$y = H.x + n \qquad (5.15)$$

Here, x is a column vector that contains the signals that are sent from the N_T transmit antennas. Similarly, n and y are column vectors containing the noise and the resulting signals at the N_R receive antennas. The *channel matrix* H has N_R rows and N_T columns, and expresses the amplitude changes and phase shifts that the air interface introduced. In the examples we considered earlier, the system had two transmit and two receive antennas, so the matrix equation above could be written as follows:

$$\begin{bmatrix} y_1 \\ y_2 \end{bmatrix} = \begin{bmatrix} H_{11} & H_{12} \\ H_{21} & H_{22} \end{bmatrix} \cdot \begin{bmatrix} x_1 \\ x_2 \end{bmatrix} + \begin{bmatrix} n_1 \\ n_2 \end{bmatrix} \qquad (5.16)$$

Now let us assume that the numbers of transmit and receive antennas are equal, so that $N_R = N_T = N$, and let us ignore the noise and interference as before. We can then invert the channel matrix and derive the following estimate of the transmitted symbols:

$$\hat{x} = \hat{H}^{-1} \cdot y \qquad (5.17)$$

Here, $\hat{H}^{-1}$ is the receiver's estimate of the inverse of the channel matrix, while $\hat{x}$ is its estimate of the transmitted signal. This is the zero-forcing detector from earlier. The detector runs into problems if the noise and interference are too great, but, in these circumstances, a *minimum mean square error* (MMSE) detector gives a more accurate answer.

If the channel matrix is well behaved, then we can measure the signals that arrive at the N receive antennas and use a suitable detector to estimate the symbols that were transmitted. As a result, we can increase the data rate by a factor N. The channel matrix may, however, be *singular* (as in Equations 5.6 and 5.11), in which case its inverse does not exist. Alternatively, the matrix may be *ill conditioned*, in which case its inverse is corrupted by noise. Either way, we need to find another solution.

The solution comes from writing the channel matrix H as follows:

$$H = P^{-1} \cdot \Lambda \cdot P \qquad (5.18)$$

Here, P is a matrix formed from the *eigenvectors* of H, while Λ is a diagonal matrix whose elements are the *eigenvalues* of H. In the two antenna example, the diagonal matrix is:

$$\Lambda = \begin{bmatrix} \lambda_1 & 0 \\ 0 & \lambda_2 \end{bmatrix} \qquad (5.19)$$

where the eigenvalues are λ_1 and λ_2.

Now let us transmit the symbols in the manner shown in Figure 5.7. At the output from the post-coding stage, the received symbol vector is:

$$r = G \cdot H \cdot F \cdot s + G \cdot n \qquad (5.20)$$

where s contains the transmitted symbols at the input to the precoding stage, F is the precoding matrix, H is the usual channel matrix, and G is the post-coding matrix. If we now choose the pre- and post-coding matrices so that they are good approximations to the matrices of eigenvectors:

$$\begin{aligned} F &\approx P^{-1} \\ G &\approx P \end{aligned} \qquad (5.21)$$

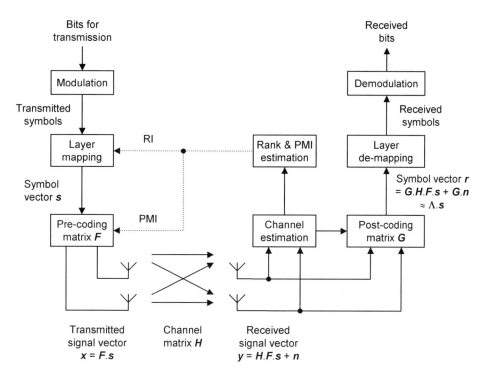

Figure 5.7 Operation of a spatial multiplexing system with an arbitrary number of antennas.

then the received symbol vector becomes the following:

$$r \approx P.H.P^{-1}.s + P.n$$
$$\approx \Lambda.s + P.n \quad (5.22)$$

Ignoring the noise, we can now write the received symbols in a two antenna spatial multiplexing system as follows:

$$\begin{bmatrix} r_1 \\ r_2 \end{bmatrix} \approx \begin{bmatrix} \lambda_1 & 0 \\ 0 & \lambda_2 \end{bmatrix} . \begin{bmatrix} s_1 \\ s_2 \end{bmatrix} \quad (5.23)$$

We therefore have two independent data streams, without any coupling between them. It is now trivial for the receiver to recover the transmitted symbols, as follows:

$$\hat{s}_i = \frac{r_i}{\lambda_i} \quad (5.24)$$

So, by a suitable choice of pre- and post-coding matrices, F and G, we can greatly simplify the design of the receiver.

If the channel matrix H is singular, then some of its eigenvalues λ_i are zero. If it is ill-conditioned, then some of the eigenvalues are very small, so that the reconstructed symbols are badly corrupted by noise. The *rank* of H is the number of usable eigenvalues and the rank indication from Section 5.2.2 equals the rank of H. In a two antenna system with a rank of 1, for example, the received symbol vector is as follows:

$$\begin{bmatrix} r_1 \\ r_2 \end{bmatrix} \approx \begin{bmatrix} \lambda_1 & 0 \\ 0 & 0 \end{bmatrix} \cdot \begin{bmatrix} s_1 \\ s_2 \end{bmatrix} \qquad (5.25)$$

The system can exploit this behaviour in the following way. The receiver estimates the channel matrix and feeds back the rank indication along with the precoding matrix F. If the rank indication is two, then the transmitter sends two symbols, s_1 and s_2, and the receiver reconstructs them from Equation (5.23). If the rank indication is one, then the transmitter just sends one symbol, s_1, and doesn't bother with s_2 at all. The receiver can then reconstruct the transmitted symbol from Equation (5.25).

In practice, the receiver does not pass a full description of F back to the transmitter, as that would require too much feedback. Instead, it selects the closest approximation to P^{-1} from a *codebook* and indicates its choice using the precoding matrix indicator, PMI.

Inspection of Equations (5.22) and (5.23) shows that the received symbols r_1 and r_2 can have different signal-to-noise ratios, which depend on the corresponding eigenvalues λ_1 and λ_2. In LTE, the transmitter can exploit this by sending the two symbols with different modulation schemes and coding rates, and also with different transmit powers.

We can also use Equation (5.15) to describe a system in which the numbers of transmit and receive antennas are different. The eigenvalue technique only works for square matrices, but it can be generalized to a technique known as *singular value decomposition* [7] that works for rectangular matrices as well. The maximum data rate is proportional to $\min(N_T, N_R)$, with any extra antennas providing additional transmit or receive diversity.

5.2.5 Implementation Issues

Spatial multiplexing is implemented in the downlink of LTE Release 8, using a maximum of four transmit antennas on the base station and four receive antennas on the mobile. There are similar implementation issues to diversity processing. Firstly, the antennas at the base station and mobile should be reasonably far apart, ideally a few wavelengths of the carrier frequency, or should handle different polarizations. If the antennas are too close together, then the channel elements H_{ij} will be very similar. This can easily take us into the situation from Section 5.2.2, where spatial multiplexing was unusable and we had to fall back to diversity processing.

A similar situation can easily arise in the case of line-of-sight transmission and reception. This leads us to an unexpected conclusion: spatial multiplexing actually works best in conditions with no direct line-of-sight and significant multipath, because, in these conditions, the channel elements H_{ij} are uncorrelated with each other. In line-of-sight conditions, we often have to fall back to diversity processing.

As in the case of closed loop transmit diversity, the PMI depends on the carrier frequency and the position of the mobile. For fast moving mobiles, delays in the feedback loop can make the PMI unreliable by the time the transmitter comes to use it, so open loop spatial multiplexing is often preferred.

5.2.6 Multiple User MIMO

Figure 5.8 shows a slightly different technique. Here, two transmit and two receive antennas are sharing the same transmission times and frequencies, in the same way as before.

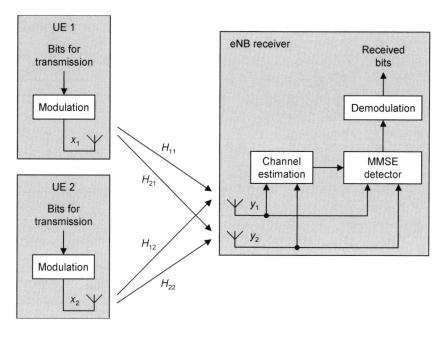

Figure 5.8 Uplink multiple user MIMO.

This time, however, the mobile antennas are on two different mobiles instead of one. This technique is known as *multiple user MIMO* (MU-MIMO), in contrast with the earlier spatial multiplexing techniques, which are sometimes known as *single user MIMO* (SU-MIMO).

Figure 5.8 specifically shows the implementation of multiple user MIMO on the uplink, which is the more common situation. Here, the mobiles transmit at the same time and on the same carrier frequency, but without using any precoding and without even knowing that they are part of a spatial multiplexing system. The base station receives their transmissions and separates them using (for example) the minimum mean square error detector that we noted earlier.

This technique only works if the channel matrix is well behaved, but we can usually guarantee this for two reasons. Firstly, the mobiles are likely to be far apart, so their ray paths are likely to be very different. Secondly, the base station can freely choose the mobiles that are taking part, so it can freely choose mobiles that lead to a well-behaved channel matrix.

Uplink multiple user MIMO does not increase the peak data rate of an individual mobile, but it is still beneficial because of the increase in cell throughput. It can also be implemented using inexpensive mobiles that just have one power amplifier and one transmit antenna, not two. For these reasons, multiple user MIMO is the standard technique in the uplink of LTE Release 8: single user MIMO is not introduced into the uplink until Release 10.

We can also apply multiple user MIMO to the downlink, as shown in Figure 5.9. This time, however, there is a problem. Mobile 1 can measure its received signal y_1 and the channel elements H_{11} and H_{12}, in the same way as before. However, it has no knowledge

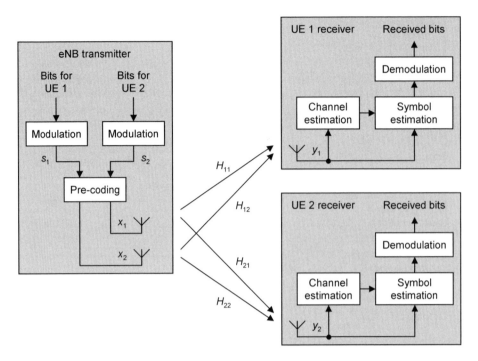

Figure 5.9 Downlink multiple user MIMO.

of the other received signal y_2, or of the other channel elements H_{21} and H_{22}. The opposite situation applies for mobile 2. Neither mobile has complete knowledge of the channel elements or of the received signals, which invalidates the techniques we have been using.

The solution is to implement downlink multiple user MIMO by adapting another multiple antenna technique, known as beamforming. We will cover beamforming in the next section and then return to downlink multiple user MIMO at the end of the chapter.

5.3 Beamforming

5.3.1 Principles of Operation

In *beamforming*, a base station uses multiple antennas in a completely different way, to increase its coverage. The principles are shown in Figure 5.10. Here, mobile 1 is a long way from the base station, on a line of sight that is at right angles to the antenna array. The signals from each antenna reach mobile 1 in phase, so they interfere constructively, and the received signal power is high. On the other hand, mobile 2 is at an oblique angle, and receives signals from alternate antennas that are 180° out of phase. These signals interfere destructively, so the received signal power is low. We have therefore created a synthetic antenna beam, which has a main beam pointing towards mobile 1 and a null pointing towards mobile 2. The beamwidth is narrower than one from a single antenna, so the transmitted power is focussed towards mobile 1. As a result, the range of the base station in the direction of mobile 1 is greater than before.

Multiple Antenna Techniques

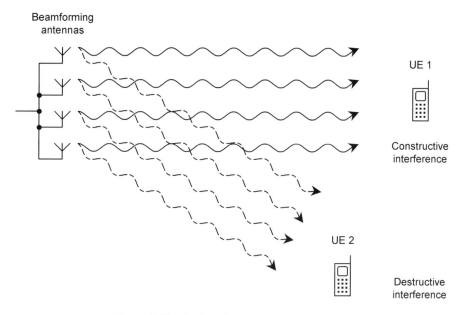

Figure 5.10 Basic principles of beamforming.

As shown in Figure 5.11, we can go a step further. By applying a phase ramp to the transmitted signals, we can change the direction at which constructive interference arises, so we can direct the beam towards any direction we choose. More generally, we can adjust the amplitudes and phases of the transmitted signals, by applying a suitable set of antenna weights. In a system with N antennas, this allows us to adjust the direction of the main beam and up to $N - 2$ nulls or sidelobes.

We can use the same technique to construct a synthetic reception beam for the uplink. By applying a suitable set of antenna weights at the base station receiver, we can ensure that the received signals add together in phase and interfere constructively. As a result, we can increase the range in the uplink as well.

In OFDMA, we can process different sub-carriers using different sets of antenna weights, so as to create synthetic antenna beams that point in different directions. We can therefore use beamforming to communicate with several different mobiles at once using different sub-carriers, even if those mobiles are in completely different locations.

Beamforming works best if the antennas are close together, with a separation comparable with the wavelength of the radio waves. This ensures that the signals sent or received by those antennas are highly correlated. This is a different situation from diversity processing or spatial multiplexing, which work best if the antennas are far apart, with uncorrelated signals. A base station is therefore likely to use two sets of antennas: a closely spaced set for beamforming and a widely spaced set for diversity and spatial multiplexing.

5.3.2 Beam Steering

We have not yet considered the question of how to calculate the antenna weights and steer the beam. How is this done?

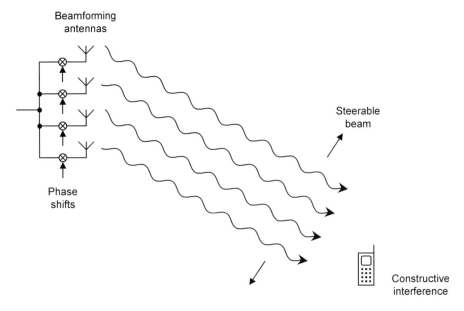

Figure 5.11 Beam steering using a set of phase shifts.

For the reception beams on the uplink, there are two main techniques [8, 9]. Using the *reference signal technique*, the base station adjusts the antenna weights so as to reconstruct the mobile's reference symbols with the correct signal phase and the greatest possible signal to interference plus noise ratio (SINR). An alternative is the *direction of arrival technique*, in which the base station measures the signals that are received by each antenna and estimates the direction of the target mobile. From this quantity, it can estimate the antenna weights that are needed for satisfactory reception.

For the transmission beams on the downlink, the answer depends on the base station's mode of operation. In TDD mode, the uplink and the downlink use the same carrier frequency, so the base station can use the same antenna weights on the downlink that it calculated for the uplink. In FDD mode, the carrier frequencies are different, so the downlink antenna weights are different and are harder to estimate. For this reason, beamforming is more common in systems that are using TDD rather than FDD.

5.3.3 Dual Layer Beamforming

Dual layer beamforming (Figure 5.12) takes the idea a step further. In this technique, the base station sends two different data streams into its antenna array, instead of just one. It then processes the data using two different sets of antenna weights and adds the results together before transmission. In doing so, it has created two separate antenna beams, which share the same sub-carriers but carry two different sets of information. The base station can then adjust the antenna weights so as to steer the beams to two different mobiles, so that the first mobile receives constructive interference from beam 1 and destructive interference from beam 2 and vice-versa. By doing this, the base station

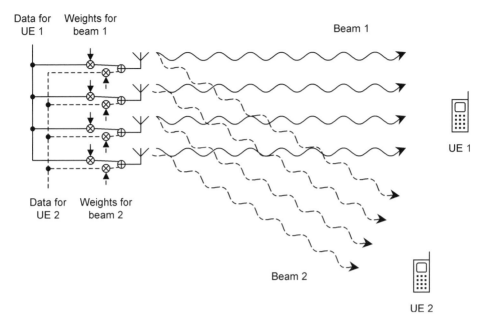

Figure 5.12 Dual layer beamforming using two parallel sets of antenna weights.

can double the capacity of the cell. Alternatively, the base station can steer the beams to two different antennas on a single mobile, so as to double that mobile's instantaneous data rate.

In ideal conditions, the maximum number of independent data streams is equal to the number of antennas in the array. LTE first supports the technique in Release 9 of the 3GPP specifications. In that release, the maximum number of data streams is limited to two, leading to the name of dual layer beamforming.

5.3.4 Downlink Multiple User MIMO Revisited

At the end of Section 5.2.6, we tried to implement downlink multiple user MIMO using the same techniques that we had previously used for spatial multiplexing. We discovered that the mobiles did not have enough information to recover the transmitted symbols, so that the previous techniques were inappropriate.

Referring back to Figure 5.9, the only reliable solution is to precode the transmitted symbols s_1 and s_2, so that s_1 is subject to constructive interference at mobile 1 and destructive interference at mobile 2, with the opposite situation applying for s_2. But that is exactly the same interpretation that we have just used for dual layer beamforming. This implies that downlink MU-MIMO is best treated as a variety of beamforming, using base station antennas that are close together rather than far apart.

The difference between downlink multiple user MIMO and dual layer beamforming lies in the calculation of the antenna weights. In multiple user MIMO, each mobile feeds back a precoding matrix from which the base station determines the antenna weights that it requires. There is no such feedback in dual layer beamforming: instead, the base station calculates the downlink antenna weights from its measurements of the mobile's uplink transmissions.

LTE first supports this implementation of downlink multiple user MIMO in Release 10 of the 3GPP specifications. There is, however, limited support for downlink multiple user MIMO in Release 8 as well. The Release 8 implementation uses the same algorithms that single user MIMO does, so it only works effectively if the codebook happens to contain a pre-coding matrix that satisfies the conditions described above. Often it does not, so the performance of downlink multiple user MIMO in Release 8 is comparatively poor.

References

1. Biglieri, E., Calderbank, R., Constantinides, A., Goldsmith, A., Paulraj, A. and Poor, H. V. (2010) *MIMO Wireless Communications*, Cambridge University Press.
2. 4G Americas (June 2009) *MIMO Transmission Schemes for LTE and HSPA Networks*.
3. 4G Americas (May 2010) *MIMO and Smart Antennas for 3G and 4G Wireless Systems: Practical Aspects and Deployment Considerations*.
4. Lee, J., Han, J. K. and Zhang, J. (2009) MIMO technologies in 3GPP LTE and LTE-Advanced, *EURASIP Journal on Wireless Communications and Networking*, **2009**, article ID 302092.
5. 3GPP TS 36.101 (October 2011) *User Equipment (UE) Radio Transmission and Reception*, Release 10, section 7.2.
6. Alamouti, S. (1998) Space block coding: A simple transmitter diversity technique for wireless communications, *IEEE Journal on Selected Areas in Communications*, **16**, 1451–1458.
7. Press, W. H., Teukolsky, S. A., Vetterling, W. T. and Flannery, B. P. (2007) *Numerical Recipes*, Section 2.6, 3rd edn. Cambridge University Press.
8. Godara, L. C. (1997) Applications of antenna arrays to mobile communications, part I: Performance improvement, feasibility, and system considerations, *Proceedings of the IEEE*, **85**, 1031–1060.
9. Godara, L. C. (1997) Application of antenna arrays to mobile communications, part II: Beam-forming and direction-of-arrival considerations, *Proceedings of the IEEE*, **85**, 1195–1245.

6

Architecture of the LTE Air Interface

Now that we have covered the principles of the air interface, we can explain how those principles are actually implemented in LTE. This task is the focus of the next five chapters.

In this chapter, we will cover the air interface's high-level architecture. We begin by reviewing the air interface protocol stack, and by listing the channels and signals that carry information between the different protocols. We then describe how the OFDMA and SC-FDMA air interfaces are organized as a function of time and frequency in a resource grid and discuss how LTE implements transmissions from multiple antennas using multiple copies of the grid. Finally, we bring the preceding material together by illustrating how the channels and signals are mapped onto the resource grids that are used in the uplink and downlink.

6.1 Air Interface Protocol Stack

Figure 6.1 reviews the protocols that are used in the air interface, from the viewpoint of the mobile. As well as the information presented in Chapter 2, the figure adds some detail to the physical layer and shows the information flows between the different levels of the protocol stack.

Let us consider the transmitter. In the user plane, the application creates data packets that are processed by protocols such as TCP, UDP and IP, while in the control plane, the radio resource control (RRC) protocol [1] writes the signalling messages that are exchanged between the base station and the mobile. In both cases, the information is processed by the packet data convergence protocol (PDCP) [2], the radio link control (RLC) protocol [3] and the medium access control (MAC) protocol [4], before being passed to the physical layer for transmission.

The physical layer has three parts. The *transport channel processor* [5] applies the error management procedures that we covered in Section 3.3, while the *physical channel processor* [6] applies the techniques of OFDMA, SC-FDMA and multiple antenna transmission from Chapters 4 and 5. Finally, the *analogue processor* [7, 8] converts the

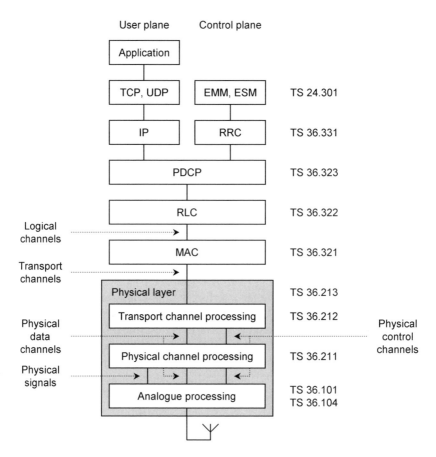

Figure 6.1 Architecture of the air interface protocol stack.

information to analogue form, filters it and mixes it up to radio frequency for transmission. A separate specification [9] describes the procedures that straddle the individual parts of the physical layer.

The information flows between the different protocols are known as channels and signals. Data and signalling messages are carried on *logical channels* between the RLC and MAC protocols, *transport channels* between the MAC and the physical layer, and *physical data channels* between the different levels of the physical layer. LTE uses several different types of logical, transport and physical channel, which are distinguished by the kind of information they carry and by the way in which the information is processed.

In the transmitter, the transport channel processor also creates *control information* that supports the low-level operation of the physical layer and sends this information to the physical channel processor in the form of *physical control channels*. The information travels as far as the transport channel processor in the receiver, but is completely invisible to higher layers. Similarly, the physical channel processor creates *physical signals*, which support the lowest-level aspects of the system. These travel as far as the physical channel processor in the receiver, but once again are invisible to higher layers.

Table 6.1 Logical channels

Channel	Release	Name	Information carried	Direction
DTCH	R8	Dedicated traffic channel	User plane data	
DCCH	R8	Dedicated control channel	Signalling on SRB 1 & 2	UL, DL
CCCH	R8	Common control channel	Signalling on SRB 0	
PCCH	R8	Paging control channel	Paging messages	
BCCH	R8	Broadcast control channel	System information	DL
MCCH	R9	Multicast control channel	MBMS signalling	
MTCH	R9	Multicast traffic channel	MBMS data	

6.2 Logical, Transport and Physical Channels

6.2.1 Logical Channels

Table 6.1 lists the logical channels that are used by LTE [10]. They are distinguished by the information they carry and can be classified in two ways. Firstly, logical traffic channels carry data in the user plane, while logical control channels carry signalling messages in the control plane. Secondly, dedicated logical channels are allocated to a specific mobile, while common logical channels can be used by more than one.

The most important logical channels are the *dedicated traffic channel* (DTCH), which carries data to or from a single mobile, and the *dedicated control channel* (DCCH), which carries the large majority of signalling messages. To be exact, the dedicated control channel carries all the mobile-specific signalling messages on signalling radio bearers 1 and 2, for mobiles that are in RRC_CONNECTED state.

The *broadcast control channel* (BCCH) carries RRC system information messages, which the base station broadcasts across the whole of the cell to tell the mobiles about how the cell is configured. These messages are divided into two unequal groups, which are handled differently by lower layers. The *master information block* (MIB) carries a few important parameters such as the downlink bandwidth, while several *system information blocks* (SIBs) carry the remainder.

The *paging control channel* (PCCH) carries *paging messages*, which the base station transmits if it wishes to contact mobiles that are in RRC_IDLE. The *common control channel* (CCCH) carries messages on signalling radio bearer 0, for mobiles that are moving from RRC_IDLE to RRC_CONNECTED in the procedure of RRC connection establishment.

Like the other tables in this chapter, Table 6.1 lists the channels that were introduced in every release of LTE. The *multicast traffic channel* (MTCH) and *multicast control channel* (MCCH) first appeared in LTE Release 9, to handle a service known as the *multimedia broadcast/multicast service* (MBMS). We will discuss these channels in Chapter 17.

6.2.2 Transport Channels

The transport channels [11] are listed in Table 6.2. They are distinguished by the ways in which the transport channel processor manipulates them.

Table 6.2 Transport channels

Channel	Release	Name	Information carried	Direction
UL-SCH	R8	Uplink shared channel	Uplink data and signalling	UL
RACH	R8	Random access channel	Random access requests	
DL-SCH	R8	Downlink shared channel	Downlink data and signalling	DL
PCH	R8	Paging channel	Paging messages	
BCH	R8	Broadcast channel	Master information block	
MCH	R8/R9	Multicast channel	MBMS	

The most important transport channels are the *uplink shared channel* (UL-SCH) and the *downlink shared channel* (DL-SCH), which carry the large majority of data and signalling messages across the air interface. The *paging channel* (PCH) carries paging messages that originated from the paging control channel. The *broadcast channel* (BCH) carries the broadcast control channel's master information block: the remaining system information messages are handled by the downlink shared channel, as if they were normal downlink data. The *multicast channel* (MCH) was fully specified in Release 8, to carry data from the multimedia broadcast/multicast service. However, it was not actually usable until the introduction of the actual service in Release 9.

The base station usually schedules the transmissions that a mobile makes, by granting it resources for uplink transmission at specific times and on specific sub-carriers. The *random access channel* (RACH) is a special channel through which the mobile can contact the network without any prior scheduling. Random access transmissions are composed by the mobile's MAC protocol and travel as far as the MAC protocol in the base station, but are completely invisible to higher layers.

The main differences between the transport channels lie in their approaches to error management. In particular, the uplink and downlink shared channels are the only transport channels that use the techniques of automatic repeat request and hybrid ARQ, and are the only channels that can adapt their coding rate to changes in the received signal to interference plus noise ratio (SINR). The other transport channels use forward error correction alone and have a fixed coding rate. The same restrictions apply to the control information that we will discuss below.

6.2.3 Physical Data Channels

Table 6.3 lists the physical data channels [12]. They are distinguished by the ways in which the physical channel processor manipulates them, and by the ways in which they are mapped onto the symbols and sub-carriers used by OFDMA.

The most important physical channels are the *physical downlink shared channel* (PDSCH) and the *physical uplink shared channel* (PUSCH). The PDSCH carries data and signalling messages from the downlink shared channel, as well as paging messages from the paging channel. The PUSCH carries data and signalling messages from the uplink shared channel and can sometimes carry the uplink control information that is described below.

The *physical broadcast channel* (PBCH) carries the master information block from the broadcast channel, while the *physical random access channel* (PRACH) carries random

Architecture of the LTE Air Interface

Table 6.3 Physical data channels

Channel	Release	Name	Information carried	Direction
PUSCH	R8	Physical uplink shared channel	UL-SCH and/or UCI	UL
PRACH	R8	Physical random access channel	RACH	
PDSCH	R8	Physical downlink shared channel	DL-SCH and PCH	
PBCH	R8	Physical broadcast channel	BCH	DL
PMCH	R8/R9	Physical multicast channel	MCH	

access transmissions from the random access channel. The *physical multicast channel* (PMCH) was fully specified in Release 8, to carry data from the multicast channel, but is not usable until Release 9.

The PDSCH and PUSCH are the only physical channels that can adapt their modulation schemes in response to changes in the received SINR. The other physical channels all use a fixed modulation scheme, usually QPSK. At least in LTE Release 8, the PDSCH is the only physical channel that uses the techniques of spatial multiplexing and beamforming from Sections 5.2 and 5.3, or the technique of closed loop transmit diversity from Section 5.1.2. The other channels are sent from a single antenna, or can use open loop transmit diversity in the case of the downlink. Once again, the same restrictions apply to the physical control channels that we will list below.

6.2.4 Control Information

The transport channel processor composes several types of control information, to support the low-level operation of the physical layer. These are listed in Table 6.4.

The *uplink control information* (UCI) contains several fields. *Hybrid ARQ acknowledgements* are the mobile's acknowledgements of the base station's transmissions on the DL-SCH. The *channel quality indicator* (CQI) describes the received SINR as a function of frequency in support of frequency-dependent scheduling, while the precoding matrix indicator (PMI) and rank indication (RI) were introduced in Chapter 5 and support the

Table 6.4 Control information

Field	Release	Name	Information carried	Direction
UCI	R8	Uplink control information	Hybrid ARQ acknowledgements Channel quality indicators (CQI) Pre-coding matrix indicators (PMI) Rank indications (RI) Scheduling requests (SR)	UL
DCI	R8	Downlink control information	Downlink scheduling commands Uplink scheduling grants Uplink power control commands	DL
CFI	R8	Control format indicator	Size of downlink control region	DL
HI	R8	Hybrid ARQ indicator	Hybrid ARQ acknowledgements	

use of spatial multiplexing. Collectively, the channel quality indicator, precoding matrix indicator and rank indication are sometimes known as *channel state information* (CSI), although this term does not actually appear in the specifications until Release 10. Finally, the mobile sends a *scheduling request* (SR) if it wishes to transmit uplink data on the PUSCH, but does not have the resources to do so.

The *downlink control information* (DCI) contains most of the downlink control fields. Using *scheduling commands* and *scheduling grants*, the base station can alert the mobile to forthcoming transmissions on the downlink shared channel and grant it resources for transmissions on the uplink shared channel. It can also adjust the power with which the mobiles are transmitting, by the use of *power control commands*.

The other sets of control information are less important. *Control format indicators* (CFIs) tell the mobiles about the organization of data and control information on the downlink, while *hybrid ARQ indicators* (HIs) are the base station's acknowledgements of the mobiles' uplink transmissions on the UL-SCH.

6.2.5 Physical Control Channels

The physical control channels are listed in Table 6.5.

In the downlink, there is a one-to-one mapping between the physical control channels and the control information listed above. As such, the *physical downlink control channel* (PDCCH), *physical control format indicator channel* (PCFICH) and *physical hybrid ARQ indicator channel* (PHICH) carry the downlink control information, control format indicators and hybrid ARQ indicators respectively. The *relay physical downlink control channel* (R-PDCCH) supports the use of relaying and was first introduced in Release 10.

The uplink control information is sent on the PUSCH if the mobile is transmitting uplink data at the same time and on the *physical uplink control channel* (PUCCH) otherwise. The PUSCH and PUCCH are transmitted on different sets of sub-carriers, so this arrangement preserves the single carrier nature of the uplink transmission, in accordance with the requirements of SC-FDMA.

6.2.6 Physical Signals

The final information streams are the physical signals, which support the lowest-level operation of the physical layer. These are listed in Table 6.6.

In the uplink, the mobile transmits the *demodulation reference signal* (DRS) at the same time as the PUSCH and PUCCH, as a phase reference for use in channel estimation. It

Table 6.5 Physical control channels

Channel	Release	Name	Information carried	Direction
PUCCH	R8	Physical uplink control channel	UCI	UL
PCFICH	R8	Physical control format indicator channel	CFI	
PHICH	R8	Physical hybrid ARQ indicator channel	HI	DL
PDCCH	R8	Physical downlink control channel	DCI	
R-PDCCH	R10	Relay physical downlink control channel	DCI	

Table 6.6 Physical signals

Signal	Release	Name	Use	Direction
DRS	R8	Demodulation reference signal	Channel estimation	UL
SRS	R8	Sounding reference signal	Scheduling	
PSS	R8	Primary synchronization signal	Acquisition	DL
SSS	R8	Secondary synchronization signal	Acquisition	
RS	R8	Cell specific reference signal	Channel estimation and scheduling	DL
	R8	UE specific reference signal	Channel estimation	
	R8/R9	MBMS reference signal	Channel estimation	
	R9	Positioning reference signal	Location services	
	R10	CSI reference signal	Scheduling	

can also transmit the *sounding reference signal* (SRS) at times configured by the base station, as a power reference in support of frequency-dependent scheduling.

The downlink usually combines these two roles in the form of the *cell specific reference signal* (RS). *UE specific reference signals* are less important and are sent to mobiles that are using beamforming in support of channel estimation. The specifications introduce other downlink reference signals as part of Releases 9 and 10. The base station also transmits two other physical signals, which help the mobile acquire the base station after it first switches on. These are known as the *primary synchronization signal* (PSS) and the *secondary synchronization signal* (SSS).

6.2.7 Information Flows

Tables 6.1 to 6.6 contain a large number of channels, but LTE uses them in just a few types of information flow. Figure 6.2 shows the information flows that are used in the uplink, with the arrows drawn from the viewpoint of the base station, so that uplink channels have arrows pointing upwards, and vice versa. Figure 6.3 shows the corresponding situation in the downlink.

6.3 The Resource Grid

6.3.1 Slot Structure

LTE maps the physical channels and physical signals onto the OFDMA symbols and sub-carriers that we introduced in Chapter 4. To understand how it does this, we first need to understand how LTE organizes its symbols and sub-carriers in the time and frequency domains [13].

First consider the time domain. The timing of the LTE transmissions is based on a time unit T_s, which is defined as follows:

$$T_s = \frac{1}{2048 \times 15000} \text{ seconds} \approx 32.6 \text{ ns} \quad (6.1)$$

T_s is the shortest time interval that is of interest to the physical channel processor. (To be exact, T_s is the sampling interval if the system uses a fast Fourier transform that contains

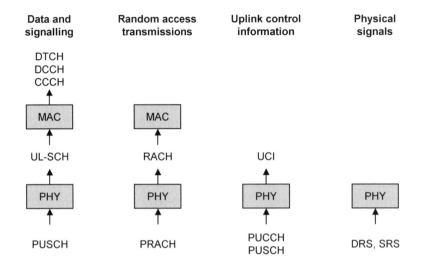

Figure 6.2 Uplink information flows used by LTE.

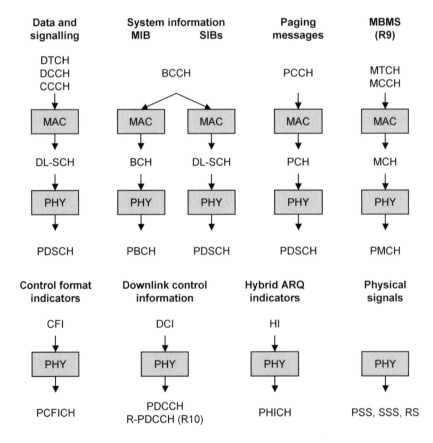

Figure 6.3 Downlink information flows used by LTE.

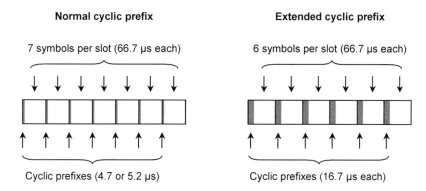

Figure 6.4 Organization of symbols into slots using the normal and extended cyclic prefix.

2048 points, which is the largest value ever likely to be used.) The 66.7 μs symbol duration is then equal to 2048 T_s.

The symbols are grouped into *slots*, whose duration is 0.5 ms (15 360 T_s). This can be done in two ways, as shown in Figure 6.4. With the *normal cyclic prefix*, each symbol is preceded by a cyclic prefix that is usually 144 T_s (4.7 μs) long. The first cyclic prefix has a longer duration of 160 T_s (5.2 μs), to tidy up the unevenness that results from fitting seven symbols into a slot.

Using the normal cyclic prefix, the receiver can remove inter-symbol interference with a delay spread of 4.7 μs, corresponding to a path difference of 1.4 km between the lengths of the longest and shortest rays. This is normally plenty, but may not be enough if the cell is unusually large or cluttered. To deal with this possibility, LTE also supports an *extended cyclic prefix*, in which the number of symbols per slot is reduced to six. This allows the cyclic prefix to be extended to 512 T_s (16.7 μs), to support a maximum path difference of 5 km.

With one exception, related to the multimedia broadcast/multicast service in Release 9, the base station sticks with either the normal or extended cyclic prefix in the downlink and does not change between the two. Mobiles generally use the same cyclic prefix duration in the uplink, but the base station can force a different choice by the use of its system information. The normal cyclic prefix is far more common, so we will use it almost exclusively.

6.3.2 Frame Structure

At a higher level, the slots are grouped into subframes and frames [14]. In FDD mode, this is done using *frame structure type 1*, which is shown in Figure 6.5.

Two slots make one *subframe*, which is 1 ms long (30 720 T_s). Subframes are used for scheduling. When a base station transmits to a mobile on the downlink, it schedules its PDSCH transmissions one subframe at a time, and maps each block of data onto a set of sub-carriers within that subframe. A similar process happens on the uplink.

In turn, 10 subframes make one *frame*, which is 10 ms long (307 200 T_s). Each frame is numbered using a *system frame number* (SFN), which runs repeatedly from 0 to 1023. Frames help to schedule a number of slowly changing processes, such as the transmission of system information and reference signals.

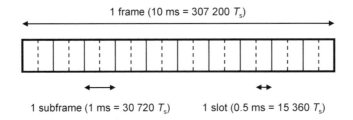

Figure 6.5 Frame structure type 1, used in FDD mode. Reproduced by permission of ETSI.

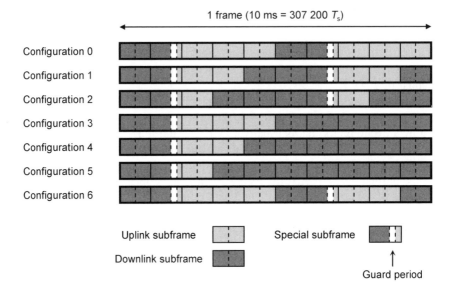

Figure 6.6 TDD configurations using frame structure type 2.

TDD mode uses *frame structure type 2*. In this structure, the slots, subframes and frames have the same duration as before, but each subframe can be allocated to either the uplink or downlink using one of the *TDD configurations* shown in Figure 6.6.

Different cells can have different TDD configurations, which are advertised as part of the cells' system information. Configuration 1 might be suitable if the data rates are similar on the uplink and downlink, for example, while configuration 5 might be used in cells that are dominated by downlink transmissions. Nearby cells should generally use the same TDD configuration, to minimize the interference between the uplink and downlink.

Special subframes are used at the transitions from downlink to uplink transmission. They contain three regions. The *special downlink region* takes up most of the subframe and is used in the same way as any other downlink region. The *special uplink region* is shorter, and is only used by the random access channel and the sounding reference signal. The two regions are separated by a *guard period* that supports the timing advance procedure described below. The cell can adjust the size of each region using a *special subframe configuration*, which again is advertised in the system information.

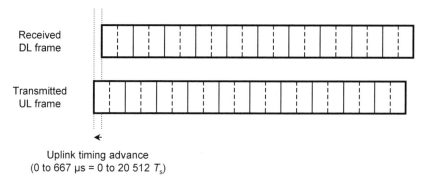

Figure 6.7 Timing relationship between the uplink and downlink in FDD mode. Reproduced by permission of ETSI.

6.3.3 Uplink Timing Advance

In LTE, a mobile starts transmitting its uplink frames at a time TA before the arrival of the corresponding frames on the downlink [15] (Figure 6.7). TA is known as the *timing advance* and is used for the following reason. Even travelling at the speed of light, a mobile's transmissions take time (typically a few microseconds) to reach the base station. However, the signals from different mobiles have to reach the base station at roughly the same time, with a spread less than the cyclic prefix duration, to prevent any risk of inter-symbol interference between them. To enforce this requirement, distant mobiles have to start transmitting slightly earlier than they otherwise would.

Because the uplink transmission time is based on the downlink arrival time, the timing advance has to compensate for the round-trip travel time between the base station and the mobile:

$$\text{TA} \approx \frac{2L}{c} \qquad (6.2)$$

Here, L is the distance between the mobile and the base station, and c is the speed of light. The timing advance does not have to be completely accurate, as the cyclic prefix can handle any remaining errors.

The specifications define the timing advance as follows:

$$\text{TA} = (N_{\text{TA}} + N_{\text{TAoffset}}) \, T_s \qquad (6.3)$$

Here, N_{TA} lies between 0 and 20 512. This gives a maximum timing advance of about 667 µs (two-thirds of a subframe), which supports a maximum cell size of 100 km. N_{TA} is initialized by the random access procedure described in Chapter 9, and updated by the timing advance procedure from Chapter 10.

N_{TAoffset} is zero in FDD mode, but 624 in TDD mode. This creates a small gap at the transition from uplink to downlink transmissions, which gives the base station time to switch from one to the other. The guard period in each special subframe creates a longer gap at the transition from downlink to uplink, which allows the mobile to advance its uplink frames without them colliding with the frames received on the downlink.

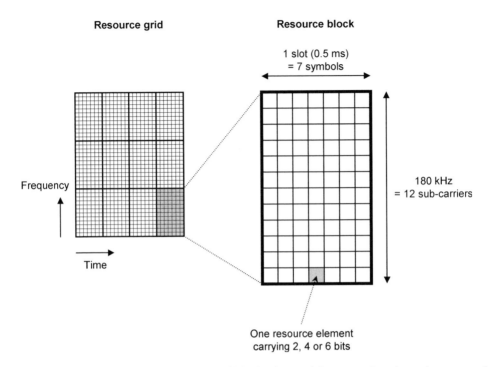

Figure 6.8 Structure of the LTE resource grid in the time and frequency domains, using a normal cyclic prefix.

6.3.4 Resource Grid Structure

In LTE, information is organized as a function of frequency as well as time, using a *resource grid* [16]. Figure 6.8 shows the resource grid for the case of a normal cyclic prefix. (There is a similar grid for the extended cyclic prefix, which uses six symbols per slot rather than seven.)

The basic unit is a *resource element* (RE), which spans one symbol by one sub-carrier. Each resource element usually carries two, four or six physical channel bits, depending on whether the modulation scheme is QPSK, 16-QAM or 64-QAM. Resource elements are grouped into *resource blocks* (RBs), each of which spans 0.5 ms (one slot), by 180 kHz (twelve sub-carriers). The base station uses resource blocks for frequency-dependent scheduling, by allocating the symbols and sub-carriers within each subframe in units of resource blocks.

6.3.5 Bandwidth Options

A cell can be configured with several different bandwidths [17], which are listed in Table 6.7. In a 5 MHz band, for example, the base station transmits using 25 resource blocks (300 sub-carriers), giving a transmission bandwidth of 4.5 MHz. That arrangement leaves room for guard bands at the upper and lower edges of the frequency band, which minimize the amount of interference with the next band along. The two guard bands are

Table 6.7 Cell bandwidths supported by LTE

Total bandwidth	Number of resource blocks	Number of sub-carriers	Occupied bandwidth	Usual guard bands
1.4 MHz	6	72	1.08 MHz	2 × 0.16 MHz
3 MHz	15	180	2.7 MHz	2 × 0.15 MHz
5 MHz	25	300	4.5 MHz	2 × 0.25 MHz
10 MHz	50	600	9 MHz	2 × 0.5 MHz
15 MHz	75	900	13.5 MHz	2 × 0.75 MHz
20 MHz	100	1200	18 MHz	2 × 1 MHz

usually the same width, but the network operator can adjust them if necessary by shifting the centre frequency in units of 100 kHz.

The existence of all these bandwidth options makes it easy for network operators to deploy LTE in a variety of spectrum management regimes. For example, 1.4 MHz is close to the bandwidths previously used by cdma2000 and TD-SCDMA, 5 MHz is the same bandwidth used by WCDMA, while 20 MHz allows an LTE base station to operate at its highest possible data rate. In FDD mode, the uplink and downlink bandwidths are usually the same. If they are different, then the base station signals the uplink bandwidth as part of its system information.

In Chapter 4, we noted that the fast Fourier transform operates most efficiently if the number of data points is an exact power of 2. This is easy to achieve, because the transmitter can simply round up the number of sub-carriers to the next highest power of 2, and can fill the extreme ones with zeros. In a 20 MHz bandwidth, for example, it will generally process the data using a 2048 point FFT, which is consistent with the value of T_s that we introduced earlier.

6.4 Multiple Antenna Transmission

6.4.1 Downlink Antenna Ports

In the downlink, multiple antenna transmissions are organized using *antenna ports*, each of which has its own copy of the resource grid that we introduced above. Table 6.8 lists the base station antenna ports that LTE uses. Ports 0 to 3 are used for single antenna transmission, transmit diversity and spatial multiplexing, while port 5 is reserved for beamforming. The remaining antenna ports are introduced in Releases 9 and 10, and will be covered towards the end of the book.

It is worth noting that an antenna port is not necessarily the same as a physical antenna: instead, it is an output from the base station transmitter that can drive one or more physical antennas. In particular, as shown in Figure 6.9, port 5 will always drive several physical antennas, which the base station uses for beamforming.

6.4.2 Downlink Transmission Modes

To support the use of multiple antennas, the base station can optionally configure the mobile into one of the *downlink transmission modes* that are listed in Table 6.9.

Table 6.8 Antenna ports used by the LTE downlink

Antenna port	Release	Application
0	R8	Single antenna transmission 2 and 4 antenna transmit diversity and spatial multiplexing
1	R8	2 and 4 antenna transmit diversity and spatial multiplexing
2	R8	4 antenna transmit diversity and spatial multiplexing
3	R8	4 antenna transmit diversity and spatial multiplexing
4	R8/R9	MBMS
5	R8	Beamforming
6	R9	Positioning reference signals
7–8	R9	Dual layer beamforming 8 antenna spatial multiplexing
9–14	R10	8 antenna spatial multiplexing
15–22	R10	CSI reference signals

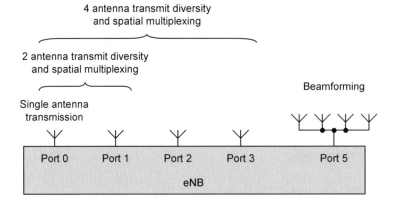

Figure 6.9 Antenna ports used by a Release 8 base station.

Table 6.9 Downlink transmission modes

			Uplink feedback required		
Mode	Release	Purpose	CQI	RI	PMI
1	R8	Single antenna transmission	✓		
2	R8	Open loop transmit diversity	✓		
3	R8	Open loop spatial multiplexing	✓	✓	
4	R8	Closed loop spatial multiplexing	✓	✓	✓
5	R8	Multiple user MIMO	✓		✓
6	R8	Closed loop transmit diversity	✓		✓
7	R8	Beamforming	✓		
8	R9	Dual layer beamforming	✓	Configurable	
9	R10	Eight layer spatial multiplexing	✓	Configurable	

Architecture of the LTE Air Interface 109

The transmission mode defines the type of multiple antenna processing that the base station will use for its transmissions on the PDSCH, and hence the type of processing that the mobile should use for PDSCH reception. It also defines the feedback that the base station will expect from the mobile, in the manner listed in the table.

If the base station does not configure the mobile in this way, then it transmits the PDSCH using either a single antenna or open loop transmit diversity, depending on the total number of antenna ports that it has.

6.5 Resource Element Mapping

6.5.1 Downlink Resource Element Mapping

The LTE physical layer transmits the physical channels and physical signals by mapping them onto the resource elements that we introduced above. The exact mapping depends on the exact configuration of the base station and mobile, so we will cover it one channel at a time as part of Chapters 7 to 9. However, it is instructive to show some example mappings for the uplink and downlink, for a typical system configuration.

Figure 6.10 shows an example resource element mapping for the downlink. The figure assumes the use of FDD mode, the normal cyclic prefix and a bandwidth of 5 MHz. Time is plotted horizontally and spans the 20 slots that make up one frame. Frequency is plotted vertically and spans the 25 resource blocks that make up the transmission band.

The cell specific reference signals are scattered across the time and frequency domains. While one antenna port is sending a reference signal, the others keep quiet, so that the mobile can measure the received reference signal from one antenna port at a time. The diagram assumes the use of two antenna ports and shows the reference signals that are sent from port 0. The exact mapping depends on the physical cell identity from Chapter 2: the one shown is suitable for a physical cell identity of 1, 7, 13, ...

Within each frame, certain resource elements are reserved for the primary and secondary synchronization signals and for the physical broadcast channel, and are read during the acquisition procedure that is described in Chapter 7. This information is only sent on the central 72 sub-carriers (1.08 MHz), which is the narrowest bandwidth ever used by LTE. This allows the mobile to read it without prior knowledge of the downlink bandwidth.

At the start of each subframe, a few symbols are reserved for the control information that the base station transmits on the PCFICH, PDCCH and PHICH. The number of control symbols can vary from one subframe to the next, depending on how much control information the base station needs to send. The rest of the subframe is reserved for data transmissions on the PDSCH and is allocated to individual mobiles in units of resource blocks within each subframe.

6.5.2 Uplink Resource Element Mapping

Figure 6.11 shows the corresponding situation on the uplink. Once again, the figure assumes the use of FDD mode, the normal cyclic prefix and a bandwidth of 5 MHz.

The outermost parts of the band are reserved for uplink control information on the PUCCH, and for the associated demodulation reference signals. The PUCCH can be sent

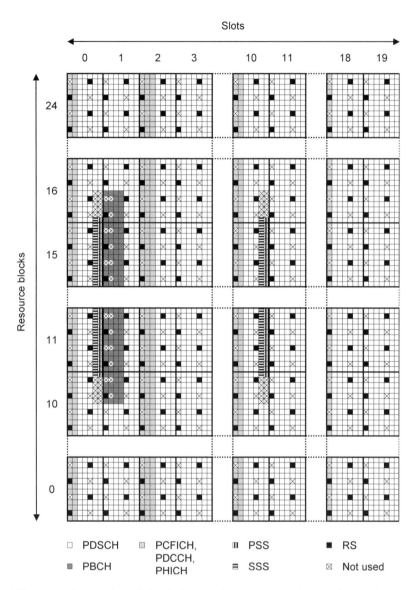

Figure 6.10 Example mapping of physical channels to resource elements in the downlink, using FDD mode, a normal cyclic prefix, a 5 MHz bandwidth, the first antenna port of two and a physical cell ID of 1.

in various different formats, depending on the information that the mobile has to transmit. In the example shown, one resource block has been reserved at each edge for PUCCH formats known as 2, 2a and 2b, which have five control symbols per slot and two reference symbols. The next resource block is being used by PUCCH formats 1, 1a and 1b, which have four control symbols per slot and three reference symbols.

Architecture of the LTE Air Interface

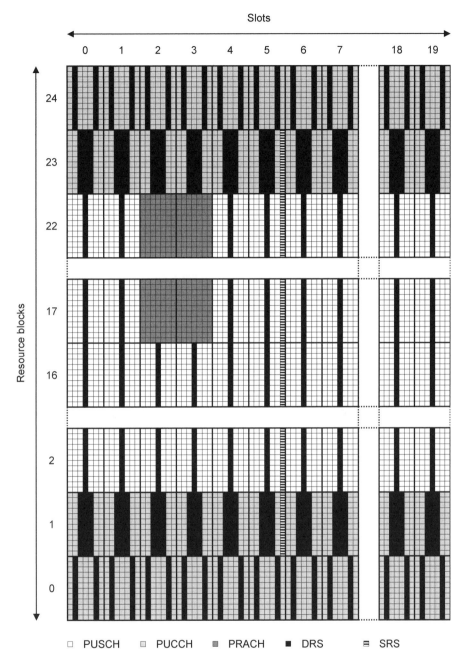

Figure 6.11 Example mapping of physical channels to resource elements in the uplink, using FDD mode, a normal cyclic prefix, a 5 MHz bandwidth and example configurations for the PUCCH, PRACH and SRS.

The rest of the band is mainly used by the PUSCH and is allocated to individual mobiles in units of resource blocks within each subframe. PUSCH transmissions contain six data symbols per slot and one reference symbol.

The base station also reserves certain resource blocks for random access transmissions on the PRACH. The PRACH has a bandwidth of six resource blocks and a duration from one to three subframes, while its locations in the resource grid are configured by the base station. In the example shown, the base station has reserved the upper end of the data region in subframe 1 (slots 2 and 3), but many other configurations are possible.

Furthermore, the base station can reserve the last symbol of certain subframes for the transmission of sounding reference signals. In the example shown, these can take place in the last symbol of subframe 2 (slot 5). Within the reserved region, the mobile transmits on alternate sub-carriers, using a mobile-specific bandwidth and frequency offset. Once again, many other configurations are possible.

References

1. 3GPP TS 36.331 (October 2011) *Radio Resource Control (RRC); Protocol Specification*, Release 10.
2. 3GPP TS 36.323 (March 2011) *Packet Data Convergence Protocol (PDCP) Specification*, Release 10.
3. 3GPP TS 36.322 (December 2010) *Radio Link Control (RLC) Protocol Specification*, Release 10.
4. 3GPP TS 36.321 (October 2011) *Medium Access Control (MAC) Protocol Specification*, Release 10.
5. 3GPP TS 36.212 (September 2011) *Evolved Universal Terrestrial Radio Access (E-UTRA); Multiplexing and Channel Coding*, Release 10.
6. 3GPP TS 36.211 (September 2011) *Evolved Universal Terrestrial Radio Access (E-UTRA); Physical Channels and Modulation*, Release 10.
7. 3GPP TS 36.101 (October 2011) *User Equipment (UE) Radio Transmission and Reception*, Release 10.
8. 3GPP TS 36.104 (October 2011) *Base Station (BS) Radio Transmission and Reception*, Release 10.
9. 3GPP TS 36.213 (September 2011) *Evolved Universal Terrestrial Radio Access (E-UTRA); Physical Layer Procedures*, Release 10.
10. 3GPP TS 36.321 (October 2011) *Medium Access Control (MAC) Protocol Specification*, Release 10, section 4.5.
11. 3GPP TS 36.212 (September 2011) *Multiplexing and Channel Coding*, Release 10, section 4.
12. 3GPP TS 36.211 (September 2011) *Physical Channels and Modulation*, Release 10, sections 5.1, 6.1.
13. 3GPP TS 36.211 (September 2011) *Physical Channels and Modulation*, Release 10, sections 5.6, 6.12.
14. 3GPP TS 36.211 (September 2011) *Physical Channels and Modulation*, Release 10, section 4.
15. 3GPP TS 36.211 (September 2011) *Physical Channels and Modulation*, Release 10, section 8.
16. 3GPP TS 36.211 (September 2011) *Physical Channels and Modulation*, Release 10, sections 5.2, 6.2.
17. 3GPP TS 36.101 (October 2011) *User Equipment (UE) Radio Transmission and Reception*, Release 10, section 5.6.

7

Cell Acquisition

After a mobile switches on, it runs a low-level acquisition procedure so as to identify the nearby LTE cells and discover how they are configured. In doing so, it receives the primary and secondary synchronization signals, reads the master information block from the physical broadcast channel and reads the remaining system information blocks from the physical downlink shared channel. It also starts reception of the downlink reference signals and the physical control format indicator channel, which it will need throughout the process of data transmission and reception later on. In this chapter, we begin by summarizing the acquisition procedure and then move to a discussion of the individual steps.

The most important specification for this chapter is the one for the physical channel processor, TS 36.211 [1]. The system information blocks form part of the radio resource control protocol and are defined in TS 36.331 [2].

7.1 Acquisition Procedure

The acquisition procedure is summarized in Table 7.1. There are several steps. The mobile starts by receiving the synchronization signals from all the nearby cells. From the primary synchronization signal (PSS), it discovers the symbol timing and gets some incomplete information about the physical cell identity. From the secondary synchronization signal (SSS), it discovers the frame timing, the physical cell identity, the transmission mode (FDD or TDD) and the cyclic prefix duration (normal or extended).

At this point, the mobile starts reception of the cell specific reference signals. These provide an amplitude and phase reference for the channel estimation process, so are essential for everything that follows. The mobile then receives the physical broadcast channel and reads the master information block. By doing so, it discovers the number of transmit antennas at the base station, the downlink bandwidth, the system frame number and a quantity called the PHICH configuration that describes the physical hybrid ARQ indicator channel.

The mobile can now start reception of the physical control format indicator channel (PCFICH), so as to read the control format indicators. These indicate how many symbols are reserved at the start of each downlink subframe for the physical control channels and how many are available for data transmissions. Finally, the mobile can start reception

Table 7.1 Steps in the cell acquisition procedure

Step	Task	Information obtained
1	Receive PSS	Symbol timing
		Cell identity within group
2	Receive SSS	Frame timing
		Physical cell identity
		Transmission mode
		Cyclic prefix duration
3	Start reception of RS	Amplitude and phase reference for demodulation
		Power reference for channel quality estimation
4	Read MIB from PBCH	Number of transmit antennas
		Downlink bandwidth
		System frame number
		PHICH configuration
5	Start reception of PCFICH	Number of control symbols per subframe
6	Read SIBs from PDSCH	System information

of the physical downlink control channel (PDCCH). This allows the mobile to read the remaining system information blocks (SIBs), which are sent on the physical downlink shared channel (PDSCH). By doing this, it discovers all the remaining details about how the cell is configured, such as the identities of the networks that it belongs to.

7.2 Synchronization Signals

7.2.1 Physical Cell Identity

The physical cell identity is a number between 0 and 503, which is transmitted on the synchronization signals [3] and used in three ways. Firstly, it determines the exact set of resource elements that are used for the cell specific reference signals and the PCFICH. Secondly, it influences a downlink transmission process known as scrambling, in a bid to minimize interference between nearby cells. Thirdly, it identifies individual cells during RRC procedures such as measurement reporting and handover. The physical cell identity is assigned during network planning or self configuration. Nearby cells should always receive different physical cell identities, to ensure that each of these roles is properly fulfilled.

It would be hard for a mobile to find the physical cell identities in one step, so they are organized into *cell identity groups* as follows:

$$N_{\text{ID}}^{\text{cell}} = 3N_{\text{ID}}^{(1)} + N_{\text{ID}}^{(2)} \tag{7.1}$$

In this equation, $N_{\text{ID}}^{\text{cell}}$ is the physical cell identity. $N_{\text{ID}}^{(1)}$ is the cell identity group, which runs from 0 to 167 and is signalled using the SSS. $N_{\text{ID}}^{(2)}$ is the cell identity within the group, which runs from 0 to 2 and is signalled using the PSS. Using this arrangement, a network planner can give each nearby base station a different cell identity group, and can distinguish its sectors using the cell identity within the group.

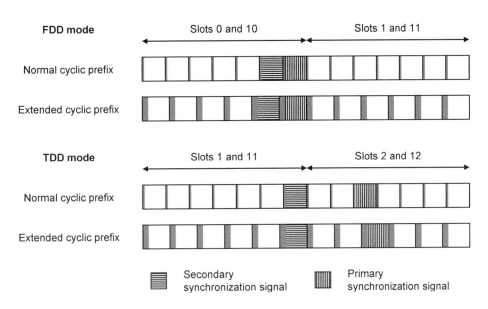

Figure 7.1 Time domain mapping of the primary and secondary synchronization signals.

7.2.2 Primary Synchronization Signal

Figure 7.1 shows the time domain mapping of the primary and secondary synchronization signals. The signals are both transmitted twice per frame. In FDD mode, the PSS is transmitted in the last symbol of slots 0 and 10, while the SSS is sent one symbol earlier. In TDD mode, the PSS is transmitted in the third symbol of slots 2 and 12, while the SSS is sent three symbols earlier.

In the frequency domain, the base station maps the synchronization signals onto the central 62 sub-carriers, and pads the resulting signal with zeros so that it occupies the central 72 sub-carriers (1.08 MHz). This second bandwidth is the smallest transmission band that LTE supports, which ensures that the mobile can receive both signals without prior knowledge of the downlink bandwidth.

The base station creates the actual signal using a *Zadoff-Chu sequence* [4, 5]. Briefly, this is a complex-valued sequence containing N_{ZC} data points. Subject to certain conditions, we can generate a maximum of N_{ZC} different *root sequences* from each value of N_{ZC}, and can then adjust each root sequence further by applying a maximum of N_{ZC} different *cyclic shifts*. The PSS simply uses three root sequences to indicate the three possible values of $N_{ID}^{(2)}$.

To receive the PSS, the mobile correlates the incoming signal with the three possible root sequences, for a time of at least 5 ms. By doing so, it measures the times at which the primary synchronization signal arrives from each of the nearby cells, and finds the cell identity within the group, $N_{ID}^{(2)}$.

Zadoff-Chu sequences are useful because they have good correlation properties. In practical terms, this means that there is little risk of the mobile making a mistake in its measurement of $N_{ID}^{(2)}$, even if the received signal to interference plus noise ratio (SINR) is low. Zadoff-Chu sequences are also used by the uplink reference signals (Chapter 8) and by the physical random access channel (Chapter 9).

7.2.3 Secondary Synchronization Signal

The base station transmits the SSS immediately before the PSS in FDD mode, or three symbols before it in TDD mode. The exact transmission time depends on the cyclic prefix duration (normal or extended), giving four possible transmission times altogether. The actual signal is created using pseudo-random sequences known as *Gold sequences* [6], the exact sequence being different for the first and second transmissions of the signal within the frame. Together, those two sequences indicate the cell identity group, $N_{ID}^{(1)}$.

To receive the secondary synchronization signal, the mobile inspects each of the four possible transmission times over two consecutive transmissions and looks for each of the possible SSS sequences. By finding the time when the signal was transmitted, it can deduce the transmission mode that the cell is using and the cyclic prefix duration. By identifying the transmitted sequences, it can deduce the cell identity group and hence the physical cell identity. Finally, by comparing the two consecutive sequences, it can deduce which is the first sequence in the frame, so can find the time at which a 10 ms frame begins.

7.3 Downlink Reference Signals

The mobile now starts reception of the downlink reference signals [7]. These are used in two ways. Their immediate role is to give the mobile an amplitude and phase reference for use in channel estimation. Later on, the mobile will use them to measure the received signal power as a function of frequency and to calculate the channel quality indicators.

Release 8 uses two types of downlink reference signal, but the cell specific reference signals are the most important. Figure 7.2 shows how these signals are mapped to resource elements, for the case of a normal cyclic prefix. As shown in the figure, the mapping depends on the number of antenna ports that the base station is using and on the antenna port number. While one antenna is transmitting a reference signal, all the others stay silent, in the manner required for spatial multiplexing.

This basic pattern is offset in the frequency domain, to minimize interference between the reference signals transmitted from nearby cells. The sub-carrier offset is:

$$v_{shift} = N_{ID}^{cell} \bmod 6 \qquad (7.2)$$

where N_{ID}^{cell} is the physical cell identity that the mobile has already found. Previously, Figure 6.10 showed the resource element mapping in the case where $v_{shift} = 1$, so that the physical cell identity could have values of 1, 7, 13 . . . .

The resource elements are filled with a Gold sequence that depends on the physical cell identity, again in a bid to minimize interference. By measuring the received reference symbols and comparing them with the ones transmitted, the mobile can measure the amplitude changes and phase shifts that the air interface has introduced. It can then estimate those quantities at intervening resource elements by interpolation.

Antenna ports 0 and 1 use four reference symbols per resource block, while antenna ports 2 and 3 use only two. This is because a cell is only likely to use four antenna ports

Cell Acquisition

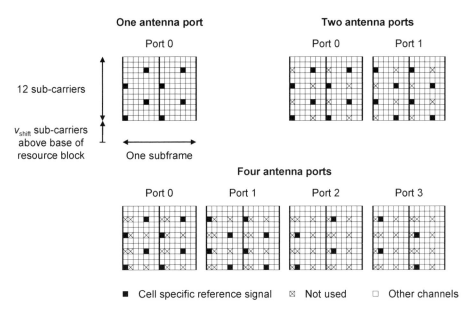

Figure 7.2 Resource element mapping for the cell-specific reference signals, using a normal cyclic prefix. Reproduced by permission of ETSI.

when it is dominated by slowly moving mobiles, for which the amplitude and phase of the received signal will only vary slowly with time.

A Release 8 base station can also transmit UE specific reference signals from antenna port 5, as an amplitude and phase reference for mobiles that are using beamforming. These differ from the cell specific reference signals in three ways. Firstly, the base station only transmits them in the resource blocks that it is using for beamforming transmissions on the PDSCH. Secondly, the Gold sequence depends on the identity of the target mobile, as well as that of the base station. Thirdly, the base station precodes the UE specific reference signals using the same antenna weights that it applies to the PDSCH. This last step ensures that the reference signals are directed towards the target mobile and also ensures that the weighting process is completely transparent: by recovering the original reference signals during channel estimation, the mobile automatically removes all the phase shifts that the weighting process introduced.

7.4 Physical Broadcast Channel

The master information block [8] contains the downlink bandwidth and the eight most significant bits of the 10-bit system frame number. It also contains a quantity known as the *PHICH configuration*, which indicates the resource elements that the base station has reserved for the physical hybrid ARQ indicator channel.

The base station transmits the master information block on the broadcast channel and the physical broadcast channel [9, 10]. It processes the information in much the same way as any other downlink channel: in the next chapter, we will illustrate these steps using the PDSCH as an example. The only significant difference is that the base station

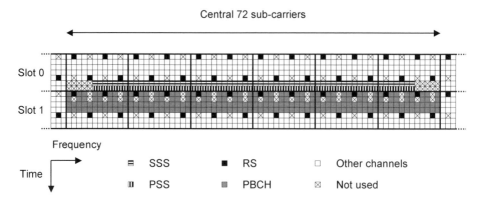

Figure 7.3 Resource element mapping for the PBCH, using FDD mode, a normal cyclic prefix, a 10 or 20 MHz bandwidth, the first antenna port of two and a physical cell ID of 1.

manipulates the broadcast channel in a manner that indicates the two least significant bits of the system frame number and the number of antenna ports. As noted in Chapter 6, the broadcast channel uses a fixed coding rate and a fixed modulation scheme (QPSK), and is transmitted using either one antenna or open loop transmit diversity.

The base station maps the master information block across four successive frames, beginning in frames where the system frame number is a multiple of four. When using a normal cyclic prefix, it transmits the physical broadcast channel on the central 72 sub-carriers, using the first four symbols of slot 1. Figure 7.3 shows an example. As shown in the figure, the mapping skips the resource elements used for cell-specific reference signals from a base station with four antenna ports, irrespective of how many ports it actually has. (Note that, in a departure from our usual convention, this diagram shows time plotted vertically and frequency horizontally.)

The mobile processes the physical broadcast channel blindly using all the possible ways in which the base station might have manipulated the information, by choosing each combination of one, two and four antenna ports and each combination of the two least significant bits of the system frame number. Only the correct choice allows the cyclic redundancy check to pass. By reading the master information block, the mobile can then discover the downlink bandwidth, the remaining bits of the system frame number and the PHICH configuration.

7.5 Physical Control Format Indicator Channel

Every downlink subframe starts with a control region that contains the PCFICH, PHICH and PDCCH and continues with a data region that contains the PDSCH. The number of control symbols per subframe can be 2, 3 or 4 in a bandwidth of 1.4 MHz and 1, 2 or 3 otherwise and can change from one subframe to the next. Every subframe, the base station indicates the number of control symbols using the control format indicator and transmits the information on the PCFICH [11, 12]. The mobile's next task is to start receiving this channel.

The PCFICH is mapped onto 16 resource elements in the first symbol of every subframe, with the precise mapping depending on the physical cell identity and the downlink

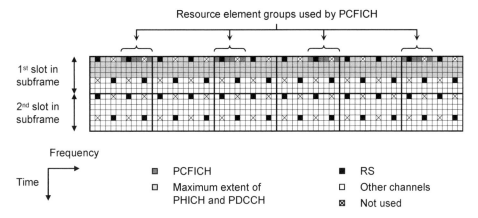

Figure 7.4 Resource element mapping for the PCFICH, using a normal cyclic prefix, a 1.4 MHz bandwidth, the first antenna port of two and a physical cell ID of 1.

bandwidth. Figure 7.4 shows an example, for a bandwidth of 1.4 MHz and a physical cell identity of 1.

As shown in the figure, the selected resource elements are organized into four *resource element groups* (REGs). Each group contains four resource elements, which are sent on nearby sub-carriers that are not required by the cell-specific reference signals. Resource element groups are used throughout the downlink control region, so are also used by the PHICH and PDCCH.

At the start of every subframe, the mobile goes to the resource elements that are occupied by the PCFICH, reads the control format indicator and determines the size of the downlink control region. Using the PHICH configuration, it can work out which of the remaining resource element groups are used by the PHICH and which by the PDCCH. It can then go on to receive downlink control information on the PDCCH and downlink data on the PDSCH. In particular, it can read the rest of the cell's system information, in the manner described below.

7.6 System Information

7.6.1 Organization of the System Information

Every cell broadcasts RRC *system information* messages [13] that indicate how it has been configured. The base station transmits these messages on the PDSCH, in a way that is almost identical to any other data transmission. The mobile's final acquisition task is to read this information.

The system information is organized into the master information block that we discussed above and into several numbered system information blocks. These are listed in Table 7.2, along with examples of the information elements that we will use in the remaining chapters.

SIB 1 defines the way in which the other system information blocks will be scheduled. It also includes the parameters that the mobile will need for network and cell selection (Chapter 11), such as the tracking area code and a list of networks that the cell belongs

Table 7.2 Organization of the system information

Block	Release	Information	Examples
MIB	R8	Master information block	Downlink bandwidth
			PHICH configuration
			System frame number/4
SIB 1	R8	Cell selection parameters	PLMN identity list
		Scheduling of other SIBs	Tracking area code
			CSG identity
			TDD configuration
			$Q_{rxlevmin}$
			SIB mapping, period, window size
SIB 2	R8	Radio resource configuration	Downlink reference signal power
			Default DRX cycle length
			Time alignment timer
SIB 3	R8	Common cell reselection data	$S_{IntraSearchP}$, $S_{NonIntraSearchP}$
		Cell independent intra frequency data	Q_{hyst}
SIB 4	R8	Cell specific intra frequency data	$Q_{offset, s, n}$
SIB 5	R8	Inter frequency reselection data	Target carrier frequency
			$Thresh_{x, LowP}$, $Thresh_{x, HighP}$
SIB 6	R8	Reselection to UMTS	UMTS neighbour list
SIB 7	R8	Reselection to GSM	GSM neighbour list
SIB 8	R8	Reselection to cdma2000	cdma2000 neighbour list
SIB 9	R8	Home eNB identifier	Name of home eNB
SIB 10	R8	ETWS primary notification	ETWS alert about natural disaster
SIB 11	R8	ETWS secondary notification	Supplementary ETWS information
SIB 12	R9	CMAS notification	CMAS emergency message
SIB 13	R9	MBMS information	Details of MBSFN areas

to. This list can identify up to six networks, which allows a base station to be easily shared amongst different network operators. SIB 2 contains parameters that describe the cell's radio resources and physical channels, such as the power that the base station is transmitting on the downlink reference signals.

SIBs 3 to 8 help to specify the cell reselection procedures that are used by mobiles in RRC_IDLE (Chapters 14 and 15). SIB 3 contains the parameters that a mobile will need for any type of cell reselection, as well as the cell-independent parameters that it will need for reselection on the same LTE carrier frequency. SIB 4 is optional and contains any cell-specific parameters that the base station might define for that process. SIB 5 covers reselection to a different LTE frequency, while SIBs 6, 7 and 8 respectively cover reselection to UMTS (both WCDMA and TD-SCDMA), GSM and cdma2000.

The remaining SIBs are more specialized. If the base station belongs to a closed subscriber group, then SIB 9 identifies its name. The mobile can then indicate this to the user, in support of closed subscriber group selection. SIBs 10 and 11 contain notifications from the earthquake and tsunami warning system. In this system, the cell broadcast centre can receive an alert about a natural disaster and can distribute it to all the base stations in the network, which then broadcast the alert through their system information. SIB 10 contains a primary notification that has to be distributed in seconds, while SIB 11 contains a secondary notification that includes less urgent supplementary information. SIBs 12 and 13 are not introduced until Release 9.

7.6.2 Transmission and Reception of the System Information

The base station can transmit its system information using two techniques. In the first technique, the base station broadcasts the system information across the whole of the cell, for use by mobiles in RRC_IDLE state and by mobiles in RRC_CONNECTED that have just handed over to a new cell. It does this in much the same way as any other downlink transmission (Chapter 8), but with a few differences. The system information transmissions do not support automatic repeat requests, which are unsuitable for one-to-many broadcast transmissions. The base station sends the system information using one antenna or open loop transmit diversity, depending on the number of antenna ports that it has, and the modulation scheme is fixed at QPSK.

There are a few rules about the timing of these system information broadcasts. The base station transmits SIB 1 in subframe 5 of frames with an even system frame number, with a full transmission taking eight frames altogether. It defines the choice of sub-carriers using its downlink scheduling command. The base station then collects the remaining SIBs into RRC *System Information* messages, and sends each message within a transmission window that has a duration of 1 to 40 ms and a period of 80 to 5120 ms. SIB 1 defines the mapping of SIBs onto messages, the period of each message and the window duration, while the downlink scheduling command defines the exact transmission time and the choice of sub-carriers.

If the base station wishes to update the system information that it is broadcasting, it first notifies the mobiles using the paging procedure (Chapter 8). It also increments a *value tag* in SIB 1, for use by mobiles that are returning from a region of poor coverage in which they may have missed a paging message. The base station then changes the system information on a pre-defined modification period boundary.

In the second technique, the base station can update the system information being used by a mobile in RRC_CONNECTED state, by sending it an explicit System Information message. It does this in the same way as any other downlink signalling transmission.

7.7 Procedures After Acquisition

After the mobile has completed the acquisition procedure, it has to run two higher-level procedures before it can exchange data with the network.

In the random access procedure (Chapter 9), the mobile acquires three pieces of information: an initial value for the uplink timing advance, an initial set of parameters for the transmission of uplink data on the physical uplink shared channel and a quantity known as the cell radio network temporary identifier (C-RNTI) that the base station will use to identify it. In the RRC connection establishment procedure (Chapter 11), the mobile acquires several other pieces of information, notably a set of parameters for the transmission of uplink control information on the physical uplink control channel and a set of protocol configurations for its data and signalling radio bearers.

Before looking at these, however, we need to cover the underlying procedure that LTE uses for data transmission and reception. That is the subject of the next chapter.

References

1. 3GPP TS 36.211 (September 2011) *Physical Channels and Modulation*, Release 10.
2. 3GPP TS 36.331 (October 2011) *Radio Resource Control (RRC); Protocol Specification*, Release 10.
3. 3GPP TS 36.211 (September 2011) *Physical Channels and Modulation*, Release 10, section 6.11.
4. Frank, R., Zadoff, S. and Heimiller, R. (1962) Phase shift pulse codes with good periodic correlation properties, *IEEE Transactions on Information Theory*, **8**, 381–382.
5. Chu, D. (1972) Polyphase codes with good periodic correlation properties, *IEEE Transactions on Information Theory* , **18**, 531–532.
6. Gold, R. (1967) Optimal binary sequences for spread spectrum multiplexing, *IEEE Transactions on Information Theory*, **13**, 619–621.
7. 3GPP TS 36.211 (September 2011) *Physical Channels and Modulation*, Release 10, section 6.10.
8. 3GPP TS 36.331 (October 2011) *Radio Resource Control (RRC); Protocol Specification*, Release 10, section 6.2.2 (*MasterInformationBlock*).
9. 3GPP TS 36.211 (September 2011) *Physical Channels and Modulation*, Release 10, section 6.6.
10. 3GPP TS 36.212 (September 2011) *Multiplexing and Channel Coding*, Release 10, section 5.3.1.
11. 3GPP TS 36.211 (September 2011) *Physical Channels and Modulation*, Release 10, sections 6.2.4, 6.7.
12. 3GPP TS 36.212 (September 2011) *Multiplexing and Channel Coding*, Release 10, section 5.3.4.
13. 3GPP TS 36.331 (October 2011) *Radio Resource Control (RRC); Protocol Specification*, Release 10, sections 5.2, 6.2.2 (*SystemInformation*, *SystemInformationBlockType1*), 6.3.1.

8

Data Transmission and Reception

Data transmission and reception is one of the more complex parts of LTE. In this chapter, we begin with an overview of the transmission and reception procedures that are used in the uplink and downlink. We then cover the three main stages of those procedures in turn, namely the delivery of scheduling messages from the base station, the actual process of data transmission and the delivery of acknowledgements and any associated control information from the receiver. We also cover the transmission of uplink reference signals as well as two associated procedures, namely uplink power control and discontinuous reception.

Several 3GPP specifications are relevant to this chapter. Data transmission and reception is defined by the physical layer specifications that we noted earlier, particularly TS 36.211 [1], TS 36.212 [2] and TS 36.213 [3], and is controlled by the MAC protocol in the manner defined by TS 36.321 [4]. In addition, the base station configures the mobile's physical layer and MAC protocols by means of RRC signalling [5].

8.1 Data Transmission Procedures

8.1.1 Downlink Transmission and Reception

Figure 8.1 shows the procedure that is used for downlink transmission and reception [6, 7]. The base station begins the procedure by sending the mobile a *scheduling command* (step 1), which is written using the downlink control information (DCI) and transmitted on the physical downlink control channel (PDCCH). The scheduling command alerts the mobile to a forthcoming data transmission and states how it will be sent, by specifying parameters such as the amount of data, the resource block allocation and the modulation scheme.

In step 2, the base station transmits the data on the downlink shared channel (DL-SCH) and the physical downlink shared channel (PDSCH). The data comprise either one or two *transport blocks*, whose duration is known as the *transmission time interval* (TTI), which equals the subframe duration of 1 millisecond. In response (step 3), the mobile composes a hybrid ARQ acknowledgement to indicate whether the data arrived correctly. It sends the acknowledgement on the physical uplink shared channel (PUSCH) if it is transmitting uplink data in the same subframe and on the physical uplink control channel (PUCCH) otherwise.

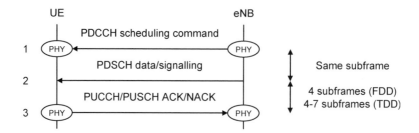

Figure 8.1 Downlink transmission and reception procedure.

Usually, the base station moves to a new transport block after a positive acknowledgement and re-transmits the original one after a negative acknowledgement. If, however, the base station reaches a certain maximum number of re-transmissions without receiving a positive response, then it moves to a new transmission anyway, on the grounds that the mobile's receive buffer may have been corrupted by a burst of interference. The radio link control (RLC) protocol then picks up the problem, for example by sending the transport block again from the beginning.

The downlink transmission timing is as follows. The scheduling command lies in the control region at the beginning of a downlink subframe, while the transport block lies in the data region of the same subframe. In FDD mode, there is a fixed time delay of four subframes between the transport block and the corresponding acknowledgement, which helps the base station to match the two pieces of information together. In TDD mode, the delay is between four and seven subframes, according to a mapping that depends on the TDD configuration. Figure 8.2 shows an example mapping, for the case of TDD configuration 1.

As described in Chapter 3, the downlink uses several parallel hybrid ARQ processes, each with its own copy of Figures 8.1 and 8.2. In FDD mode, the maximum number of hybrid ARQ processes is eight. In TDD mode, the maximum number depends on the TDD configuration, up to an absolute maximum of 15 for TDD configuration 5. The LTE downlink uses a technique known as *asynchronous hybrid ARQ*, in which the base station explicitly specifies the hybrid ARQ process number in every scheduling command. As a

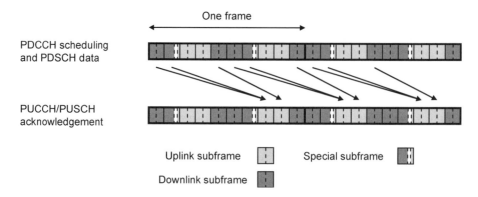

Figure 8.2 Relationship between the timing of downlink data and uplink acknowledgements, for TDD configuration 1.

result, there is no need to define the timing delay between a negative acknowledgement and a re-transmission: instead, the base station schedules a re-transmission whenever it likes and simply states the hybrid ARQ process number that it is using.

8.1.2 Uplink Transmission and Reception

Figure 8.3 shows the corresponding procedure for the uplink [8, 9]. As in the downlink, the base station starts the procedure by sending the mobile a *scheduling grant* on the PDCCH (step 1). This grants permission for the mobile to transmit and states all the transmission parameters that it should use, for example the transport block size, the resource block allocation and the modulation scheme. In response, the mobile carries out an uplink data transmission on the uplink shared channel (UL-SCH) and the PUSCH (step 2).

If the base station receives the data correctly, then it sends the mobile a positive acknowledgement on the physical hybrid ARQ indicator channel (PHICH). If it does not, then there are two ways for it to respond. In one technique, the base station can trigger a *non adaptive re-transmission* by sending the mobile a negative acknowledgement on the PHICH. The mobile then re-transmits the data with the same parameters that it used first time around. Alternatively, the base station can trigger an *adaptive re-transmission* by explicitly sending the mobile another scheduling grant. It can do this to change the parameters that the mobile uses for the re-transmission, such as the resource block allocation or the uplink modulation scheme. If the mobile receives a PHICH acknowledgement and a PDCCH scheduling grant in the same subframe, then the scheduling grant takes priority.

In the diagram, steps 3 to 5 assume that the base station fails to decode the mobile's first transmission, but succeeds with the second. If the mobile reaches a maximum number of re-transmissions without receiving a positive reply, then it moves to a new transmission anyway and leaves the RLC protocol to solve the problem.

Once again, the uplink uses several hybrid ARQ processes, each with its own copy of Figure 8.3. In FDD mode, the maximum number of hybrid ARQ processes is eight. In TDD mode, the absolute maximum is seven in TDD configuration 0.

The uplink uses a technique known as *synchronous hybrid ARQ*, in which the hybrid ARQ process number is not signalled explicitly, but is instead defined using the transmission

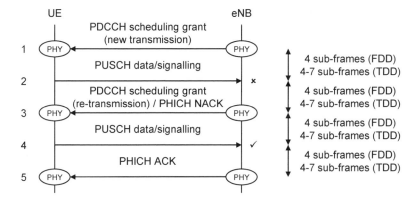

Figure 8.3 Uplink transmission and reception procedure.

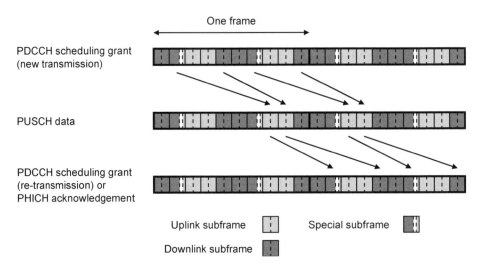

Figure 8.4 Relationship between the timing of scheduling grants, uplink data and downlink acknowledgements, for TDD configuration 1.

timing. In FDD mode, there is a delay of four subframes between the scheduling grant and the corresponding uplink transmission, and another four subframe delay before any re-transmission request on the same hybrid ARQ process. This gives all the information that the devices need to match up the scheduling grants, transmissions, acknowledgements and re-transmissions.

As in the downlink, TDD mode uses a variable set of delays, according to a mapping that depends on the TDD configuration. Figure 8.4 shows an example mapping, for the case of TDD configuration 1.

8.1.3 Semi Persistent Scheduling

When using *semi persistent scheduling* (SPS) [10–12], the base station can schedule several transmissions spanning several subframes, by sending the mobile a single scheduling message that contains just one resource allocation. Semi persistent scheduling is designed for services such as voice over IP. For these services, the data rate is low, so the overhead of the scheduling message can be high. However, the data rate is also constant, so the base station can confidently use the same resource allocation from one transmission to the next.

The base station configures a mobile for semi persistent scheduling by means of a mobile-specific RRC signalling message. As part of the message, it specifies the interval between transmissions, which lies between 10 and 640 subframes. (A 20 subframe interval is often suitable for voice over IP, consistent with the 20 ms block duration of the *adaptive multi rate* (AMR) voice codec.) Later on, the base station can activate semi persistent scheduling by sending the mobile a specially formatted scheduling command or scheduling grant.

In the downlink, the base station sends new transmissions on the PDSCH at the interval defined by the mobile's SPS configuration, in the manner indicated by the original scheduling command. The mobile cycles the hybrid ARQ process number for every new transmission, because the base station has no opportunity to specify it. However, the base station continues to schedule all its re-transmissions explicitly, in the manner shown in Figure 8.1. It can therefore specify different transmission parameters for them such as different resource block allocations or different modulation schemes. A similar situation applies in the uplink: the mobile sends new transmissions on the PUSCH at the interval defined by its SPS configuration, but the base station continues to schedule any re-transmissions in the manner shown in Figure 8.2.

Eventually, the base station can release the SPS assignment by sending the mobile another specially formatted scheduling message. In addition, the mobile can implicitly release an uplink SPS assignment if it has reached a maximum number of transmission opportunities without having any data to send.

8.2 Transmission of Scheduling Messages on the PDCCH

8.2.1 Downlink Control Information

In looking at the details of the transmission and reception procedures, we will begin with the transmission of downlink control information on the PDCCH. The base station uses its downlink control information to send downlink scheduling commands, uplink scheduling grants and uplink power control commands to the mobile. The DCI can be written using several different formats, which are listed in Table 8.1 [13]. Each format contains a specific set of information and has a specific purpose.

Table 8.1 List of DCI formats and their applications

DCI format	Release	Purpose		Resource allocation	DL mode
0	R8	UL scheduling grants	1 antenna	-	-
1	R8	DL scheduling commands	1 antenna, open loop diversity, beamforming	Type 0, 1	1, 2, 7
1A	R8		1 antenna, open loop diversity	Type 2	Any
1B	R8		Closed loop diversity	Type 2	6
1C	R8		System information, paging, random access responses	Type 2	Any
1D	R8		Multiple user MIMO	Type 2	5
2	R8	DL scheduling commands	Closed loop MIMO	Type 0, 1	4
2A	R8		Open loop MIMO	Type 0, 1	3
2B	R9		Dual layer beamforming	Type 0, 1	8
2C	R10		8 layer MIMO	Type 0, 1	9
3	R8	UL power control	2 bit power adjustments	-	-
3A	R8		1 bit power adjustments	-	-
4	R10	UL scheduling grants	Closed loop MIMO	-	-

DCI format 0 contains scheduling grants for the mobile's uplink transmissions. The scheduling commands for downlink transmissions are more complicated, and are handled in Release 8 by DCI formats 1 to 1D and 2 to 2A.

DCI format 1 schedules data that the base station will transmit using one antenna, open loop diversity or beamforming, for mobiles that already have been configured into one of the downlink transmission modes 1, 2 or 7. When using this format, the base station can allocate the downlink resource blocks in a flexible way, by means of two resource allocation schemes known as type 0 and type 1 that we will describe below.

Format 1A is similar, but the base station uses a compact form of resource allocation known as type 2. Format 1A can also be used in any downlink transmission mode. If the mobile has previously been configured into one of transmission modes 3 to 7, then it receives the data by falling back to single antenna reception if the base station has one antenna port, or open loop transmit diversity otherwise.

Skipping a line, format 1C uses a very compact format that only specifies the resource allocation and the amount of data that the base station will send. In the ensuing data transmission, the modulation scheme is fixed at QPSK and hybrid ARQ is not used. Format 1C is only used to schedule system information messages, paging messages and random access responses, for which this very compact format is appropriate.

Formats 1B, 1D, 2 and 2A are respectively used for closed loop transmit diversity, the Release 8 implementation of multiple user MIMO, and closed and open loop spatial multiplexing. They include extra fields to signal information such as the precoding matrix that the base station will apply to the PDSCH and the number of layers that the base station will transmit.

Unlike the others, DCI formats 3 and 3A do not schedule any transmissions: instead, they control the power that the mobile transmits on the uplink by means of embedded power control commands. We will cover this procedure later in the chapter. Formats 2B, 2C and 4 are introduced in Releases 9 and 10, and are covered towards the end of the book.

8.2.2 Resource Allocation

The base station has various ways of allocating the resource blocks to individual mobiles in the uplink and downlink [14, 15]. In the downlink, as noted above, it can use two flexible resource allocation formats that are known as types 0 and 1, and a compact format known as type 2.

When using downlink resource allocation type 0, the base station collects the resource blocks into *resource block groups* (RBGs), which it assigns individually using a bitmap. With resource allocation type 1, it can assign individual resource blocks within a group, but has less flexibility over the assignment of the groups themselves. Allocation type 1 might be suitable in environments with severe frequency-dependent fading, in which the frequency resolution of type 0 might be too coarse.

When using resource allocation type 2, the base station gives the mobile a contiguous allocation of *virtual resource blocks* (VRBs). In the downlink, these come in two varieties: localized and distributed. Localized virtual resource blocks are identical to the *physical resource blocks* (PRBs) that we have been considering elsewhere, so, when using these, the mobile simply receives a contiguous resource block allocation. Distributed virtual resource blocks are related to physical resource blocks by a mapping operation, which is different in the first and second slots of a subframe. The use of distributed virtual resource blocks gives the mobile extra frequency diversity and is suitable in environments that are subject to frequency-dependent fading.

The mobile also receives a contiguous allocation of virtual resource blocks for its uplink transmissions. Their meaning depends on whether the base station has requested the use of *frequency hopping* in DCI format 0. If frequency hopping is disabled, then the uplink virtual resource blocks map directly onto physical resource blocks. If frequency hopping is enabled, then the virtual and physical resource blocks are related using a mapping that is either explicitly signalled (type 1 hopping) or follows a pseudo-random pattern (type 2 hopping). A mobile can also change transmission frequency in every subframe or in every slot, depending on a hopping mode that is configured using RRC signalling.

In the uplink, the number of resource blocks per mobile must be either 1, or a number whose prime factors are 2, 3 or 5. The reason lies in the extra Fourier transform used by SC-FDMA, which runs quickly if the number of sub-carriers is a power of 2 or a product of small prime numbers alone, but slowly if a large prime number is involved.

8.2.3 Example: DCI Format 1

To illustrate the DCI formats, Table 8.2 shows the contents of DCI format 1 in Release 8. The other formats are not that different: the details can be found in the specifications.

The base station indicates whether the mobile should use resource allocation type 0 or 1 using the resource allocation header and carries out the allocation using the resource block assignment. In a bandwidth of 1.4 MHz, allocation type 0 is not supported, so the header field is omitted.

The modulation and coding scheme is a 5 bit number, from which the mobile can look up the modulation scheme that the PDSCH will use (QPSK, 16-QAM or 64-QAM) and the number of bits in the transport block. By comparing the transport block size with the number of resource elements in its allocation, the mobile can calculate the coding rate for the DL-SCH.

As noted earlier, the base station explicitly signals the hybrid ARQ process number in every downlink scheduling command. The base station also toggles the new data indicator for every new transmission, while leaving it unchanged for a re-transmission.

Table 8.2 Contents of DCI format 1 in 3GPP Release 8

Field	Number of bits
Resource allocation header	0 (1.4 MHz) or 1 (otherwise)
Resource block assignment	6 (1.4 MHz) to 25 (20 MHz)
Modulation and coding scheme	5
HARQ process number	3 (FDD) or 4 (TDD)
New data indicator	1
Redundancy version	2
TPC command for PUCCH	2
Downlink assignment index	2 (TDD only)
Padding	0 or 1

The redundancy version indicates which of the turbo coded bits will be transmitted after the rate matching stage and which will be punctured.

The base station uses the *transmit power control* (TPC) command for PUCCH to adjust the power that the mobile will use when sending uplink control information on the PUCCH. (This is an alternative technique to the adjustment of transmit power using DCI formats 3 and 3A.) In TDD mode, it uses the downlink assignment index to assist the mobile's transmission of uplink acknowledgements, in the manner that we will describe later on.

There is a significant omission from Table 8.2: there is no header field to indicate what the DCI format actually is. Although some of the other formats do contain such headers, the mobile usually distinguishes the different DCI formats by the fact that they contain different numbers of bits. The base station occasionally adds a padding bit to the end of the scheduling command, to ensure that format 1 contains a different number of bits from all the others.

8.2.4 Radio Network Temporary Identifiers

The base station transmits a PDCCH scheduling message by addressing it to a *radio network temporary identifier* (RNTI) [16]. In LTE, an RNTI defines two things: the identity of the mobile(s) that should read the scheduling message and the type of information that is being scheduled. Table 8.3 lists the RNTIs that are used by LTE, along with the hexadecimal values that they can use.

The *cell RNTI* (C-RNTI) is the most important. The base station assigns a unique C-RNTI to a mobile as part of the random access procedure. Later on, it can schedule a transmission that extends over one subframe, by addressing a scheduling message to the mobile's C-RNTI.

The *SPS C-RNTI* is used for semi persistent scheduling. The base station first assigns an SPS C-RNTI to a mobile using mobile-specific RRC signalling. Later on, it can schedule a transmission that extends over several subframes by addressing a specially formatted scheduling message to the SPS C-RNTI.

The *paging RNTI* (P-RNTI) and *system information RNTI* (SI-RNTI) are fixed values, which are used to schedule the transmission of paging and system information messages to

Table 8.3 List of radio network temporary identifiers and their applications. Reproduced by permission of ETSI

Type of RNTI	Release	Information scheduled	Hex value
RA-RNTI	R8	Random access response	0001 - 003C
Temporary C-RNTI	R8	Random access contention resolution	
C-RNTI	R8	One UL or DL transmission	
SPS C-RNTI	R8	Several UL or DL transmissions	003D - FFF3
TPC-PUCCH-RNTI	R8	Embedded PUCCH TPC command	
TPC-PUSCH-RNTI	R8	Embedded PUSCH TPC command	
M-RNTI	R9	MBMS change notification	FFFD
P-RNTI	R8	Paging message	FFFE
SI-RNTI	R8	System information message	FFFF

all the mobiles in the cell. The *temporary C-RNTI* and *random access RNTI* (RA-RNTI) are temporary fields during the random access procedure (Chapter 9), while the *MBMS RNTI* (M-RNTI) is used by the multimedia broadcast/multicast service (Chapter 17). Finally, the TPC-PUCCH-RNTI and TPC-PUSCH-RNTI are used to send embedded uplink power control commands using DCI formats 3 and 3A.

8.2.5 Transmission and Reception of the PDCCH

We are now in a position to discuss how the PDCCH is transmitted and received, a process that is summarized in Figure 8.5. In its transport channel processor, the base station first manipulates the DCI by attachment of a cyclic redundancy check (CRC) and error correction coding [17], in a manner that depends on the RNTI of the target mobile. It then processes the PDCCH using QPSK modulation and either single antenna transmission or open loop transmit diversity, depending on the number of antenna ports that it has [18]. Finally, the base station maps the PDCCH onto the chosen resource elements.

The resource element mapping for the PDCCH is organized using *control channel elements* (CCEs) [19], each of which contains nine resource element groups that have not already been assigned to the physical control format indicator channel (PCFICH) or the PHICH. Depending on the length of the DCI message, the base station can transmit a PDCCH scheduling message by mapping it onto one, two, four or eight consecutive CCEs; in other words onto 36, 72, 144 or 288 resource elements.

In turn, the control channel elements are organized into *search spaces*. These come in two types. *Common search spaces* are available to all the mobiles in the cell and are located at fixed positions within the downlink control region. *UE-specific search spaces* are assigned to groups of mobiles and have locations that depend on the mobiles' RNTIs. Each search space contains up to 16 control channel elements, so it contains several

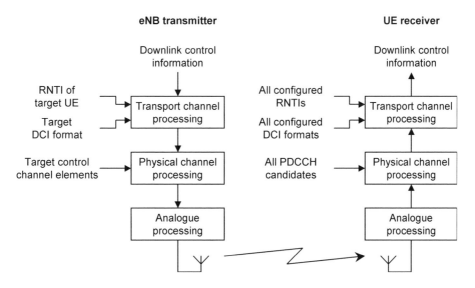

Figure 8.5 Transmission and reception of the PDCCH.

locations where the base station might transmit downlink control information. The base station can therefore use these search spaces to send several PDCCH messages to several different mobiles at the same time.

A mobile then receives the PDCCH as follows. Every subframe, the mobile reads the control format indicator, and establishes the size of the downlink control region and the locations of the common and UE-specific search spaces. Within each search space, it identifies the possible *PDCCH candidates*, which are control channel elements where the base station might have transmitted downlink control information. The mobile then attempts to process each PDCCH candidate, using all the combinations of RNTI and DCI format that it has been configured to look for. If the observed CRC bits match the ones expected, then it concludes that the message was sent using the RNTI and DCI format that it was looking for. It then reads the downlink control information and acts upon it.

The cyclic redundancy check can fail for several reasons: the base station may not have sent a scheduling message in those control channel elements, or it may have sent a scheduling message using a different DCI format or a different RNTI, or the mobile may have failed to read the message due to an uncorrected bit error. Whichever situation applies, the mobile's response is the same: it moves on to the next combination of PDCCH candidate, RNTI and DCI format, and tries again.

8.3 Data Transmission on the PDSCH and PUSCH

8.3.1 Transport Channel Processing

After the base station has sent the mobile a scheduling command, it can transmit the DL-SCH in the way that the scheduling command defined. After reception of an uplink scheduling grant, the mobile can transmit the UL-SCH in a similar way. Figure 8.6 shows the steps that the transport channel processor uses to send the data [20].

At the top of the figure, the medium access control (MAC) protocol sends information to the physical layer in the form of transport blocks. The size of each transport block is defined by the downlink control information, while its duration is the 1 ms transmission time interval.

In the uplink, the mobile sends one transport block at a time. In the downlink, the base station usually sends one transport block to each mobile, but can send two when using spatial multiplexing (DCI formats 2 and 2A). The two transport blocks can have different modulation schemes and coding rates, are mapped to different layers and are separately acknowledged. This increases the amount of signalling, so it adds some overhead to the transmission. As noted in Chapter 5, however, different layers can reach the mobile with different values of the signal to interference plus noise ratio, so we can improve the performance of the air interface by transmitting a high SINR layer using a fast modulation scheme and coding rate, and vice versa. By limiting the maximum number of transport blocks to two rather than four, we reach a compromise between these two conflicting criteria.

In the downlink (Figure 8.6a), the base station adds a 24 bit CRC to each DL-SCH transport block, segments it into code blocks with a maximum size of 6144 bits and adds another CRC to each one. It then passes the data through a $1/3$ rate turbo coder. The rate matching stage stores the resulting bits in a circular buffer and then selects bits from the

Data Transmission and Reception 133

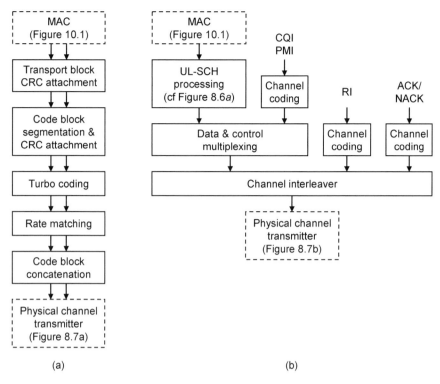

Figure 8.6 Transport channel processing in Releases 8 and 9, for the (a) DL-SCH (b) UL-SCH. Reproduced by permission of ETSI.

buffer for transmission. The number of transmitted bits is determined by the size of the resource allocation and the exact choice is determined by the redundancy version. Finally, the base station reassembles the coded transport blocks and sends them to the physical channel processor in the form of *codewords*.

The mobile processes the received data in the manner described in Chapter 3. The turbo decoding algorithm is an iterative one, which continues until the code block CRC is passed. The receiver then reassembles each transport block and uses the transport block CRC for error detection.

In the uplink (Figure 8.6b), the mobile transmits the UL-SCH by means of the same steps that the base station used on the downlink. If the mobile is sending uplink control information in the same subframe, then it processes the control bits using forward error correction and multiplexes them into the UL-SCH, in the manner indicated by the diagram.

8.3.2 Physical Channel Processing

The transport channel processor passes the outgoing codeword(s) to the physical channel processor, which transmits them in the manner shown in Figure 8.7 [21].

In the downlink (Figure 8.7a), the scrambling stage mixes each codeword with a pseudo-random sequence that depends on the physical cell ID and the target RNTI, to reduce

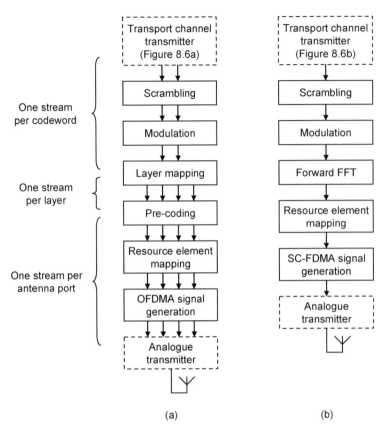

Figure 8.7 Physical channel processing in Releases 8 and 9, for the (a) PDSCH (b) PUSCH. Reproduced by permission of ETSI.

the interference between transmissions from nearby cells. The modulation mapper takes the resulting bits in groups of two, four or six and maps them onto the in-phase and quadrature components using QPSK, 16-QAM or 64-QAM.

The next two stages implement the multiple antenna transmission techniques from Chapter 5. The layer mapping stage takes the codewords and maps them onto one to four independent layers, while the precoding stage applies the chosen precoding matrix and maps the layers onto the different antenna ports.

The resource element mapper carries out a serial-to-parallel conversion and maps the resulting sub-streams onto the chosen sub-carriers, along with the sub-streams resulting from all the other data transmissions, control channels and physical signals. The PDSCH occupies resource elements in the data region of each subframe that have not been assigned to other channels or signals, in the manner shown in Figure 6.10. Finally, the OFDMA signal generator applies an inverse fast Fourier transform and a parallel-to-serial conversion and inserts the cyclic prefix. The result is a digital representation of the time-domain data that will be transmitted from each antenna port.

There are just a few differences in the uplink (Figure 8.7b). Firstly, the process includes the forward FFT that is the distinguishing feature of SC-FDMA. Secondly, there is no layer mapping or precoding, because the uplink does not use single user MIMO in LTE

Data Transmission and Reception 135

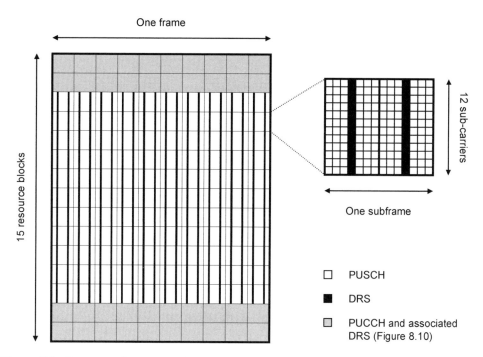

Figure 8.8 Resource element mapping for the PUSCH and its demodulation reference signal, using FDD mode, a normal cyclic prefix, a 3 MHz bandwidth and two pairs of resource blocks for the PUCCH.

Release 8. Thirdly, the PUSCH occupies a contiguous set of resource blocks towards the centre of the uplink band, with the edges reserved for the PUCCH. Each subframe contains six PUSCH symbols and one demodulation reference symbol, in the manner illustrated in Figure 8.8.

8.4 Transmission of Hybrid ARQ Indicators on the PHICH

8.4.1 Introduction

We can now start to discuss the feedback that the receiver sends back to the transmitter. The base station's feedback is easier to understand than the mobile's and is a better place to begin.

During the procedure for uplink transmission and reception, the base station sends acknowledgements to the mobiles in the form of hybrid ARQ indicators and transmits them on the physical hybrid ARQ indicator channel [22–25]. The exact transmission technique depends on the cell's PHICH configuration, which contains two parameters: the *PHICH duration* (normal or extended) and a parameter N_g that can take values of $1/6$, $1/2$, 1 or 2. The transmission technique also depends on the cyclic prefix duration.

In the discussion that follows, we will generally assume that the base station is using the normal PHICH duration and the normal cyclic prefix. The details of the other techniques are rather different, but the underlying principles remain the same.

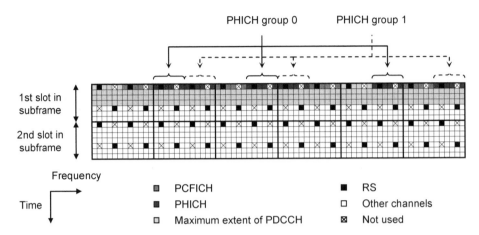

Figure 8.9 Resource element mapping for the PHICH, using a normal PHICH duration, a normal cyclic prefix, a 1.4 MHz bandwidth, the first antenna port of two, a physical cell ID of 1 and two PHICH groups.

8.4.2 Resource Element Mapping of the PHICH

The base station transmits each hybrid ARQ indicator in the downlink control region, using a set of three resource element groups (12 resource elements) that is known as a *PHICH group*. The number of PHICH groups depends on the cell bandwidth and the value of N_g. It is constant in FDD mode, but can vary from one subframe to the next in TDD mode because the base station has to send more acknowledgements in some TDD subframes than in others.

Each PHICH group is mapped to resource element groups that have not already been assigned to the PCFICH. These lie in the first symbol of a subframe when using the normal PHICH duration, but can cover two or three symbols when using the extended PHICH duration. Figure 8.9 shows an example mapping, for a base station that is using the normal PHICH duration and two PHICH groups.

A PHICH group is not dedicated to a single mobile: instead, it is shared amongst eight mobiles, by assigning each mobile a different *orthogonal sequence index*. A mobile determines the PHICH group number and orthogonal sequence index that it should inspect using two parameters from its original scheduling grant, namely the first physical resource block that it used for the uplink transmission and a parameter called the cyclic shift that we will see in Section 8.6 later. Together, the PHICH group number and orthogonal sequence index are known as a *PHICH resource*.

8.4.3 Physical Channel Processing of the PHICH

To transmit a hybrid ARQ indicator, the base station modulates it by means of BPSK, using symbols of +1 and −1 for positive and negative acknowledgements respectively. It then spreads each indicator across the four symbols in a resource element group, by multiplying it by the chosen orthogonal sequence. There are four basic sequences available to the base station, namely [+1 +1 +1 +1], [+1 −1 +1 −1], [+1 +1 −1 −1] and [+1 −1 −1 +1], but each can be applied to the in-phase and quadrature components of the signal, making a total of eight orthogonal sequences in all. The base station can send simultaneous acknowledgements to the eight mobiles in a PHICH group, by assigning them different orthogonal sequence indexes and adding the resulting symbols. This technique will be familiar to those with experience of code division multiple access and is one of the few uses that LTE makes of CDMA.

The PHICH symbols are then repeated across three resource element groups to increase the received symbol energy and are transmitted in a similar manner to the other downlink physical channels.

8.5 Uplink Control Information

8.5.1 Hybrid ARQ Acknowledgements

The mobile sends three types of uplink control information to the base station [26–28]: hybrid ARQ acknowledgements of the base station's downlink transmissions, uplink scheduling requests and channel state information. In turn, the channel state information comprises the channel quality indicator (CQI), the precoding matrix indicator (PMI) and the rank indication (RI).

First, let us consider the hybrid ARQ acknowledgements. In FDD mode, the mobile computes one or two acknowledgements per subframe, depending on the number of transport blocks that it received. It then transmits them four subframes later.

In TDD mode, things are more complicated. If the mobile is acknowledging one downlink subframe at a time, then it does so in the same way as in FDD mode. There are two ways for it to acknowledge multiple subframes. Using *ACK/NACK bundling*, the mobile sends a maximum of two acknowledgements, one for each parallel stream of transport blocks. Each acknowledgement is positive if it successfully received the corresponding transport block in all the downlink subframes and negative otherwise. Using *ACK/NACK multiplexing*, the mobile computes one acknowledgement for each downlink subframe. Each acknowledgement is positive if it successfully received both transport blocks in that subframe and negative otherwise. When using ACK/NACK multiplexing, the specifications only require the mobile to transmit a maximum of four acknowledgements at the same time, for the data received in four downlink subframes. To achieve this, the technique is not supported in TDD configuration 5.

In TDD mode, the scheduling command included a quantity known as the downlink assignment index. This indicates the total number of downlink transmissions that the mobile should be acknowledging at the same time as the scheduled data. It reduces the risk of a wrongly formatted acknowledgment if the mobile missed an earlier scheduling command, so reduces the overall error rate on the air interface.

8.5.2 Channel Quality Indicator

The channel quality indicator is a 4-bit quantity, which indicates the maximum data rate that the mobile can handle with a block error ratio of 10% or below. The CQI mainly depends on the received signal to interference plus noise ratio, because a high data rate can only be received successfully at a high SINR. However, it also depends on the

Table 8.4 Interpretation of the channel quality indicator, in terms of the modulation scheme and coding rate that a mobile can successfully receive. Reproduced by permission of ETSI

CQI	Modulation scheme	Coding rate (units of 1/1 024)	Information bits per symbol
0	n/a	0	0.00
1	QPSK	78	0.15
2	QPSK	120	0.23
3	QPSK	193	0.38
4	QPSK	308	0.60
5	QPSK	449	0.88
6	QPSK	602	1.18
7	16-QAM	378	1.48
8	16-QAM	490	1.91
9	16-QAM	616	2.41
10	64-QAM	466	2.73
11	64-QAM	567	3.32
12	64-QAM	666	3.90
13	64-QAM	772	4.52
14	64-QAM	873	5.12
15	64-QAM	948	5.55

implementation of the mobile receiver, because an advanced receiver can successfully process the incoming data at a lower SINR than a more basic one.

Table 8.4 shows how the CQI is interpreted, in terms of the downlink modulation scheme and coding rate. The last column shows the number of information bits per symbol and is calculated by multiplying the coding rate by 2, 4 or 6.

Because of frequency-dependent fading, the channel quality can often vary across the downlink band. To reflect this, the base station can configure the mobile to report the CQI in three different ways. *Wideband reporting* covers the whole of the downlink band. For *higher layer configured sub-band reporting*, the base station divides the downlink band into sub-bands, and the mobile reports one CQI value for each. For *UE selected sub-band reporting*, the mobile selects the sub-bands that have the best channel quality and reports their locations, together with one CQI that spans them and a separate wideband CQI. If the mobile is receiving more than one transport block, then it can also report a different CQI value for each, to reflect the fact that different layers can reach the mobile with different values of the SINR.

The base station uses the received CQI in its calculation of the modulation scheme and coding rate, and in support of frequency-dependent scheduling. Despite the frequency dependence of the CQI, however, the base station only uses one frequency-independent modulation scheme and coding rate per transport block when it comes to transmit the downlink data.

8.5.3 Rank Indication

The mobile reports a rank indication when it is configured for spatial multiplexing in transmission mode 3 or 4. The rank indication lies between 1 and the number of base station antenna ports and indicates the maximum number of layers that the mobile can successfully receive.

The mobile reports a single rank indication, which applies across the whole of the downlink band. The rank indication can be calculated jointly with the PMI, by choosing the combination that maximizes the expected downlink data rate.

8.5.4 Precoding Matrix Indicator

The mobile reports a precoding matrix indicator when it is configured for closed loop spatial multiplexing, multiple user MIMO or closed loop transmit diversity, in transmission modes 4, 5 or 6. The PMI indicates the precoding matrix that the base station should apply before transmitting the signal.

The PMI can vary across the downlink band, in a similar way to the CQI. To reflect this, there are two options for PMI reporting. The mobile can report a *single PMI* spanning the whole downlink band or spanning all of the UE-selected sub-bands. When using *multiple PMIs*, it either reports both of these quantities, or reports one PMI for each higher layer configured sub-band.

The base station uses the received PMI to calculate the precoding matrix that it should apply to its next downlink transmission. Once again, the base station actually transmits the data using one frequency-independent precoding matrix, despite the frequency dependence of the PMI.

8.5.5 Channel State Reporting Mechanisms

The mobile can return channel state information to the base station in two ways. *Periodic reporting* is carried out at regular intervals, which lie between 2 and 160 ms for the CQI and PMI and are up to 32 times greater for the RI. The information is usually carried by the PUCCH, but is transferred to the PUSCH if the mobile is sending uplink data in the same subframe. The maximum number of bits in each periodic report is 11, to reflect the low data rate that is available on the PUCCH.

Aperiodic reporting is carried out at the same time as a PUSCH data transmission and is requested using a field in the mobile's scheduling grant. If both types of report are scheduled in the same subframe, then the aperiodic report takes priority.

For both techniques, the base station can configure the mobile into a *channel quality reporting mode* using RRC signalling. The reporting mode defines the type of channel quality information that the base station requires, in the manner defined by Tables 8.5 and 8.6. In each mode, the first number describes the type of CQI feedback that the

Table 8.5 Channel quality reporting modes for periodic reporting on the PUCCH or PUSCH. Reproduced by permission of ETSI

PMI feedback type	Downlink transmission modes	CQI feedback type	
		Wideband	UE selected sub-bands
None	1, 2, 3, 7	Mode 1-0	Mode 2-0
Single	4, 5, 6	Mode 1-1	Mode 2-1

Table 8.6 Channel quality reporting modes for aperiodic reporting on the PUSCH. Reproduced by permission of ETSI

PMI feedback type	Downlink transmission modes	CQI feedback type		
		Wideband	UE selected sub-bands	Higher layer configured sub-bands
None	1, 2, 3, 7	-	Mode 2-0	Mode 3-0
Single	4, 5, 6	-	-	Mode 3-1
Multiple	4, 6	Mode 1–2	Mode 2-2	-

base station requires, while the second describes the type of PMI feedback. The precise definitions of each reporting mode are covered in the specifications and are different for periodic and aperiodic reporting, because of the need to limit the amount of data transferred on the PUCCH. In particular, periodic mode 2-0 is defined differently from aperiodic mode 2-0.

8.5.6 Scheduling Requests

If the mobile has data waiting for transmission on the PUSCH, then the data eventually trigger a scheduling request. Normally, the mobile composes a one-bit scheduling request for transmission on the PUCCH. However, it does not send the request right away, because it has to share the PUCCH with other mobiles. Instead, it transmits the scheduling request in a subframe that is configured by RRC signalling, which recurs with a period between 5 and 80 ms. The mobile never sends channel state information at the same time as a scheduling request: instead, the scheduling request takes priority.

A well behaved base station should reply to a scheduling request by giving the mobile a scheduling grant. However, it is not obliged to do so. If the mobile reaches a maximum number of scheduling requests without receiving a reply, then it triggers the random access procedure that is covered in Chapter 9. The base station is obliged to give the mobile a scheduling grant as part of that procedure, which solves the problem.

A mobile in RRC_IDLE state cannot transmit on the PUCCH, so it cannot send a scheduling request at all. Instead, it uses the random access procedure right away.

8.6 Transmission of Uplink Control Information on the PUCCH

8.6.1 PUCCH Formats

If the mobile wishes to send uplink control information and is not carrying out a PUSCH transmission in the same subframe, then it transmits the information on the physical uplink control channel [29–32]. The PUCCH can be transmitted using several different formats. Table 8.7 shows how these formats are used, for the case of a normal cyclic prefix.

As indicated towards the right of the table, the transport channel processor applies error correction coding to the channel state information, which increases the number of CSI bits to 20. However, it sends the scheduling request and acknowledgement bits directly down to the physical layer, without any coding at all.

Data Transmission and Reception 141

Table 8.7 List of PUCCH formats and their applications in the case of a normal cyclic prefix. Reproduced by permission of ETSI

PUCCH format	Release	Application	Number of UCI bits	Number of PUCCH bits
1	R8	SR	1	1
1a	R8	1 bit HARQ-ACK and optional SR	1 or 2	1 or 2
1b	R8	2 bit HARQ-ACK and optional SR	2 or 3	2 or 3
2	R8	CQI, PMI, RI	≤ 11	20
2a	R8	CQI, PMI, RI and 1 bit HARQ-ACK	≤ 12	21
2b	R8	CQI, PMI, RI and 2 bit HARQ-ACK	≤ 13	22
3	R10	20 bit HARQ-ACK and optional SR	≤ 21	48

The mobile transmits the PUCCH at the edges of the uplink band (Figure 8.10), to keep it separate from the PUSCH. The base station reserves resource blocks at the extreme edges of the band for PUCCH formats 2, 2a and 2b, with the exact number of blocks advertised in SIB 2. Formats 1, 1a and 1b use resource blocks further in, with the number of blocks varying dynamically from one subframe to the next depending on the

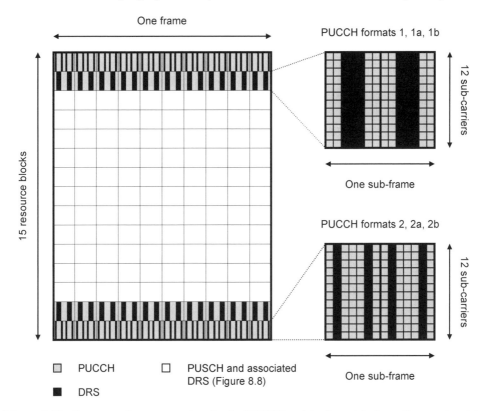

Figure 8.10 Resource element mapping for the PUCCH and its demodulation reference signal in Releases 8 and 9, using FDD mode, a normal cyclic prefix, a 3 MHz bandwidth and one pair of resource blocks for both types of PUCCH format.

number of acknowledgements that the base station is expecting. The base station can also share an intermediate pair of resource blocks amongst all the PUCCH formats, which can be useful if the bandwidth is small. When using the normal cyclic prefix, formats 1, 1a and 1b use four PUCCH symbols per slot and three demodulation reference symbols, while formats 2, 2a and 2b use five PUCCH symbols per slot and two demodulation reference symbols.

An individual mobile transmits the PUCCH using two resource blocks, which are in the first and second slots of a subframe and at opposite sides of the frequency band. However, a mobile does not have these resource blocks to itself. In PUCCH formats 2, 2a and 2b, each pair of resource blocks is shared amongst 12 mobiles, using a mobile-specific parameter known as the *cyclic shift* that runs from 0 to 11. In PUCCH formats 1, 1a and 1b, the resource blocks are shared amongst 36 mobiles, using the cyclic shift and another mobile-specific parameter, the *orthogonal sequence index*, which runs from 0 to 2.

8.6.2 PUCCH Resources

A *PUCCH resource* is a number that determines three things: the resource blocks on which the mobile should transmit the PUCCH, and the orthogonal sequence index and cyclic shift that it should use. The base station can assign three types of PUCCH resource to each mobile.

The first PUCCH resource, denoted $n^{(1)}_{PUCCH}$, is used for stand-alone hybrid ARQ acknowledgements in formats 1a and 1b. The mobile calculates $n^{(1)}_{PUCCH}$ dynamically, using the index of the first control channel element that the base station used for its downlink scheduling command.

The second PUCCH resource, denoted $n^{(1)}_{PUCCH, SRI}$, is used for scheduling requests in format 1. The third, denoted $n^{(2)}_{PUCCH}$, is used for channel state information and optional acknowledgements in formats 2, 2a and 2b. The mobile receives both of these resources by means of mobile-specific RRC signalling messages.

If a mobile wishes to send hybrid ARQ acknowledgements at the same time as a scheduling request, then it processes the acknowledgements in the usual way, but transmits them using $n^{(1)}_{PUCCH, SRI}$. The base station is already expecting the acknowledgements, so it knows how to process them, while it recognizes the scheduling request from the mobile's use of $n^{(1)}_{PUCCH, SRI}$.

If the mobile is using ACK/NACK multiplexing in TDD mode, then it may have to send more than two acknowledgements in one subframe. It usually does this by transmitting on a maximum of four PUCCH resources, denoted $n^{(1)}_{PUCCH, 0}$ to $n^{(1)}_{PUCCH, 3}$, which it calculates from the first CCE in a similar way to $n^{(1)}_{PUCCH}$. If, however, it wishes to send a scheduling request or channel state information at the same time, then it compresses the hybrid ARQ acknowledgements down to two bits, and sends them on $n^{(1)}_{PUCCH, SRI}$ or $n^{(2)}_{PUCCH}$ in the usual way.

8.6.3 Physical Channel Processing of the PUCCH

We now have enough information to describe the physical channel processing for the PUCCH.

When using PUCCH formats 1, 1a and 1b, the mobile modulates the bits onto one symbol, using on-off modulation for a scheduling request, BPSK for a one-bit acknowledgement and QPSK for a two-bit acknowledgement. It then spreads the information in the time domain using the orthogonal sequence index, usually across four symbols, but across three symbols in slots which support a sounding reference signal that is taking priority over these PUCCH formats (Section 8.7.2). The spreading process follows the same technique that the base station used for the PHICH, and allows the symbols to be shared amongst three different mobiles.

The mobile then spreads the information across 12 sub-carriers in the frequency domain using the cyclic shift. This technique is implemented differently from the one above but has the same objective, namely sharing the sub-carriers amongst 12 different mobiles. Finally, the mobile repeats its transmission in the first and second slots of the subframe.

When using PUCCH format 2, the mobile modulates the channel state information bits onto 10 symbols using QPSK and spreads the information in the frequency domain using the cyclic shift. It can also send simultaneous acknowledgements in formats 2a and 2b, by modulating the second reference symbol in each subframe using BPSK or QPSK.

8.7 Uplink Reference Signals

8.7.1 Demodulation Reference Signal

The mobile transmits the demodulation reference signal [33] along with the PUSCH and PUCCH, to help the base station carry out channel estimation. As shown in Figures 8.8 and 8.10, the signal occupies three symbols per slot when the mobile is using PUCCH formats 1, 1a and 1b, two when using PUCCH formats 2, 2a and 2b, and one when using the PUSCH.

The demodulation reference signal can contain 12, 24, 36, ... data points, corresponding to transmission bandwidths of 1, 2, 3, ... resource blocks. To generate the signal, each cell is assigned to one of 30 *sequence groups*. With one exception, described at the end of this section, each sequence group contains one *base sequence* of every possible length, which is generated either from a Zadoff-Chu sequence, or, in the case of the very shortest sequences, from a look-up table. The base sequence is then modified by one of 12 cyclic shifts, to generate the reference signal itself.

There are two ways to assign the sequence groups. In *sequence group planning*, each cell is permanently assigned to one of the sequence groups during radio network planning. Nearby cells should lie in different sequence groups, so as to minimize the interference between them. In *sequence group hopping*, the sequence group changes from one slot to the next according to one of 510 pseudo-random hopping patterns. The hopping pattern depends on the physical cell identity and can be calculated without the need for any further planning.

When sending PUSCH reference signals, the mobile calculates the cyclic shift from a field that the base station supplied in its scheduling grant. In the case of uplink multiple user MIMO, the base station can distinguish different mobiles that are sharing the same resource blocks by giving them different cyclic shifts. The remaining cyclic shifts can be used to distinguish nearby cells that share the same sequence group.

When sending PUCCH reference signals, the mobile applies the same cyclic shift that it used for the PUCCH transmission itself, and modifies the demodulation reference signal further in the case of formats 1, 1a and 1b using the orthogonal sequence index. This process allows the base station to distinguish the reference signals from all the mobiles that are sharing each pair of resource blocks.

There are two other complications. Firstly, each sequence group actually contains two base sequences for every transmission bandwidth of six resource blocks or more. In *sequence hopping*, a mobile can be configured to switch between the two sequences according to a pseudo-random pattern. Secondly, *cyclic shift hopping* makes the cyclic shift change in a pseudo-random manner from one slot to the next. Both techniques reduce the interference between nearby cells that share the same sequence group.

8.7.2 Sounding Reference Signal

The mobile transmits the sounding reference signal (SRS) [34–36] to help the base station measure the received signal power across a wide transmission bandwidth. The base station then uses the information for frequency dependent scheduling.

The base station controls the timing of sounding reference signals in two ways. Firstly, it tells the mobiles which subframes support sounding, using a parameter in SIB 2 called the *SRS subframe configuration*. Secondly, it configures each mobile with a sounding period of 2 to 320 subframes and an offset within that period, using a mobile-specific parameter called the *SRS configuration index*. A mobile transmits the sounding reference signal whenever the resulting transmission times coincide with a subframe that supports sounding.

The mobile usually sends the sounding reference signal in the last symbol of the subframe, as shown in Figure 8.11. In TDD mode, it can also send the signal in the uplink region of a special subframe. The mobile generates the signal in a similar way to the demodulation reference signal described above. The main difference is that the sounding reference signal uses eight cyclic shifts rather than 12, so that eight mobiles can share the same set of resource elements.

In the frequency domain, the base station controls the starting position and transmission bandwidth using cell- and mobile-specific parameters called the SRS bandwidth configuration, SRS bandwidth, frequency domain position and SRS hopping bandwidth. As shown

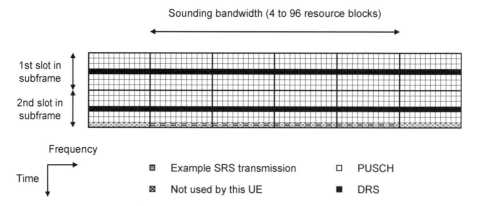

Figure 8.11 Example resource element mapping for the sounding reference signal, using a normal cyclic prefix.

in the figure, an individual mobile transmits on alternate sub-carriers, as configured by a transmission comb.

There are various ways to prevent clashes between the sounding reference signal and the mobile's other transmissions. A mobile does not transmit the PUSCH in the last symbol of subframes that support sounding, so it can always send the PUSCH and SRS in the same subframe. PUCCH formats 2, 2a and 2b take priority over the sounding reference signal, as they have reserved frequencies at the edge of the transmission band that are of no interest to the sounding procedure. The base station can configure PUCCH formats 1, 1a and 1b to use either technique by means of RRC signalling.

8.8 Uplink Power Control

8.8.1 Uplink Power Calculation

The uplink power control procedure [37, 38] sets the mobile's transmit power to the smallest value that is consistent with satisfactory reception of the signal. This reduces the interference between mobiles that are transmitting on the same resource elements in nearby cells and increases the mobile's battery life. In LTE, the mobile estimates its transmit power as well as it can and the base station adjusts this estimate using power control commands. The mobile uses slightly different calculations for the PUSCH, PUCCH and SRS, so, to illustrate the principles, we will just look at the PUSCH.

The PUSCH transmit power is calculated as follows:

$$P_{\text{PUSCH}}(i) = \min(P(i), P_{\text{CMAX}}) \quad (8.1)$$

In this equation, $P_{\text{PUSCH}}(i)$ is the power transmitted on the PUSCH in subframe i, measured in decibels relative to one milliwatt (dBm). P_{CMAX} is the mobile's maximum transmit power, while $P(i)$ is calculated as follows:

$$P(i) = P_{\text{O\_PUSCH}} + 10\log_{10}(M_{\text{PUSCH}}(i)) + \Delta_{\text{TF}}(i) + \alpha.\text{PL} + f(i) \quad (8.2)$$

Here, $P_{\text{O\_PUSCH}}$ is the power that the base station expects to receive over a bandwidth of one resource block. It has two components, a cell-specific baseline $P_{\text{O\_NOMINAL\_PUSCH}}$ and a mobile-specific adjustment $P_{\text{O\_UE\_PUSCH}}$, which are sent to the mobile using RRC signalling. $M_{\text{PUSCH}}(i)$ is the number of resource blocks on which the mobile is transmitting in subframe i. $\Delta_{\text{TF}}(i)$ is an optional adjustment for the data rate in subframe i, which ensures that the mobile uses a higher transmit power for a larger coding rate or a faster modulation scheme such as 64-QAM.

PL is the downlink path loss. The base station advertises the power transmitted on the downlink reference signals as part of SIB 2, so the mobile can estimate PL by reading this quantity and subtracting the power received. α is a weighting factor that reduces the impact of changes in the path loss, in a technique known as *fractional power control*. By setting α to a value between zero and one, the base station can ensure that mobiles at the cell edge transmit a weaker signal than would otherwise be expected. This reduces the interference that they send into nearby cells and can increase the capacity of the system.

Using the parameters covered so far, the mobile can make its own estimate of the PUSCH transmit power. However, this estimate may not be accurate, particularly in FDD mode, where the fading patterns are likely to be different on the uplink and downlink. The base station therefore adjusts the mobile's power using power control commands, which are handled by the last parameter, $f(i)$.

8.8.2 Uplink Power Control Commands

The base station can send power control commands for the PUSCH in two ways. Firstly, it can send stand-alone power control commands to groups of mobiles using DCI formats 3 and 3A. When using these formats, the base station addresses the PDCCH message to a radio network identity known as the TPC-PUSCH-RNTI, which is shared amongst all the mobiles in the group. The message contains a power control command for each of the group's mobiles, which is found using an offset that has previously been configured by means of RRC signalling.

The mobile then accumulates its power control commands in the following way:

$$f(i) = f(i-1) + \delta_{PUSCH}(i - K_{PUSCH}) \quad (8.3)$$

Here, the mobile receives a power adjustment of δ_{PUSCH} in subframe $i - K_{PUSCH}$, and applies it in subframe i. K_{PUSCH} is four in FDD mode, while in TDD mode, it can lie between four and seven in the usual way. When using DCI format 3, the power control command contains two bits and causes power adjustments of -1, 0, 1 and 3 decibels. When using DCI format 3A, the command only contains one bit and causes power adjustments of -1 and 1 decibels.

The base station can also send two bit power control commands to one mobile as part of an uplink scheduling grant. Usually, the mobile interprets them in the manner described above. However, the base station can also disable the accumulation of power control commands using RRC signalling, in which case the mobile interprets them as follows:

$$f(i) = \delta_{PUSCH}(i - K_{PUSCH}) \quad (8.4)$$

In this case, the power adjustment δ_{PUSCH} can take values of -4, -1, 1 and 4 decibels.

8.9 Discontinuous Reception

8.9.1 Discontinuous Reception and Paging in RRC_IDLE

When a mobile is in a state of *discontinuous reception* (DRX), the base station only sends it downlink control information on the PDCCH in certain subframes. Between those subframes, the mobile can stop monitoring the PDCCH and can enter a low-power state known as *sleep mode*, so as to maximize its battery life. Discontinuous reception is implemented using two different mechanisms, which support paging in RRC_IDLE and low data rate transmission in RRC_CONNECTED.

In RRC_IDLE state, discontinuous reception is defined using a *DRX cycle* [39, 40], which lies between 32 and 256 frames (0.32 and 2.56 seconds). The base station specifies a default DRX cycle length in SIB 2, but the mobile can request a different cycle length during an attach request or a tracking area update (Chapters 11 and 14).

Data Transmission and Reception 147

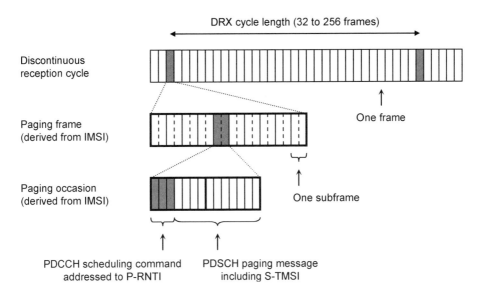

Figure 8.12 Operation of discontinuous reception and paging in RRC_IDLE.

As shown in Figure 8.12, the mobile wakes up once every DRX cycle frames, in a *paging frame* whose system frame number depends on the mobile's international mobile subscriber identity. Within that frame, the mobile inspects a subframe known as a *paging occasion*, which also depends on the IMSI. If the mobile finds downlink control information addressed to the P-RNTI, then it goes on to receive an RRC *Paging* message on the PDSCH.

Several mobiles can share the same paging occasion. To resolve this conflict, the Paging message contains the identity of the target mobile, using the S-TMSI (if available) or the IMSI (otherwise). If the mobile detects a match, then it responds to the paging message using an EPS mobility management procedure known as a service request, in the manner described in Chapter 14.

8.9.2 Discontinuous Reception in RRC_CONNECTED

In RRC_CONNECTED state, the base station configures a mobile's discontinuous reception parameters by means of mobile-specific RRC signalling. During discontinuous reception (Figure 8.13), the mobile wakes up every *DRX cycle* subframes, in a subframe defined by a *DRX start offset*. It monitors the PDCCH continuously for a duration known as the *active time* and then goes back to sleep [41, 42].

Several timers contribute to the active time. Initially, the mobile stays awake for a time of *on duration* (1 to 200 subframes), waiting for a scheduling message on the PDCCH. If one arrives, then the mobile stays awake for a time of *DRX inactivity timer* (1 to 2560 subframes) after every PDCCH command. Other timers ensure that the mobile stays awake while waiting for information such as hybrid ARQ re-transmissions, but, if all the timers expire, then the mobile goes back to sleep. The base station can also send the

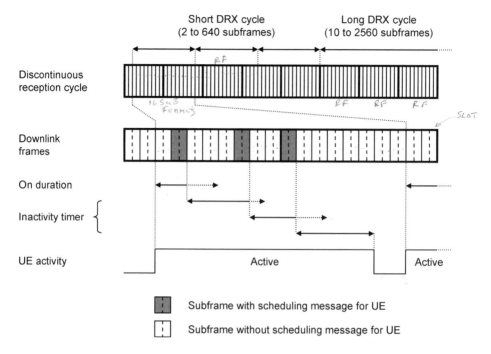

Figure 8.13 Operation of discontinuous reception in RRC_CONNECTED.

mobile to sleep explicitly, by sending it a MAC control element (Chapter 10) known as a *DRX command*.

There are actually two discontinuous reception cycles, the *long DRX cycle* (10 to 2560 subframes) and the optional *short DRX cycle* (2 to 640 subframes). If both are configured, then the mobile starts by using the short cycle, but moves to the long cycle if it goes for *DRX short cycle timer* (1 to 16) cycles without receiving a PDCCH command.

References

1. 3GPP TS 36.211 (September 2011) *Physical Channels and Modulation*, Release 10.
2. 3GPP TS 36.212 (September 2011) *Multiplexing and Channel Coding*, Release 10.
3. 3GPP TS 36.213 (September 2011) *Physical Layer Procedures*, Release 10.
4. 3GPP TS 36.321 (October 2011) *Medium Access Control (MAC) Protocol Specification*, Release 10.
5. 3GPP TS 36.331 (October 2011) *Radio Resource Control (RRC); Protocol Specification*, Release 10.
6. 3GPP TS 36.321 (October 2011) *Medium Access Control (MAC) Protocol Specification*, Release 10, section 5.3.
7. 3GPP TS 36.213 (September 2011) *Physical Layer Procedures*, Release 10, section 7.
8. 3GPP TS 36.321 (October 2011) *Medium Access Control (MAC) Protocol Specification*, Release 10, section 5.4.
9. 3GPP TS 36.213 (September 2011) *Physical Layer Procedures*, Release 10, section 8.
10. 3GPP TS 36.321 (October 2011) *Medium Access Control (MAC) Protocol Specification*, Release 10, section 5.10.
11. 3GPP TS 36.213 (September 2011) *Physical Layer Procedures*, Release 10, section 9.2.
12. 3GPP TS 36.331 (October 2011) *Radio Resource Control (RRC); Protocol Specification*, Release 10, section 6.3.2 (*SPS-Config*).

Data Transmission and Reception 149

13. 3GPP TS 36.212 (September 2011) *Multiplexing and Channel Coding*, Release 10, section 5.3.3.1.
14. 3GPP TS 36.213 (September 2011) *Physical Layer Procedures*, Release 10, sections 7.1.6, 8.1, 8.4.
15. 3GPP TS 36.211 (September 2011) *Physical Channels and Modulation*, Release 10, section 5.2.3, 5.3.4, 6.2.3.
16. 3GPP TS 36.321 (October 2011) *Medium Access Control (MAC) Protocol Specification*, Release 10, section 7.1.
17. 3GPP TS 36.212 (September 2011) *Multiplexing and Channel Coding*, Release 10, sections 5.3.3.2, 5.3.3.3, 5.3.3.4.
18. 3GPP TS 36.211 (September 2011) *Physical Channels and Modulation*, Release 10, section 6.8.
19. 3GPP TS 36.213 (September 2011) *Physical Layer Procedures*, Release 10, section 9.1.1.
20. 3GPP TS 36.212 (September 2011) *Multiplexing and Channel Coding*, Release 10, sections 5.1, 5.2.2, 5.2.4, 5.3.2.
21. 3GPP TS 36.211 (March 2010) *Physical Channels and Modulation*, Release 9, sections 5.3, 5.6, 5.8, 6.3, 6.4, 6.12, 6.13.
22. 3GPP TS 36.211 (September 2011) *Physical Channels and Modulation*, Release 10, section 6.9.
23. 3GPP TS 36.212 (September 2011) *Multiplexing and Channel Coding*, Release 10, section 5.3.5.
24. 3GPP TS 36.213 (September 2011) *Physical Layer Procedures*, Release 10, section 9.1.2.
25. 3GPP TS 36.331 (October 2011) *Radio Resource Control (RRC); Protocol Specification*, Release 10, section 6.3.2 (*PHICH-Config*).
26. 3GPP TS 36.213 (September 2011) *Physical Layer Procedures*, Release 10, sections 7.2, 7.3, 10.1.
27. 3GPP TS 36.321 (October 2011) *Medium Access Control (MAC) Protocol Specification*, Release 10, section 5.4.4.
28. 3GPP TS 36.331 (October 2011) *Radio Resource Control (RRC); Protocol Specification*, Release 10, section 6.3.2 (*CQI-ReportConfig*, *SchedulingRequestConfig*).
29. 3GPP TS 36.211 (September 2011) *Physical Channels and Modulation*, Release 10, section 5.4.
30. 3GPP TS 36.212 (September 2011) *Multiplexing and Channel Coding*, Release 10, section 5.2.3.
31. 3GPP TS 36.213 (September 2011) *Physical Layer Procedures*, Release 10, section 10.1.
32. 3GPP TS 36.331 (October 2011) *Radio Resource Control (RRC); Protocol Specification*, Release 10, section 6.3.2 (*PUCCH-Config*).
33. 3GPP TS 36.211 (September 2011) *Physical Channels and Modulation*, Release 10, sections 5.5.1, 5.5.2.
34. 3GPP TS 36.211 (September 2011) *Physical Channels and Modulation*, Release 10, section 5.5.3.
35. 3GPP TS 36.213 (September 2011) *Physical Layer Procedures*, Release 10, section 8.2.
36. 3GPP TS 36.331 (October 2011) *Radio Resource Control (RRC); Protocol Specification*, Release 10, section 6.3.2 (*SoundingRS-UL-Config*).
37. 3GPP TS 36.213 (September 2011) *Physical Layer Procedures*, Release 10, section 5.
38. 3GPP TS 36.331 (October 2011) *Radio Resource Control (RRC); Protocol Specification*, Release 10, section 6.3.2 (*Uplink Power Control*, *TPC-PDCCH-Config*).
39. 3GPP TS 36.304 (October 2011) *Evolved Universal Terrestrial Radio Access (E-UTRA); User Equipment (UE) Procedures in Idle Mode*, Release 10, section 7.
40. 3GPP TS 36.331 (October 2011) *Radio Resource Control (RRC); Protocol Specification*, Release 10, sections 5.3.2, 6.2.2 (*Paging*).
41. 3GPP TS 36.321 (October 2011) *Medium Access Control (MAC) Protocol Specification*, Release 10, section 5.7.
42. 3GPP TS 36.331 (October 2011) *Radio Resource Control (RRC); Protocol Specification*, Release 10, section 6.3.2 (*MAC-MainConfig*).

9
Random Access

As described in the previous chapter, the base station explicitly schedules all the transmissions that the mobile carries out on the physical uplink shared channel. If the mobile wishes to transmit on the PUSCH but does not have the resources to do so, then it usually sends a scheduling request on the physical uplink control channel. If it does not have the resources to do that, then it initiates the random access procedure. This can happen in a few different situations, primarily during the establishment of an RRC connection, during a handover, or if the mobile has lost timing synchronization with the base station. The base station can also trigger the random access procedure, if it wishes to transmit to the mobile after a loss of timing synchronization.

The procedure begins when the mobile transmits a random access preamble on the physical random access channel (PRACH). This initiates an exchange of messages between the mobile and the base station that has two main variants, non contention based and contention based. As a result of the procedure, the mobile receives three quantities: resources for an uplink transmission on the PUSCH, an initial value for the uplink timing advance and, if it does not already have one, a C-RNTI.

The random access procedure is defined by the same specifications that were used for data transmission and reception. The most important are TS 36.211 [1], TS 36.213 [2], TS 36.321 [3] and TS 36.331 [4].

9.1 Transmission of Random Access Preambles on the PRACH

9.1.1 Resource Element Mapping

The best place to start discussing the PRACH [5–7] is with the resource element mapping. In the frequency domain, a PRACH transmission has a bandwidth of six resource blocks. In the time domain, the transmission is usually one subframe long, but it can be longer or shorter. Figure 9.1 shows an example, for the case of FDD mode and a bandwidth of 3 MHz.

Looking in more detail, the PRACH transmission comprises a cyclic prefix, a preamble sequence and a guard period. In turn, the preamble sequence contains one or two PRACH symbols, which are usually 800 μs long. The mobile transmits the PRACH without any

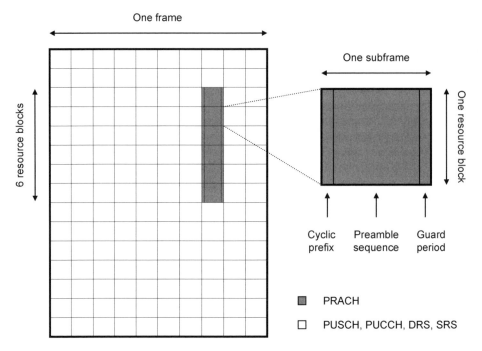

Figure 9.1 Resource element mapping for the physical random access channel, using FDD mode, a normal cyclic prefix, a 3 MHz bandwidth, a PRACH configuration index of 5 and a PRACH frequency offset of 7.

Table 9.1 Random access preamble formats. Reproduced by permission of ETSI

	Approximate duration (µs)				
Format	Cyclic prefix	Preamble	Guard period	Total	Application
0	103	800	97	1000	Normal cells
1	684	800	516	2000	Large cells
2	203	1600	197	2000	Weak signals
3	684	1600	716	3000	Large cells and weak signals
4	15	133	9	157	Small TDD cells

timing advance, but the guard period prevents it from colliding at the base station with the symbols that follow.

The base station can specify the duration of each component using the preamble formats listed in Table 9.1. The most common is format 0, in which the transmission takes up one subframe. Formats 1 and 3 have long guard periods, so are useful in large cells, while formats 2 and 3 have two PRACH symbols, so are useful if the received signal is weak. Format 4 is only used by small TDD cells, and maps the PRACH onto the uplink part of a special subframe.

The base station reserves specific resource blocks for the PRACH using two parameters that it advertises in SIB 2, namely the *PRACH configuration index* and the *PRACH frequency offset*. These parameters have different meanings in FDD and TDD modes.

In FDD mode, the PRACH configuration index specifies the preamble format and the subframes in which random access transmissions can begin, while the PRACH frequency offset specifies their location in the frequency domain. In Figure 9.1, for example, the PRACH configuration index is 5, which supports PRACH transmissions in subframe 7 of every frame using preamble format 0. The PRACH frequency offset is 7, so the transmissions occupy resource blocks 7 to 12 in the frequency domain.

In TDD mode, random access transmissions cannot take place during downlink subframes, so there are fewer opportunities for them. To compensate for this, mobiles can transmit the PRACH at a maximum of six locations in the frequency domain, instead of one. Together, the PRACH configuration index and frequency offset define a mapping into the resource grid, from which the mobile can discover the times and frequencies that support the PRACH.

9.1.2 Preamble Sequence Generation

Each cell supports 64 different preamble sequences. The mobile generates these from the Zadoff-Chu sequences that we introduced in Chapter 7, using 64 different combinations of the root sequence and the cyclic shift. The base station tells the mobile which combinations to use, by means of a parameter called the *root sequence index* that it advertises in SIB 2. Nearby cells use different root sequence indexes, which are assigned either during network planning or by the self optimization functions that we will discuss in Chapter 19.

The base station can distinguish mobiles that are transmitting on the same set of resource blocks, provided that their preamble sequences are different. To help achieve this, it reserves some of the 64 preamble sequences for the non contention based random access procedure that we will discuss next and assigns them to individual mobiles by means of RRC signalling. The remainder are available for the contention based procedure and are chosen at random by the mobile.

9.1.3 Signal Transmission

To transmit the physical random access channel, the mobile simply generates the appropriate time domain preamble sequence and passes it to the forward Fourier transform in its physical layer. There are, however, some differences from the usual techniques for resource element mapping that we introduced in Chapter 6. In particular, the PRACH symbol duration is 800 μs in formats 0 to 3, instead of the usual value of 66.7 μs. That implies that the sub-carrier spacing is 1250 Hz, instead of the usual value of 15 kHz. In format 4, the symbol duration is 133 μs, so the sub-carrier spacing is 7500 Hz. The use of a smaller sub-carrier spacing means that the PRACH sub-carriers are not orthogonal to the sub-carriers used by the PUCCH and PUSCH. Because of this, the PRACH transmission band contains small guard bands at its upper and lower edges, to minimize the amount of interference that occurs.

Power control works differently on the random access channel from the other uplink channels. The mobile first transmits a random access preamble with the following power:

$$P_{PRACH} = \min\left(P_{PREAMBLE,\ INITIAL} + PL, P_{CMAX}\right) \quad (9.1)$$

Here, P_{CMAX} is the mobile's maximum transmit power, PL is its estimate of the downlink path loss and $P_{PREAMBLE,\ INITIAL}$ is a parameter that the base station advertises in SIB 2, which describes the power that it expects to receive.

The mobile then awaits a response from the base station, in a random access window whose duration lies between 2 and 10 subframes. If it does not receive a response within this time, then it assumes that the transmit power was too low for the base station to hear it, so it increases the transmit power by a value that lies between 0 and 6 decibels and repeats the transmission. This process continues until the mobile receives a response, or until it reaches a maximum number of re-transmissions.

9.2 Non Contention Based Procedure

When the mobile sends a PRACH transmission to the base station, it initiates the random access procedure [8–10]. There are two variants of this procedure, namely non contention based and contention based.

If the network can reserve a preamble sequence for a mobile, then it can guarantee that no other mobile will be using that sequence in the same set of resource blocks. This idea is the basis of the non contention based random access procedure, which is typically used as part of a handover as shown in Figure 9.2.

Before the procedure begins, the old base station sends the mobile an RRC message known as *RRC Connection Reconfiguration*. This tells the mobile how to reconfigure itself for communication with the new base station and identifies a preamble sequence that the new base station has reserved for it. The mobile reads the RRC message and reconfigures itself as instructed. However, it does not yet have timing synchronization, so it triggers the random access procedure.

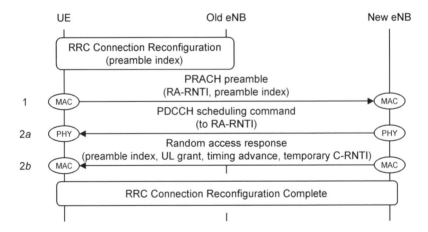

Figure 9.2 Non contention based random access procedure, as used during a handover.

The mobile reads the new cell's random access configuration from SIB 2, chooses the next available PRACH transmission time and sends a preamble using the requested sequence (step 1). The transmission frequency is fixed in FDD mode, while in TDD mode it is chosen at random. Together, the transmission time and frequency determine a mobile identity known as the random access RNTI (RA-RNTI). If necessary, the mobile repeats the transmission in the manner described above, until it receives a response.

Once the base station receives the preamble, it measures the arrival time and calculates the required timing advance. It replies first with a PDCCH scheduling command (step 2a), which it writes using DCI format 1A or 1C and addresses to the mobile's RA-RNTI. It follows this with a *random access response* (step 2b), which identifies the preamble sequence that the mobile used, and gives the mobile an uplink scheduling grant and an initial value for the uplink timing advance. (The base station also gives the mobile an identity known as the temporary C-RNTI, but the mobile does not actually use it in this version of the procedure.) The base station can identify several preamble sequences in one response, so it can simultaneously reply to all the mobiles that transmitted on the same resource blocks but with different preambles.

The mobile receives the base station's response and initializes its timing advance. It can then reply to the base station's signalling message, using an *RRC Connection Reconfiguration Complete*.

A base station can also initiate the non contention based random access procedure if it wishes to transmit to the mobile on the downlink, but has lost timing synchronization with it. To do this, it triggers the procedure using a variant of DCI format 1A known as a *PDCCH order* [11]. The procedure then continues in the manner described above.

9.3 Contention Based Procedure

A mobile uses the contention based random access procedure if it has not been allocated a preamble index. This typically happens as part of a procedure known as RRC connection establishment, in the manner shown in Figure 9.3.

In this example, the mobile wishes to send the base station an RRC message known as an *RRC Connection Request*, in which it asks to move from RRC_IDLE to RRC_CONNECTED. It has no PUSCH resources on which to send the message and no PUCCH resources on which to send a scheduling request, so it triggers the random access procedure.

The mobile reads the cell's random access configuration from SIB 2 and chooses a preamble sequence at random from the ones available. It then transmits the preamble in the usual way (step 1). There is a risk of contention, if two or mobiles transmit on the same resource blocks using the same preamble sequence. As before, the base station sends the mobile a scheduling command followed by a random access response (steps 2a and 2b).

Using the uplink grant, the mobile sends its RRC message in the usual way (step 3a). As part of the message, the mobile uniquely identifies itself using either its S-TMSI or a random number (Section 11.3.1). There is still a risk of contention between the mobiles that initiated the procedure, but if one of the transmissions is much stronger than the others, then the base station will be able to decode it. The other transmissions will only

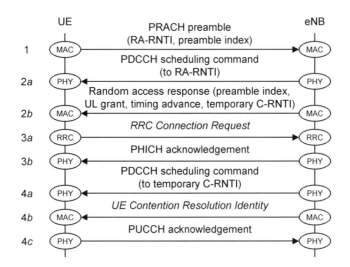

Figure 9.3 Contention based random access procedure, as used during RRC connection establishment.

cause interference. The base station sends an acknowledgement using the PHICH resource that was indicated by the scheduling grant (step 3b).

The base station now sends the mobile another scheduling command (step 4a), which it addresses to the temporary C-RNTI that it allocated earlier. It follows the command with a MAC control element called the *UE contention resolution identity* (step 4b). This echoes back the RRC message that the mobile transmitted in step 3, so it includes the identity of the successful mobile.

If a mobile receives an echo of the message that it originally transmitted, then it sends an acknowledgement using the PUCCH resource indicated by the scheduling command (step 4c). It then promotes the temporary C-RNTI to a full C-RNTI and continues the RRC procedure. If the message does not match, then the mobile discards the temporary C-RNTI and tries the random access procedure again at a later time. As a result, the base station has selected one of the mobiles that were originally competing for its attention and has told the others to back off.

A mobile can also initiate the contention based procedure in RRC_CONNECTED state, if it wishes to transmit to the base station but has lost timing synchronization, or if it has reached a maximum number of scheduling requests without receiving a reply. In this situation, however, the mobile already has a C-RNTI. In step 3 of the procedure, it replaces the RRC message with a C-RNTI MAC control element (Chapter 10) and the base station then uses the C-RNTI as the basis for contention resolution.

References

1. 3GPP TS 36.211 (September 2011) *Physical Channels and Modulation*, Release 10.
2. 3GPP TS 36.213 (September 2011) *Physical Layer Procedures*, Release 10.
3. 3GPP TS 36.321 (October 2011) *Medium Access Control (MAC) Protocol Specification*, Release 10.
4. 3GPP TS 36.331 (October 2011) *Radio Resource Control (RRC); Protocol Specification*, Release 10.

5. 3GPP TS 36.211 (September 2011) *Physical Channels and Modulation*, Release 10, section 5.7.
6. 3GPP TS 36.212 (September 2011) *Multiplexing and Channel Coding*, Release 10, section 5.2.1.
7. 3GPP TS 36.331 (October 2011) *Radio Resource Control (RRC); Protocol Specification*, Release 10, section 6.3.2 (*PRACH-Config, RACH-ConfigCommon, RACH-ConfigDedicated*).
8. 3GPP TS 36.300 (October 2011) *Evolved Universal Terrestrial Radio Access (E-UTRA) and Evolved Universal Terrestrial Radio Access Network (E-UTRAN); Overall Description; Stage 2*, Release 10, section 10.1.5.
9. 3GPP TS 36.321 (October 2011) *Medium Access Control (MAC) Protocol Specification*, Release 10, sections 5.1, 6.1.3.2, 6.1.3.4, 6.1.5, 6.2.2, 6.2.3.
10. 3GPP TS 36.213 (September 2011) *Physical Layer Procedures*, Release 10, section 6.
11. 3GPP TS 36.212 (September 2011) *Multiplexing and Channel Coding*, Release 10, section 5.3.3.1.3.

10

Air Interface Layer 2

We have now completed our survey of the air interface's physical layer. In this chapter, we round off our discussion of the LTE air interface by describing the three protocols in the data link layer, layer 2 of the OSI model. The medium access control protocol schedules all the transmissions that are made on the LTE air interface and controls the low-level operation of the physical layer. The radio link control protocol maintains the data link across the air interface, if necessary by re-transmitting packets that the physical layer has not delivered correctly. Finally, the packet data convergence protocol maintains the security of the air interface, compresses the headers of IP packets and ensures the reliable delivery of packets after a handover.

10.1 Medium Access Control Protocol

10.1.1 Protocol Architecture

The medium access control (MAC) protocol [1, 2] schedules the transmissions that are carried out on the air interface and controls the low-level operation of the physical layer. There is one MAC entity in the base station and one in the mobile. To illustrate its architecture, Figure 10.1 is a high-level block diagram of the mobile's MAC protocol.

In the transmitter, the *logical channel prioritization* function determines how much data the mobile should transmit from each incoming logical channel in every transmission time interval. In response, the mobile grabs the required data from transmit buffers in the RLC protocol, in the form of MAC service data units (SDUs). The *multiplexing* function combines the service data units together, attaches a header and sends the resulting data on a transport channel to the physical layer. It does this by way of the *hybrid ARQ transmission* function, which controls the operation of the physical layer's hybrid ARQ protocol. The output data packets are known as MAC protocol data units (PDUs) and are identical to the transport blocks that we saw in Chapter 8. The functions are reversed in the mobile's receiver.

The principles are much the same in the base station, with the downlink channels transmitted and the uplink channels received. However, there are two main differences. Firstly, the base station's MAC protocol has to carry out transmissions to different mobiles

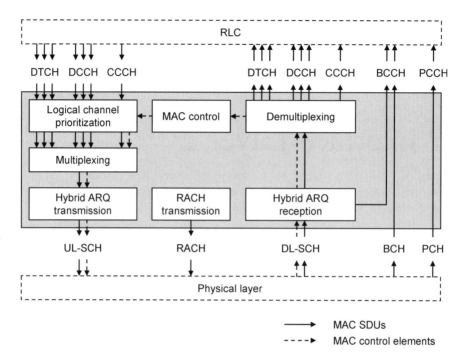

Figure 10.1 High-level architecture of the mobile's MAC protocol. Reproduced by permission of ETSI.

on the downlink and receive transmissions from different mobiles on the uplink. Secondly, the protocol includes a scheduling function, which organizes the base station's transmissions on the downlink and the mobiles' transmissions on the uplink, and which ultimately determines the contents of the PDCCH scheduling commands and scheduling grants.

The MAC protocol also sends and receives a number of *MAC control elements*, which control the low-level operation of the physical layer. There are several types of control element, which are listed in Table 10.1. We have already seen three of them. During discontinuous reception in RRC_CONNECTED, the base station can send the mobile to sleep using a *DRX command*, while the *UE contention resolution identity* and *C-RNTI* control elements are both used by the contention based random access procedure. We will discuss the remaining Release 8 control elements below, before moving on to the remaining blocks in the diagram.

10.1.2 Timing Advance Commands

After initializing a mobile's timing advance using the random access procedure, the base station updates it using MAC control elements known as *Timing advance commands*. Each command adjusts the timing advance by an amount ranging from $-496T_s$ to $+512T_s$, with a resolution of $16T_s$ [3]. This corresponds to a change of -2.4 to $+2.5$ km in the distance between the mobile and the base station, with a resolution of 80 metres.

Table 10.1 List of MAC control elements

MAC control element	Release	Application	Direction
Buffer status report	R8	UE transmit buffer occupancy	
C-RNTI	R8	UE identification option during random access	UL
Power headroom	R8	UE transmit power headroom	
Extended power headroom	R10	Power headroom during carrier aggregation	
DRX command	R8	Sends UE to sleep during DRX	
Timing advance command	R8	Adjusts UE timing advance	
UE contention resolution identity	R8	Resolves contention during random access	DL
MCH scheduling information	R9	Informs UE about scheduling of MBMS	
Activation/deactivation	R10	Activates/deactivates secondary cells	

The mobile expects to receive timing advance commands from the base station at regular intervals. The maximum permitted interval is a quantity known as *timeAlignmentTimer*, which can take a value from 500 to 10240 subframes (0.5 to 10.24 seconds), or can be infinite if the cell size is small [4]. If the time elapsed since the previous timing advance command exceeds this value, then the mobile concludes that it has lost timing synchronization with the base station. In response, it releases all its PUSCH and PUCCH resources, notably the parameters $n^{(1)}_{PUCCH}$, $n^{(1)}_{PUCCH, SRI}$ and $n^{(2)}_{PUCCH}$ from Chapter 8. Any subsequent attempt to transmit will trigger the random access procedure, through which the mobile can recover its timing synchronization.

10.1.3 Buffer Status Reporting

The mobile transmits *Buffer status report* (BSR) MAC control elements to tell the base station about how much data it has available for transmission. There are three types of buffer status report, of which the most important is the *regular BSR*. A mobile sends this in three situations: if data become ready for transmission when the transmit buffers were previously empty, or if data become ready for transmission on a logical channel with a higher priority than the buffers were previously storing, or if a timer expires while data are waiting for transmission. The mobile expects the base station to reply with a scheduling grant.

If the mobile wishes to send a regular BSR, but does not have the PUSCH resources on which to do so, then it instead sends the base station a scheduling request on the PUCCH. (In fact a scheduling request is always triggered in this way, by an inability to send a regular BSR.) If, however, the mobile is in RRC_IDLE or has lost timing synchronization with the base station, then it has no PUCCH resources either. In that situation, it runs the random access procedure instead.

There are two other types of buffer status report. The mobile transmits *periodic BSRs* at regular intervals during data transmission on the PUSCH, and *padding BSRs* if it has enough spare room during a normal PUSCH transmission.

10.1.4 Power Headroom Reporting

A mobile's power headroom is the difference between its maximum transmit power and the power requested for its PUSCH transmission [5]. The power headroom is usually positive, but it can be negative if the requested power exceeds the power available.

The mobile reports its power headroom to the base station using a *Power headroom* MAC control element. It can do so in two situations: periodically, or if the downlink path loss has changed significantly since the last report. The base station uses the information to support its uplink scheduling procedure, typically by limiting the data rate at which it asks the mobile to transmit.

10.1.5 Multiplexing and De-Multiplexing

We can now discuss the internal structure of a MAC protocol data unit (PDU) and the way in which it is assembled. In its most general form (Figure 10.2), a MAC protocol data unit contains several MAC service data units and several control elements, which collectively make up the MAC payload. Each service data unit contains data received from the RLC on a single logical channel, while each control element is one of the elements listed in Table 10.1. The payload can also contain padding, which rounds the protocol data unit up to one of the permitted transport block sizes.

Each item in the MAC payload is associated with a sub-header. The SDU sub-header identifies the size of the service data unit and the logical channel from which it originated, while the control element sub-header identifies the control element's size and type.

In the uplink, the mobile discovers the required PDU size from the base station's scheduling grant. Using the prioritization algorithm described below, the mobile decides how it will fill the available space in the protocol data unit, grabs service data units from the buffers in the overlying RLC protocol and grabs control elements from the MAC control unit. The multiplexing function then writes the corresponding sub-headers and assembles the PDU. The same technique is used in the downlink, except that, as we will see, the prioritization process is rather different.

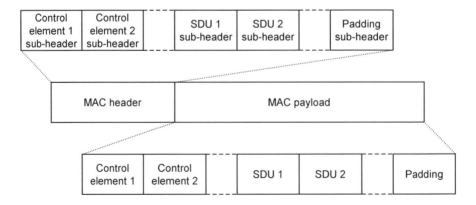

Figure 10.2 Structure of a MAC PDU. Reproduced by permission of ETSI.

10.1.6 Logical Channel Prioritization

We saw above that the base station tells the mobile about the size of every uplink MAC protocol data unit using its scheduling grant. However the scheduling grant says nothing about what the protocol data unit should contain. The mobile therefore runs a prioritization algorithm, to decide how to fill it.

To support the algorithm, each logical channel is associated with a priority from 1 to 16, where a small number corresponds to a high priority. The logical channel is also associated with a *prioritized bit rate* (PBR), which runs from zero to 256 kbps and is a target for the long-term average bit rate. The specifications also support an infinite prioritized bit rate, with the interpretation 'as fast as possible'. Ultimately, these parameters are all derived from the quality of service parameters that we will discuss in Chapter 13.

The algorithm is fully defined by the MAC protocol specification and the principles are as follows. First, the MAC runs through the logical channels in order of priority and grabs data from channels that have fallen behind their long-term average bit rates. It then runs through the channels in priority order once again and fills any space in the PDU that remains. The algorithm also prioritizes the control elements, and the resulting priority order is as follows: data on the common control channel together with any associated C-RNTI control elements, regular or periodic buffer status reports, power headroom reports, data on other logical channels and finally padding buffer status reports.

Prioritization in the downlink is rather different, because the base station is free to fill up the PDUs in any way that it likes. In practice, the downlink prioritization algorithm will form part of the proprietary scheduling algorithm that we discuss below.

10.1.7 Scheduling of Transmissions on the Air Interface

The base station's scheduling algorithm has to decide the contents of every downlink scheduling command and uplink scheduling grant, on the basis of all the information available to it at the time. The specifications say nothing about how it should work, but to illustrate its operation, Figure 10.3 shows some of the main inputs and outputs.

The information available to the downlink scheduler includes the following. Each bearer is associated with a buffer occupancy, as well as information about its quality of service such as the priority and prioritized bit rate that we introduced earlier. To support the scheduling function, the mobile returns hybrid ARQ acknowledgements, channel quality indicators and rank indications. The base station also knows the discontinuous reception pattern for every mobile in the cell and can receive load information from nearby cells about their own use of the downlink sub-carriers.

Using this information, the scheduler has to decide how many information bits to send to each mobile, whether to send a new transmission or a re-transmission and how to divide new transmissions amongst the available bearers. It also has to decide the modulation schemes and coding rates to use, the number of layers in the case of spatial multiplexing, and the allocation of resource blocks to every mobile.

The uplink scheduler follows the same principles, although some of the inputs and outputs are different. For example, the base station does not have complete knowledge of the uplink buffer occupancy and does not tell the mobiles which logical channels they

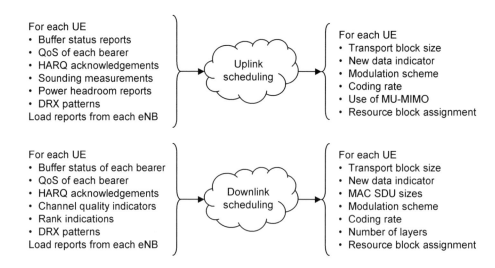

Figure 10.3 Inputs and outputs for the uplink and downlink scheduling algorithms.

should use for their uplink transmissions. In addition, the base station derives its channel quality information from the sounding procedure, instead of from the mobiles' channel quality indicators.

At a basic level, two extreme scheduling algorithms are available. A *maximum rate scheduler* allocates resources to the mobiles with the highest signal-to-noise ratios, which can transmit or receive at the highest data rates. This maximizes the throughput of the cell but is grossly unfair, as distant mobiles may not get a chance to transmit or receive at all. At the other extreme, a *round-robin scheduler* gives the same data rate to every mobile. This is fair but grossly inefficient, as distant mobiles with low signal-to-noise ratios will dominate the cell's use of resources. In practice, techniques such as *proportional fair scheduling* try to strike a balance between the two extremes.

To summarize, the scheduler is a complex piece of software, and one of the harder parts of the system to implement effectively. It is likely to be an important differentiator between equipment manufacturers and network operators.

10.2 Radio Link Control Protocol

10.2.1 Protocol Architecture

The radio link control (RLC) protocol [6, 7] maintains the layer 2 data link between the mobile and the base station, for example by ensuring reliable delivery for data streams that have to reach the receiver correctly. Figure 10.4 shows the high-level architecture of the RLC. The transmitter receives service data units from higher layers in the form of modified IP packets or signalling messages, and sends PDUs to the MAC protocol on the logical channels. The process is reversed in the receiver.

The RLC has three modes of operation, namely *transparent mode* (TM), *unacknowledged mode* (UM) and *acknowledged mode* (AM). These are set up on a channel-by-channel basis, so that each logical channel is associated with an RLC object that is configured in one of

Air Interface Layer 2   165

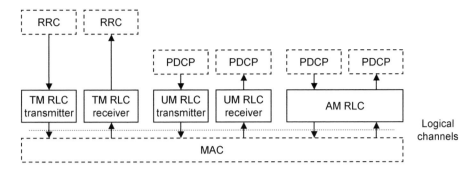

Figure 10.4 High-level architecture of the RLC protocol. Reproduced by permission of ETSI.

these modes. The transparent and unacknowledged mode RLC objects are uni-directional, while the acknowledged mode object is bi-directional.

10.2.2 Transparent Mode

Transparent mode handles three types of signalling message: system information messages on the broadcast control channel, paging messages on the paging control channel and RRC connection establishment messages on the common control channel. Its architecture (Figure 10.5) is very simple.

In the transmitter, the RLC receives signalling messages directly from the RRC protocol, and stores them in a buffer. The MAC protocol grabs the messages from the buffer as RLC PDUs, without any modification. (The messages are short enough to fit into a single transport block, without segmenting them.) In the receiver, the RLC passes the received messages directly up to the RRC.

10.2.3 Unacknowledged Mode

Unacknowledged mode handles data streams on the dedicated traffic channel for which timely delivery is more important than reliability, such as voice over IP and streaming video. Its architecture is shown in Figure 10.6.

The RLC transmitter receives service data units from the PDCP in the form of modified IP packets and stores them in a buffer in the same way as before. This time, however,

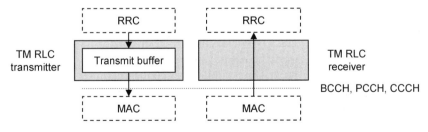

Figure 10.5 Internal architecture of the RLC protocol in transparent mode. Reproduced by permission of ETSI.

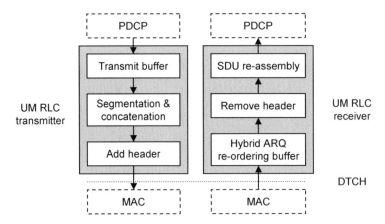

Figure 10.6 Internal architecture of the RLC protocol in unacknowledged mode. Reproduced by permission of ETSI.

the MAC protocol tells the RLC to send it a PDU with a specific size. In response, the segmentation and concatenation function cuts up the buffered IP packets and splices their ends together, so as to deliver a PDU with the correct size down to the MAC. As a result, the output PDU size does not bear any resemblance to the size of the input SDU. Finally, the RLC adds a header that contains two important pieces of information: a PDU sequence number, and a description of any segmentation and concatenation that it has done.

The PDUs can reach the receiver's RLC protocol in a different order, because of the underlying hybrid ARQ processes. To deal with this problem, the hybrid ARQ re-ordering function stores the received PDUs in a buffer, and uses their sequence numbers to send them upwards in the correct order. The receiver can then remove the header from every PDU, use the header information to undo the segmentation and concatenation process and reconstruct the original packets.

10.2.4 Acknowledged Mode

Acknowledged mode handles two types of information: data streams on the dedicated traffic channel such as web pages and emails, for which reliability is more important than speed of delivery, and mobile-specific signalling messages on the dedicated control channel. It is similar to unacknowledged mode, but also re-transmits any packets that have not reached the receiver correctly. The architecture (Figure 10.7) is bi-directional, in the sense that the same acknowledged mode object handles transmission and reception. There are also two types of PDU: data PDUs carry higher-layer data and signalling messages, and control PDUs carry RLC-specific control information.

The transmitter sends data packets in a similar way to the unacknowledged mode RLC. This time, however, it stores the transmitted PDUs in a re-transmission buffer, until it knows that they have reached the receiver correctly. At regular intervals, the transmitter also sets a *polling bit* in one of the data PDU headers. This tells the receiver to return a type of control PDU known as a *status PDU*, which lists the data PDUs it has received and the ones it has missed. In response, the transmitter discards the data PDUs that have arrived correctly and re-transmits the ones that have not.

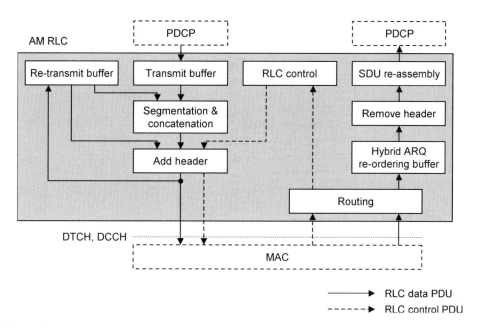

Figure 10.7 Internal architecture of the RLC protocol in acknowledged mode. Reproduced by permission of ETSI.

There is one problem. If the SINR is falling, then the MAC protocol may request a smaller PDU size for the re-transmission than it did first time around. The PDUs in the re-transmission buffer will then be too large to send. To solve the problem, the RLC protocol can cut a previously transmitted PDU into smaller segments. To support this process, the data PDU header includes extra fields, which describe the position of a re-transmitted segment within a previously transmitted PDU. The receiver can acknowledge each segment individually, using similar fields in the status PDU.

Figure 10.8 shows an example. At the start of the sequence, the transmitter sends four data PDUs to the receiver and labels each one with a sequence number. PDUs 1 and 4 reach the receiver correctly, but PDUs 2 and 3 are lost. (To be exact, the hybrid ARQ transmitter reaches its maximum number of re-transmissions, and moves on to the next PDU.)

The transmitter sets a polling bit in PDU 4 and the receiver replies by returning a status PDU. The transmitter can re-transmit PDU 2, but a fall in the SINR means that PDU 3 is now too large. In response, the transmitter cuts PDU 3 into two segments and re-transmits them individually. The first segment of PDU 3 arrives correctly, but the second is lost. In response to another status PDU, the transmitter can discard the first segment and can re-transmit the second.

10.3 Packet Data Convergence Protocol

10.3.1 Protocol Architecture

The packet data convergence protocol (PDCP) [8, 9] supports data streams that are using RLC acknowledged mode, by ensuring that none of their packets are lost during a handover. It also manages three air interface functions, namely header compression,

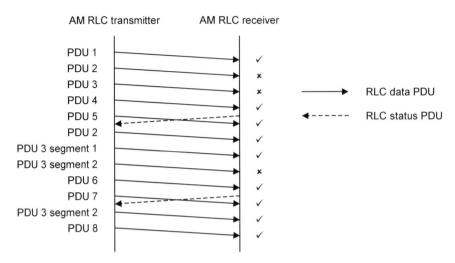

Figure 10.8 Operation of transmission, re-segmentation and re-transmission in RLC acknowledged mode.

ciphering and integrity protection. We will cover header compression here, while leaving ciphering and integrity protection until our discussion of security in Chapter 12.

The PDCP is only used by the dedicated traffic and control channels, for which the underlying RLC protocol is operating in unacknowledged or acknowledged mode. As shown in Figure 10.9, there are different architectures for data and signalling.

In the user plane, the transmitter receives PDCP service data units in the form of IP packets, adds a PDCP sequence number and stores any packets that are using RLC acknowledged mode in a re-transmission buffer. It then compresses the IP headers, ciphers the information, adds a PDCP header and outputs the resulting PDU. The receiver reverses the process, stores incoming packets in the receive buffer and uses the sequence number to deliver them in the correct order to higher layers.

In the control plane, the signalling messages are protected by an extra security procedure, known as integrity protection. There is no re-transmission buffer, because the re-transmission function is only used for data and there are no IP headers to compress.

10.3.2 Header Compression

Headers can make up a large proportion of a slow packet data stream. In the case of voice over IP, for example, the narrowband adaptive multi rate codec has a maximum bit rate of 12.2 kbps and a packet duration of 20 ms, giving a typical packet size of 31 bytes [10]. However, the header typically contains 40 or 60 bytes, comprising 12 bytes from the *real time protocol* (RTP), 8 bytes from UDP and either 20 bytes from IP version 4 or 40 bytes from IP version 6. Such an overhead is inappropriate across the bottleneck of the air interface.

To solve the problem, the PDCP includes an IETF protocol known as *robust header compression* (ROHC) [11]. The principle is that the transmitter sends the full header in

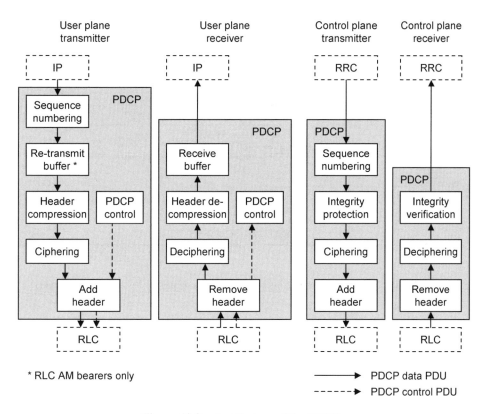

Figure 10.9 Architecture of the PDCP.

the first packet, but only sends differences in subsequent packets. Most of the header stays the same from one packet to the next, so the difference fields are considerably smaller. The protocol can compress the original 40 and 60 byte headers to as little as 1 and 3 bytes respectively, which greatly reduces the overhead.

Robust header compression has an advantage over other header compression protocols, in that it is designed to work well even if the underlying rate of packet loss is high. This makes it suitable for an unreliable data link such as the LTE air interface, particularly for real-time data streams, such as voice over IP, that are using RLC unacknowledged mode.

10.3.3 Prevention of Packet Loss During Handover

When transmitting data streams that are using RLC acknowledged mode, the PDCP stores each service data unit in a re-transmission buffer, until the RLC tells it that the SDU has been successfully received. During a handover, the processes of transmission and reception are briefly interrupted, so there is a risk of packet loss. On completion of the handover, the PDCP solves the problem by re-transmitting any service data units that it is still storing.

There is, however, a secondary problem: some of those service data units may actually have reached the receiver, but the acknowledgements may have been lost instead.

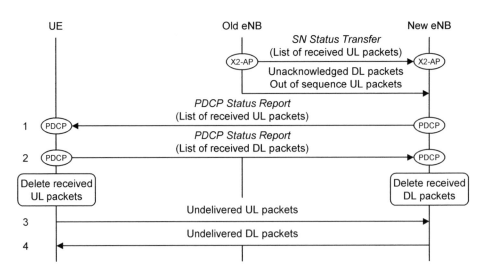

Figure 10.10 PDCP status reporting and prevention of packet loss after a handover.

To prevent them from being transmitted twice, the system can use a second procedure known as *PDCP status reporting*. Figure 10.10 shows the combined effect of the two procedures. Note that the messages in this figure apply only to bearers that are using RLC acknowledged mode.

As part of the handover procedure described in Chapter 14, the old base station sends the new base station an X2-AP message known as *SN Status Transfer*, in which it lists the PDCP sequence numbers that it has received on the uplink. It also forwards any downlink PDCP service data units that the mobile has not yet acknowledged, as well as any uplink SDUs that it has received out of sequence.

The new base station can now send the mobile a PDCP control PDU known as a *PDCP Status Report* (step 1), in which it lists the sequence numbers that it has just received from the old base station. The mobile can delete these from its re-transmission buffer, and only has to re-transmit the remainder (step 3). At the same time, the mobile can send a PDCP Status Report to the new base station (step 2), in which it lists the PDCP sequence numbers that it has received on the downlink. The new base station can delete these in the same way, before beginning its own re-transmission (step 4).

References

1. 3GPP TS 36.321 (October 2011) *Medium Access Control (MAC) Protocol Specification*, Release 10.
2. 3GPP TS 36.331 (October 2011) *Radio Resource Control (RRC); Protocol Specification*, Release 10, section 6.3.2 (*Logical Channel Config, MAC-Main Config*).
3. 3GPP TS 36.213 (September 2011) *Physical Layer Procedures*, Release 10, section 4.2.3.
4. 3GPP TS 36.331 (October 2011) *Radio Resource Control (RRC); Protocol Specification*, Release 10, section 6.3.2 (*Time Alignment Timer*).
5. 3GPP TS 36.133 (October 2011) *Evolved Universal Terrestrial Radio Access (E-UTRA); Requirements for Support of Radio Resource Management*, Release 10, section 9.1.8.
6. 3GPP TS 36.322 (December 2010) *Radio Link Control (RLC) Protocol Specification*, Release 10.

7. 3GPP TS 36.331 (October 2011) *Radio Resource Control (RRC); Protocol Specification*, Release 10, section 6.3.2 (*RLC-Config*).
8. 3GPP TS 36.323 (March 2011) *Packet Data Convergence Protocol (PDCP) Specification*, Release 10.
9. 3GPP TS 36.331 (October 2011) *Radio Resource Control (RRC); Protocol Specification*, Release 10, section 6.3.2 (*PDCP-Config*).
10. 3GPP TS 26.101 (April 2011) *Mandatory Speech Codec Speech Processing Functions; Adaptive Multi-Rate (AMR) Speech Codec Frame Structure*, Release 10, annex A.
11. IETF RFC 4995 (July 2007) *The RObust Header Compression (ROHC) Framework*.

11

Power-On and Power-Off Procedures

We have now completed our discussion of the LTE air interface. In the course of the next five chapters, we will cover the signalling procedures that govern the high-level operation of LTE.

In this chapter, we describe the procedures that a mobile follows after it switches on, to select a cell and register its location with the network. We begin by reviewing the procedure at a high level and then continue with the three main steps of network and cell selection, RRC connection establishment and registration with the evolved packet core. A final section describes the detach procedure, through which the mobile switches off. As part of the chapter, we will refer to several of the low-level procedures that we have already encountered, notably the ones for cell acquisition and random access.

Network and cell selection are covered by two specifications, which describe the idle mode procedures for the non access stratum [1] and the access stratum [2]. The signalling procedures in the rest of the chapter are summarized by the usual stage 2 specifications for LTE [3, 4]. Readers who require more detail about these procedures can find it by digging down into the relevant stage 3 signalling specifications [5–9], which define both the procedures on each individual interface and the contents of every signalling message.

11.1 Power-On Sequence

Figure 11.1 summarizes the procedures that the mobile follows after it switches on. The mobile begins by running the procedure for network and cell selection, which has three steps. In the first step, the mobile selects a public land mobile network (PLMN) that it will register with. In the second step, the mobile can optionally ask the user to select a closed subscriber group (CSG) for registration. In the third, the mobile selects a cell that belongs to the selected network and if necessary to the selected CSG. In doing so, it is said to *camp* on the cell.

The mobile then contacts the corresponding base station using the contention based random access procedure from Chapter 9 and initiates the procedure for RRC connection establishment. During the RRC procedure, the mobile establishes a signalling

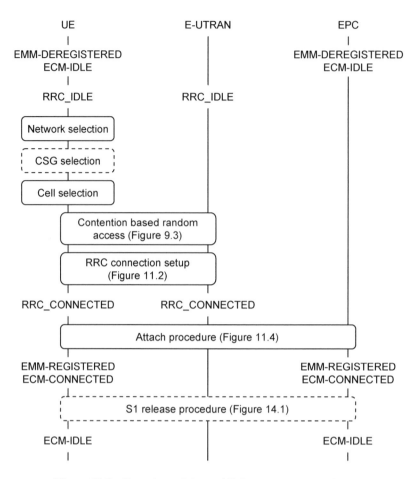

Figure 11.1 Overview of the mobile's power-on procedures.

connection with the selected base station, configures signalling radio bearer 1 and moves from RRC_IDLE into RRC_CONNECTED. It also acquires a set of parameters through which it can communicate with the base station, such as a set of resources for the transmission of uplink control information on the physical uplink control channel (PUCCH).

In the final step, the mobile uses the attach procedure to contact the evolved packet core. As a result of that procedure, the mobile registers its location with a mobility management entity (MME) and moves to the states EMM-REGISTERED and ECM-CONNECTED. It also configures signalling radio bearer 2, acquires an IP address and establishes a default bearer through which it can communicate with the outside world.

The mobile is now in the states EMM-REGISTERED, ECM-CONNECTED and RRC_CONNECTED, and will stay in those states for as long as it is exchanging data with the network. If the user does nothing, then the network can transfer the mobile into ECM-IDLE and RRC_IDLE, using a procedure known as S1 release that is covered in Chapter 14.

11.2 Network and Cell Selection

11.2.1 Network Selection

In the network selection procedure [10–12], the mobile selects a public land mobile network (PLMN) that it will register with. To start the procedure, the mobile equipment interrogates the USIM and retrieves the globally unique temporary identity (GUTI) that it was using when last switched on, as well as the tracking area identity in which it was registered. From these quantities, it can identify the corresponding network, which is known as the *registered PLMN*. The mobile runs the CSG and cell selection procedures described below, in the hope of finding a suitable cell that belongs to the registered PLMN.

If the mobile cannot find the registered PLMN, then it scans all the LTE carrier frequencies that it supports and identifies the networks that it can actually find. To do this, the mobile uses the acquisition procedure from Chapter 7 to find the strongest LTE cell on each frequency, reads SIB 1 from its system information and identifies the network or networks that the cell belongs to. If the mobile also supports UMTS, GSM or cdma2000, then it runs a similar procedure to find networks that are using those radio access technologies.

There are then two network selection modes, automatic and manual. In *automatic mode*, the mobile runs in priority order through a list of networks that it should treat as home PLMNs, together with an associated list of radio access technologies. (These lists are both stored on the USIM.) When it encounters a network that it has previously found, the mobile runs the CSG and cell selection procedures in the manner described below.

If the mobile cannot find a home PLMN, then it repeats the procedure using first any user-defined list of networks and radio access technologies, and then any operator-defined list. If it cannot find any of those networks, then the mobile tries to select a cell from any network that is available. In this last case, it enters a limited service state, in which it can only make emergency calls and receive warnings from the earthquake and tsunami warning system.

In *manual mode*, the mobile presents the user with the list of networks that it has found, using the same priority order as in automatic mode. The user selects a preferred network and the mobile proceeds to the CSG and cell selection procedures as before.

11.2.2 Closed Subscriber Group Selection

A home base station is a base station that is controlling a femtocell, which can only be selected by registered subscribers. To support this restriction, the base station is associated with a closed subscriber group and a home eNB name, which it advertises in SIB 1 and SIB 9 respectively. Each USIM lists any closed subscriber groups that the subscriber is allowed to use [13], together with the identities of the corresponding networks.

If the USIM contains any closed subscriber groups, then the mobile has to run an additional procedure, known as *CSG selection* [14, 15]. The procedure has two modes of operation, automatic and manual, which are distinct from the network selection modes described above. In *automatic mode*, the mobile sends the list of allowed closed subscriber groups to the cell selection procedure, which selects either a non-CSG cell, or a cell whose CSG is in the list. *Manual mode* is more restrictive. Here, the mobile identifies the CSG cells that it can find in the selected network. It presents this list to the user, indicates the corresponding home eNB names and indicates whether each CSG is in the list of allowed CSGs. The user selects a preferred closed subscriber group and the mobile selects a cell belonging to that CSG.

11.2.3 Cell Selection

During the *cell selection* procedure [16], the mobile selects a *suitable cell* that belongs to the selected network and, if necessary, to the selected closed subscriber group. It can do this in two ways. Usually, it has access to stored information about potential LTE carrier frequencies and cells, either from when it was last switched on, or from the network selection procedure described above. If this information is unavailable, then the mobile scans all the LTE carrier frequencies that it supports and identifies the strongest cell on each carrier that belongs to the selected network.

A suitable cell is a cell that satisfies several criteria. The most important is the cell selection criterion:

$$S_{rxlev} > 0 \tag{11.1}$$

During initial network selection, the mobile calculates S_{rxlev} as follows:

$$S_{rxlev} = Q_{rxlevmeas} - Q_{rxlevmin} - P_{compensation} \tag{11.2}$$

In this equation, $Q_{rxlevmeas}$ is the cell's *reference signal received power* (RSRP), which is the average power per resource element that the mobile is receiving on the cell specific reference signals [17]. $Q_{rxlevmin}$ is a minimum value for the RSRP, which the base station advertises in SIB 1. These quantities ensure that a mobile will only select the cell if it can hear the base station's transmissions on the downlink. The final parameter, $P_{compensation}$, is calculated as follows:

$$P_{compensation} = \max(P_{EMAX} - P_{PowerClass}, 0) \tag{11.3}$$

Here, P_{EMAX} is an upper limit on the transmit power that a mobile is allowed to use, which the base station advertises as part of SIB 1. $P_{PowerClass}$ is the mobile's intrinsic maximum power. By combining these quantities, $P_{compensation}$ reduces the value of S_{rxlev} if the mobile cannot reach the power limit that the base station is assuming. It therefore ensures that a mobile will only select the cell if the base station can hear it on the uplink.

The cell selection procedure is enhanced in release 9 of the 3GPP specifications, so that a suitable cell also has to satisfy the following criterion:

$$S_{qual} > 0 \tag{11.4}$$

where:

$$S_{qual} = Q_{qualmeas} - Q_{qualmin} \tag{11.5}$$

In this equation, $Q_{qualmeas}$ is the *reference signal received quality* (RSRQ), which is the signal to interference plus noise ratio of the cell specific reference signals. $Q_{qualmin}$ is a minimum value for the RSRQ, which the base station advertises in SIB 1 as before. This condition prevents a mobile from selecting a cell on a carrier frequency that is subject to high levels of interference.

A suitable cell must also satisfy a number of other criteria. If the USIM contains a list of closed subscriber groups, then the cell has to meet the criteria for automatic or manual CSG selection that we defined above. If the USIM does not, then the cell must lie outside any closed subscriber groups. In addition, the network operator can bar a cell to all users or reserve it for operator use, by means of flags in SIB 1.

11.3 RRC Connection Establishment

11.3.1 Basic Procedure

Once the mobile has selected a network and a cell to camp on, it runs the contention based random access procedure from Chapter 9. In doing so, it obtains a C-RNTI, an initial value for the timing advance and resources on the physical uplink shared channel (PUSCH) through which it can send a message to the network.

The mobile can then begin a procedure known as *RRC connection establishment* [18]. Figure 11.2 shows the message sequence. In step 1, the mobile's RRC protocol composes a message known as an *RRC Connection Request*. In this message, it specifies two parameters. The first is a unique non access stratum (NAS) identity, either the S-TMSI (if the mobile was registered in the cell's tracking area when last switched on), or a randomly chosen value (otherwise). The second is the establishment cause, which can be mobile originated signalling (as in this example), mobile originated data, mobile terminated access (a response to paging), high priority access, or an emergency call.

The mobile transmits the message using signalling radio bearer 0, which has a simple configuration that the base station advertises in SIB 2. The message is sent on the common control channel, the uplink shared channel and the physical uplink shared channel.

The base station reads the message, takes on the role of serving eNB and composes a reply known as an *RRC Connection Setup* (step 2). In this message, it configures the mobile's physical layer and MAC protocols, as well as SRB 1. These configurations include several parameters that we have already seen. For example, the physical layer parameters include the PUCCH resources $n^{(1)}_{PUCCH,SRI}$ and $n^{(2)}_{PUCCH}$, the CQI reporting mode and the radio network temporary identities TPC-PUCCH-RNTI and TPC-PUSCH-RNTI. Similarly, the MAC parameters include the time alignment timer, the timer for periodic buffer status reports and the maximum number of hybrid ARQ transmissions on

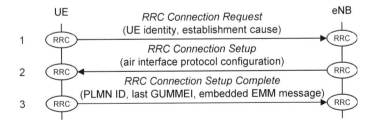

Figure 11.2 RRC connection establishment procedure. Reproduced by permission of ETSI.

the uplink. Finally, the parameters for SRB 1 include the priorities and prioritized bit rates of the corresponding logical channels and the parameters that govern polling and status reporting in the RLC. To reduce the size of the message, the base station can set several parameters to default values that are defined in the specifications. The base station transmits its message on SRB 0, as before, because the mobile does not yet understand the configuration of SRB 1.

The mobile reads the message, configures its protocols in the manner required and moves into RRC_CONNECTED. It then writes a confirmation message known as *RRC Connection Setup Complete* (step 3) and transmits it on SRB 1. In the message, the mobile includes three information elements. The first identifies the PLMN that it would like to register with. The second is the globally unique identity of the MME that was previously serving the mobile, which the mobile has extracted from its GUTI. The third is an embedded EPS mobility management message, which in this example is an attach request, but can also be a detach request, a service request or a tracking area update request.

The RRC connection establishment procedure is also used later on, whenever a mobile in RRC_IDLE wishes to communicate with the network. We will see several examples in the chapters that follow.

11.3.2 Relationship with Other Procedures

As shown in Figure 11.3, the RRC connection establishment procedure overlaps with two other procedures, namely the random access procedure that precedes it and the EPS mobility management (EMM) procedure that follows.

The mobile sends its RRC Connection Request in the third step of the contention based random access procedure. The base station therefore uses the message in two ways: it echoes back the message during contention resolution and it replies to the message with its RRC Connection Setup. Similarly, the message RRC Connection Setup Complete is also the first step of the EMM procedure that follows. The base station accepts the RRC

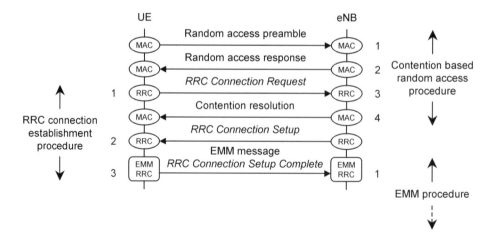

Figure 11.3 Relationships between RRC connection establishment and other procedures.

message as an acknowledgement of its RRC Connection Setup and forwards the embedded EMM message to a suitable MME. Overlapping the procedures makes it harder to follow what is going on, but it brings a big advantage: it makes the signalling delays lower than in earlier systems, which helps the system to meet the latency requirements that were laid out in Chapter 1.

There is one final point to make. In Figures 11.2 and 11.3, we have only shown the high-level signalling messages that are transmitted on the PUSCH and PDSCH. We have omitted the lower-level control information on the PUCCH, PDCCH and PHICH, as well as the possibility of re-transmissions. We will follow this convention throughout the rest of the book, but it is worth remembering that the full sequence of air interface messages may be considerably longer than the resulting figures imply.

11.4 Attach Procedure

11.4.1 IP Address Allocation

During the attach procedure, the mobile acquires an IP version 4 address and/or an IP version 6 address, which it will subsequently use to communicate with the outside world. Before looking at the attach procedure itself, it is useful to discuss the methods that the network can use for IP address allocation [19].

IPv4 addresses are 32 bits long. In the usual technique, the PDN gateway allocates a dynamic IPv4 address to the mobile as part of the attach procedure. It can either allocate the IP address by itself, or acquire a suitable IP address from a *dynamic host configuration protocol version 4* (DHCPv4) server. As an alternative, the mobile can itself use DHCPv4 to acquire a dynamic IP address after the attach procedure has completed. To do this, it contacts the PDN gateway over the user plane, with the PDN gateway acting as a DHCPv4 server towards the mobile. As before, the PDN gateway can obtain a suitable IP address from elsewhere, by acting as a DHCPv4 client towards another DHCPv4 server.

IPv6 addresses are 128 bits long and have two parts, namely a 64 bit network prefix and a 64 bit interface identifier. They are allocated using a procedure known as *IPv6 stateless address auto-configuration* [20]. In LTE's implementation of the procedure, the PDN gateway assigns the mobile a globally unique IPv6 prefix during the attach procedure, as well as a temporary interface identifier. It passes the interface identifier back to the mobile, which uses it to construct a temporary link-local IPv6 address. After the attach procedure has completed, the mobile uses the temporary address to contact the PDN gateway over the user plane and retrieve the IPv6 prefix, in a process known as router solicitation. It then uses the prefix to construct a full IPv6 address. Because the prefix is globally unique, the mobile can actually do this using any interface ID that it likes.

A mobile can also use a static IPv4 address or IPv6 prefix. The mobile does not store these permanently, however: instead, the network stores them in the home subscriber server or a DHCP server and sends them to the mobile during the attach procedure. Static IP addresses are unusual in the case of IPv4, due to the chronic shortage of IPv4 addresses.

Subsequently, the mobile will use the same IP address for any dedicated bearers that it sets up with the same packet data network. If it establishes communications with another packet data network, then it will acquire another IP address using the same technique.

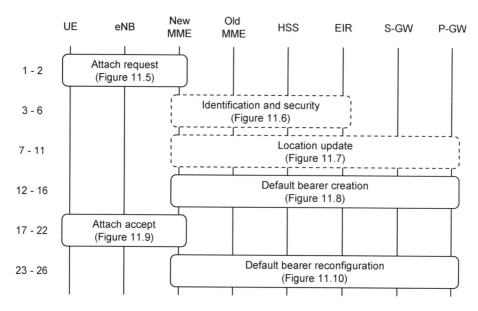

Figure 11.4 Overview of the attach procedure.

11.4.2 Overview of the Attach Procedure

The *attach procedure* has four main objectives. The mobile uses the procedure to register its location with a serving MME. The network configures signalling radio bearer 2, which carries subsequent non access stratum signalling messages across the air interface. The network also gives the mobile an IP version 4 address and/or an IP version 6 address, using either or both of the techniques described above, and sets up a default EPS bearer, which provides the mobile with always-on connectivity to a default PDN.

Figure 11.4 summarizes the attach procedure. We will run through the individual steps of the procedure in the following sections, for the case where the S5/S8 interface is using the GPRS tunnelling protocol (GTP). In this figure and the ones that follow, solid lines show mandatory messages, while dashed lines indicate messages that are optional or conditional. The message numbers are the same as in TS 23.401 [21], a convention that we will follow for most of the other procedures in the book.

11.4.3 Attach Request

Figure 11.5 shows the first two steps of the procedure, which cover the mobile's attach request. The mobile starts by running the contention based random access procedure and the first two steps of RRC connection establishment, in the manner described earlier.

The mobile then composes an EPS session management (ESM) message, *PDN Connectivity Request*, which asks the network to establish a default EPS bearer. The message includes a PDN type, which indicates whether the mobile supports IPv4, IPv6 or both. It can also include a set of protocol configuration options, which list any parameters that

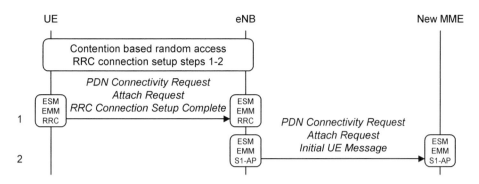

Figure 11.5 Attach procedure. (1) Attach request. Reproduced by permission of ETSI.

relate to the external network, such as a preferred access point name, or a request to receive an IPv4 address over the user plane by means of DHCPv4. The mobile can either list its configuration options here or can set an ESM information transfer flag, which indicates a wish to send the options securely later on, after security activation. The mobile always uses the latter option if it wishes to indicate a preferred APN.

The mobile embeds the PDN connectivity request into an EMM *Attach Request*, in which it asks for registration with a serving MME. The message includes the globally unique temporary identity that the mobile was using when last switched on and the identity of the tracking area in which the mobile was last located. It also includes the mobile's non access stratum capabilities, primarily the security algorithms that it supports.

In turn, the mobile embeds the Attach Request into the last message from the RRC connection establishment procedure, RRC Connection Setup Complete. As noted earlier, the RRC message also identifies the PLMN that the mobile would like to register with and the identity of its last serving MME. In step 1 of the attach procedure, the mobile sends this message to the serving eNB.

As described in Chapter 12, the mobile and MME can store their LTE security keys after the mobile switches off. If the mobile has a valid set of security keys, then it uses these to secure the attach request using a process known as integrity protection. This assures the MME that the request is coming from a genuine mobile, and not from an intruder.

The base station extracts the EMM and ESM messages and embeds them into an S1-AP *Initial UE Message*, which requests the establishment of an S1 signalling connection for the mobile. As part of this message, the base station specifies the RRC establishment cause and the requested PLMN, which it received from the mobile during the RRC procedure.

The base station can now forward the message to a suitable MME (step 2). Usually, the chosen MME is the same one that the mobile was previously registered with. This can be done if two conditions are met: the base station has to lie in one of the old MME's pool areas and the old MME has to lie in the requested PLMN. If the mobile has changed pool area since it was last switched on, or if it is asking to register with a different network, then the base station selects another MME. It does so by choosing at random from the ones in its pool area, according to a load balancing algorithm [22].

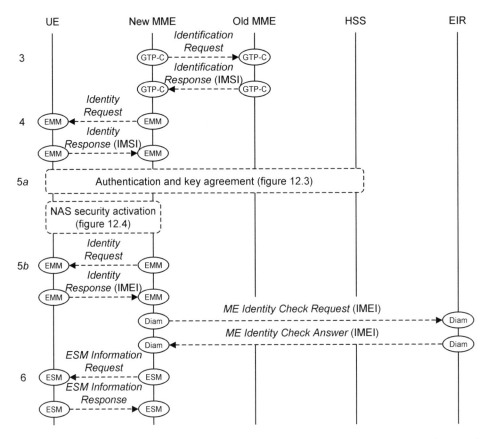

Figure 11.6 Attach procedure. (2) Identification and security procedures. Reproduced by permission of ETSI.

11.4.4 Identification and Security Procedures

The MME receives the messages from the base station, and can now run some procedures that relate to identification and security (Figure 11.6).

If the mobile has moved to a new MME since it was last switched on, then the MME has to find out the mobile's identity. To do this, it extracts the identity of the old MME from the mobile's GUTI and sends the GUTI to the old MME in a GTP-C *Identification Request* (3). The old MME's response includes the international mobile subscriber identity (IMSI) and the mobile's security keys. In exceptional cases, however, the mobile may be unknown to the old MME. If this happens, then the new MME asks the mobile for its IMSI using an EMM *Identity Request* (4), a message that is transported using the NAS information transfer procedure from Chapter 2.

The network can now run two security procedures (5a). In *authentication and key agreement*, the mobile and network confirm each other's identities and set up a new set of security keys. In *NAS security activation*, the MME activates those keys and initiates the secure protection of all subsequent EMM and ESM messages. These steps are mandatory if there was any problem with the integrity protection of the attach request, and are optional otherwise. If the integrity check succeeded, then the MME can implicitly re-activate the mobile's old keys by sending it a signalling message that it has secured using those keys, thus skipping both of these procedures.

The MME then retrieves the international mobile equipment identity (IMEI) (5b). It can combine this message with NAS security activation to reduce the amount of signalling, but it is mandatory for the MME to retrieve the IMEI somehow. As a protection against stolen mobiles, the MME can optionally send the IMEI to the equipment identity register, which responds by either accepting or rejecting the device.

If the mobile set the ESM information transfer flag in its PDN Connectivity Request, then the MME can now send it an *ESM Information Request* (6). The mobile sends its protocol configuration options in response. Now that the network has activated NAS security, the mobile can send the message securely.

11.4.5 Location Update

The MME can now update the network's record of the mobile's location (Figure 11.7). If the mobile is re-attaching to its previous MME without having properly detached (for example, if its battery ran out), then the MME may still have some EPS bearers that are associated with the mobile. If this is the case then the MME deletes them (7), by following steps from the detach procedure that we will see later on.

If the MME has changed, then the new MME sends the mobile's IMSI to the home subscriber server (HSS), in a Diameter *Update Location Request* (8). The HSS updates its record of the mobile's location, and tells the old MME to forget about the mobile (9). If the old MME has any EPS bearers that are associated with the mobile, then it deletes these as before (10).

In step 11, the HSS sends an *Update Location Answer* to the new MME, which includes the user's *subscription data* [23]. The subscription data list all the access point names (APNs) that the user has subscribed to and define each one using an *APN configuration*.

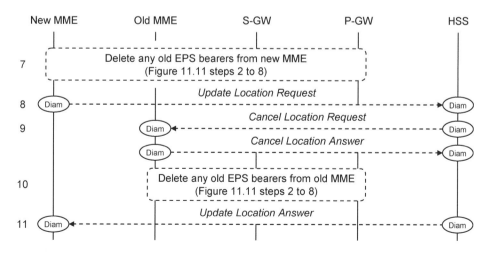

Figure 11.7 Attach procedure. (3) Location update. Reproduced by permission of ETSI.

Each APN configuration identifies the access point name and states whether the corresponding packet data network supports IPv4, IPv6 or both. It also includes the user's maximum bit rate from all non-GBR bearers on that APN, known as the *per APN aggregate maximum bit rate* (APN-AMBR), and the parameters that describe the default EPS bearer's quality of service. Optionally, it can also indicate a static IPv4 address or IPv6 prefix for the mobile to use when connecting to that APN.

In addition, the subscription data identify one of the APN configurations as the default and define the user's maximum total bit rate from all non-GBR bearers, known as the *per UE aggregate maximum bit rate* (UE-AMBR).

11.4.6 Default Bearer Creation

The MME now has all the information that it needs to set up the default EPS bearer (Figure 11.8). It begins by selecting a suitable PDN gateway, using the mobile's preferred APN if it supplied one and the subscription data support it, or the default APN otherwise. It then selects a serving gateway, and sends it a GTP-C *Create Session Request* (12). In this message, the MME includes the relevant subscription data and identifies the mobile's IMSI and the destination PDN gateway.

The serving gateway receives the message and forwards it to the PDN gateway (13). In the message, the serving gateway includes a GTP-U tunnel endpoint identifier (TEID), which the PDN gateway will eventually use to send it downlink packets across the S5/S8 interface.

If the message does not contain a static IP address, then the PDN gateway can allocate a dynamic IPv4 and/or IPv6 address for the mobile, using the methods we covered earlier. Alternatively, it can defer the allocation of an IPv4 address until later, if the mobile requested that in its protocol configuration options. The PDN gateway also runs a procedure known as *IP connectivity access network* (IP-CAN) *session establishment* (14). This procedure sets up the quality of service of the default EPS bearer and can also trigger the

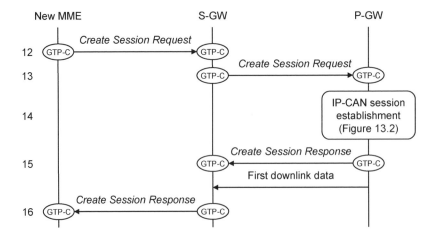

Figure 11.8 Attach procedure. (4) Default bearer creation. Reproduced by permission of ETSI.

establishment of one or more dedicated EPS bearers. We will describe it along with the other procedures for managing quality of service, as part of Chapter 13.

The PDN gateway now acknowledges the serving gateway's request by means of a GTP-C *Create Session Response* (15). In the message, it includes any IP address that the mobile has been allocated, as well as the quality of service of the default EPS bearer. The PDN gateway also includes a TEID of its own, which the serving gateway will eventually use to route uplink packets across S5/S8. The serving gateway forwards the message to the MME (16), except that it replaces the PDN gateway's tunnel endpoint identifier with an uplink TEID for the base station to use across S1-U.

11.4.7 Attach Accept

The MME can now reply to the mobile's attach request, as shown in Figure 11.9. It first initiates an ESM procedure known as *Default EPS bearer context activation*, which is a response to the mobile's PDN Connectivity Request and which starts with a message known as *Activate Default EPS Bearer Context Request*. The message includes the EPS bearer identity, the access point name, the quality of service and any IP address that the network has allocated to the mobile.

The MME embeds the ESM message into an EMM *Attach Accept*, which is a response to the mobile's original attach request. The message includes a list of tracking areas in which the MME has registered the mobile and a new globally unique temporary identity.

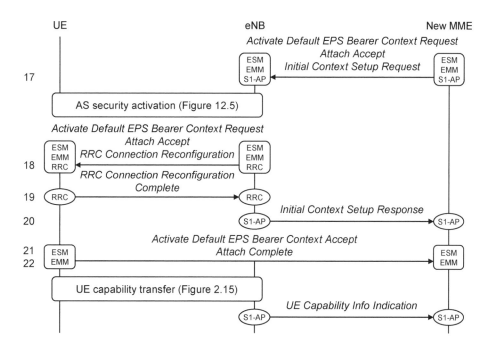

Figure 11.9 Attach procedure. (5) Attach accept. Reproduced by permission of ETSI.

In turn, the MME embeds both messages into an S1-AP *Initial Context Setup Request*. This is the start of a procedure known as *Initial context setup*, which was triggered by the base station's Initial UE Message. The procedure tells the base station to set up an S1 signalling connection for the mobile, and S1 and radio bearers that correspond to the default EPS bearer. The message includes the bearers' quality of service, the uplink TEID that the MME received from the serving gateway and a key for the activation of access stratum security. The MME sends all three messages to the base station, in step 17.

The base station now activates access stratum security, using the secure key that it has just received. From this point, all the data and RRC signalling messages on the air interface are secured. It then composes an *RRC Connection Reconfiguration* message, in which it modifies the mobile's RRC connection so as to set up two new radio bearers: a radio bearer that will carry the default EPS bearer, and SRB 2. It sends this message to the mobile, along with the EMM and ESM messages that it has just received from the MME (18).

The mobile reconfigures its RRC connection as instructed and sets up the default EPS bearer. It then sends its acknowledgements to the network in two stages. Using SRB 1, the mobile first sends the base station an acknowledgement known as *RRC Connection Reconfiguration Complete*, which triggers an S1-AP *Initial Context Setup Response* to the MME (20). The S1-AP message includes a downlink TEID for the serving gateway to use across S1-U.

The mobile then composes an ESM *Activate Default EPS Bearer Context Accept* and embeds it into an EMM *Attach Complete*, to acknowledge the ESM and EMM parts of message 18. It sends these messages to the base station on SRB 2 (21), using the NAS information transfer procedure, and the base station forwards the messages to the MME (22).

At about this point, the base station retrieves the mobile's radio access capabilities, using the procedure we covered in Chapter 2 [24]. It sends the capabilities back to the MME using an S1-AP *UE Capability Info Indication*, which stores them until the mobile detaches from the network.

11.4.8 Default Bearer Update

The mobile can now send uplink data as far as the PDN gateway. However, we still need to tell the serving gateway about the identity of the selected base station and send it the tunnel endpoint identifier that the base station has just provided. To do this (Figure 11.10), the MME sends a GTP-C *Modify Bearer Request* to the serving gateway (23) and the serving gateway responds (24). From this point, downlink data packets can flow to the mobile.

The MME can also notify the HSS about the chosen PDN gateway and APN (25). It does this if the chosen PDN gateway is different from the one in the default APN configuration, for example, if the mobile requested an access point name of its own to connect to. The HSS stores the chosen PDN gateway, for use in any future handovers to non 3GPP systems, and responds (26).

Finally, the mobile may have to contact the PDN gateway across the user plane, to complete the allocation of its IP addresses. It does this when obtaining an IPv6 prefix using stateless auto-configuration, and also when obtaining an IPv4 address using DHCPv4.

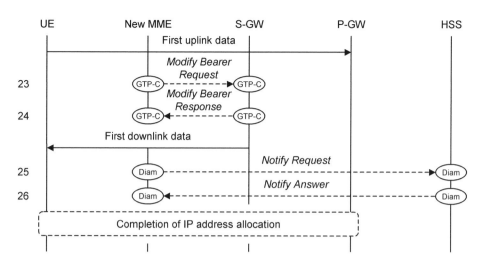

Figure 11.10 Attach procedure. (6) Default bearer update. Reproduced by permission of ETSI.

The mobile is now in the states EMM-REGISTERED, ECM-CONNECTED and RRC_CONNECTED and will stay in these states for as long as the user is actively communicating with the outside world. If the user does nothing, the network can transfer the mobile into ECM-IDLE and RRC_IDLE using a procedure known as S1 release. We will cover this procedure later, as part of Chapter 14.

11.5 Detach Procedure

The last process to consider in this chapter is the *Detach procedure* [25]. This cancels the mobile's registration with the evolved packet core and is normally used when the mobile switches off, as shown in Figure 11.11.

We will assume that the mobile starts in ECM-CONNECTED and RRC_CONNECTED, consistent with its state at the end of the previous section. The user triggers the procedure by telling the mobile to shut down. In response, the mobile composes an EMM *Detach Request*, in which it specifies its GUTI, and sends the message to the MME (1). After sending the message, the mobile can switch off without waiting for a reply.

The MME now has to tear down the mobile's EPS bearers. To do this, it looks up the mobile's serving gateway and sends it a GTP-C *Delete Session Request* (2). The serving gateway forwards the message to the PDN gateway (3), which runs a procedure known as *IP-CAN session termination* (4) that undoes the earlier effect of IP-CAN session establishment. The PDN gateway then tears down all the mobile's bearers and replies to the serving gateway (5), which tears down its bearers in the same way and replies to the MME (6). If necessary, these steps are repeated for any other network that the mobile is connected to.

To finish the procedure, the MME tells the base station to tear down all the resources that are related to the mobile and indicates that the cause is a detach request (7). The base station does so and responds (8). The MME can now delete most of the information that

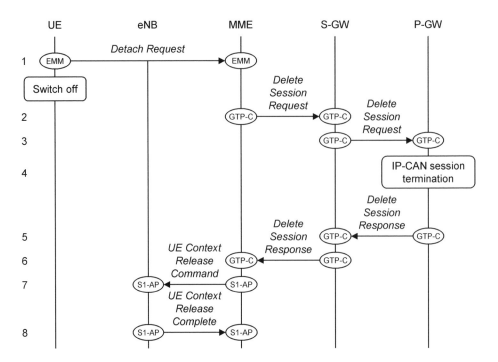

Figure 11.11 Detach procedure, triggered by the mobile switching off. Reproduced by permission of ETSI.

it associated with the mobile. However, it keeps a record of the mobile's IMSI, GUTI and security keys, as it will need these next time the mobile switches on.

If the mobile starts in ECM-IDLE and RRC_IDLE, then it cannot send the detach request right away. Instead, it starts by running the contention based random access procedure, followed by steps 1 and 2 of RRC connection establishment. It then embeds the detach request into the message RRC Connection Setup Complete, and the detach procedure continues as before.

References

1. 3GPP TS 23.122 (September 2011) *Non-Access-Stratum (NAS) Functions Related to Mobile Station (MS) in Idle Mode*, Release 10.
2. 3GPP TS 36.304 (October 2011) *User Equipment (UE) Procedures in Idle Mode*, Release 10.
3. 3GPP TS 23.401 (September 2011) *General Packet Radio Service (GPRS) Enhancements for Evolved Universal Terrestrial Radio Access Network (E-UTRAN) Access*, Release 10.
4. 3GPP TS 36.300 (October 2011) *Evolved Universal Terrestrial Radio Access (E-UTRA) and Evolved Universal Terrestrial Radio Access Network (E-UTRAN); Overall Description; Stage 2*, Release 10.
5. 3GPP TS 24.301 (September 2011) *Non-Access-Stratum (NAS) protocol for Evolved Packet System (EPS); Stage 3*, Release 10.
6. 3GPP TS 29.272 (September 2011) *Evolved Packet System (EPS); Mobility Management Entity (MME) and Serving GPRS Support Node (SGSN) Related Interfaces Based on Diameter Protocol*, Release 10.
7. 3GPP TS 29.274 (September 2011) *3GPP Evolved Packet System (EPS); Evolved General Packet Radio Service (GPRS) Tunnelling Protocol for Control Plane (GTPv2-C); Stage 3*, Release 10.

8. 3GPP TS 36.331 (October 2011) *Radio Resource Control (RRC); Protocol Specification*, Release 10.
9. 3GPP TS 36.413 (September 2011) *Evolved Universal Terrestrial Radio Access Network (E-UTRAN); S1 Application Protocol (S1AP)*, Release 10.
10. 3GPP TS 23.122 (September 2011) *Non-Access-Stratum (NAS) Functions Related to Mobile Station (MS) in Idle Mode*, Release 10, sections 3.1, 4.3.1, 4.4.
11. 3GPP TS 36.304 (October 2011) *User Equipment (UE) Procedures in Idle Mode*, Release 10, sections 4, 5.1.
12. 3GPP TS 31.102 (October 2011) *Characteristics of the Universal Subscriber Identity Module (USIM) Application*, Release 10, sections 4.2.2, 4.2.5, 4.2.53, 4.2.54, 4.2.84, 4.2.91.
13. 3GPP TS 31.102 (October 2011) *Characteristics of the Universal Subscriber Identity Module (USIM) Application*, Release 10, section 4.4.6.
14. 3GPP TS 23.122 (September 2011) *Non-Access-Stratum (NAS) Functions Related to Mobile Station (MS) in Idle Mode*, Release 10, sections 3.1A, 4.4.3.1.3.
15. 3GPP TS 36.304 (October 2011) *User Equipment (UE) Procedures in Idle Mode*, Release 10, section 5.5.
16. 3GPP TS 36.304 (October 2011) *User Equipment (UE) Procedures in Idle Mode*, Release 10, sections 5.2.1, 5.2.2, 5.2.3, 5.3.
17. 3GPP TS 36.214 (March 2011) *Evolved Universal Terrestrial Radio Access (E-UTRA); Physical Layer; Measurements*, Release 10, section 5.1.1.
18. 3GPP TS 36.331 (October 2011) *Radio Resource Control (RRC); Protocol Specification*, Release 10, sections 5.3.3, 6.2.2 (*RRCConnectionRequest, RRCConnectionSetup, RRCConnectionSetupComplete*).
19. 3GPP TS 23.401 (September 2011) *General Packet Radio Service (GPRS) Enhancements for Evolved Universal Terrestrial Radio Access Network (E-UTRAN) Access*, Release 10, section 5.3.1.
20. IETF RFC 4862 (September 2007) *IPv6 Stateless Address*.
21. 3GPP TS 23.401 (September 2011) *General Packet Radio Service (GPRS) Enhancements for Evolved Universal Terrestrial Radio Access Network (E-UTRAN) Access*, Release 10, section 5.3.2.1.
22. 3GPP TS 23.401 (September 2011) *General Packet Radio Service (GPRS) Enhancements for Evolved Universal Terrestrial Radio Access Network (E-UTRAN) Access*, Release 10, sections 4.3.7, 4.3.8.
23. 3GPP TS 29.272 (September 2011) *Evolved Packet System (EPS); Mobility Management Entity (MME) and Serving GPRS Support Node (SGSN) Related Interfaces Based on Diameter Protocol*, Release 10, sections 7.3.2, 7.3.34, 7.3.35.
24. 3GPP TS 36.300 (October 2011) *Evolved Universal Terrestrial Radio Access (E-UTRA) and Evolved Universal Terrestrial Radio Access Network (E-UTRAN); Overall Description; Stage 2*, Release 10, section 18.
25. 3GPP TS 23.401 (September 2011) *General Packet Radio Service (GPRS) Enhancements for Evolved Universal Terrestrial Radio Access Network (E-UTRAN) Access*, Release 10, section 5.3.8.2.

12

Security Procedures

In this chapter, we review the security techniques that protect LTE against attacks from intruders. The most important issue is network access security, which protects the mobile's communications with the network across the air interface. In the first part of this chapter, we cover the architecture of network access security, the procedures that establish secure communications between the network and mobile, and the security techniques that are subsequently used. The system must also secure certain types of communication within the radio access network and the evolved packet core. This issue is known as network domain security and is the subject of the second part.

The 3GPP security procedures are covered by the 33 series specifications: those for LTE are summarized in TS 33.401 [1]. As in the last chapter, the details of the individual messages are in the specifications for the relevant signalling protocols [2–5]. For a detailed account of security in LTE, see Reference [6].

12.1 Network Access Security

12.1.1 Security Architecture

Network access security (Figure 12.1) protects the mobile's communications with the network across the air interface, which is the most vulnerable part of the system. It does this using four main techniques.

During *authentication*, the network and mobile confirm each other's identities. The evolved packet core (EPC) confirms that the user is authorized to use the network's services and is not using a cloned device. Similarly, the mobile confirms that the network is genuine and is not a spoof network set up to steal the user's personal data.

Confidentiality protects the user's identity. The international mobile subscriber identity (IMSI) is one of the quantities that an intruder needs to clone a mobile, so LTE avoids broadcasting it across the air interface wherever possible. Instead, the network identifies the user by means of temporary identities. If the EPC knows the MME pool area that the mobile is in (for example, during paging), then it uses the 40 bit S-TMSI. Otherwise (for example, during the attach procedure), it uses the longer GUTI. Similarly, the radio access network uses the radio network temporary identifiers (RNTIs) that we introduced in Chapter 8.

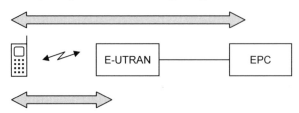

Figure 12.1 Network access security architecture.

Ciphering, also known as *encryption*, ensures that intruders cannot read the data and signalling messages that the mobile and network exchange. *Integrity protection* detects any attempt by an intruder to replay or modify signalling messages. It protects the system against problems such as man-in-the-middle attacks, in which an intruder intercepts a sequence of signalling messages and modifies and re-transmits them, in an attempt to take control of the mobile.

GSM and UMTS only implemented ciphering and integrity protection in the air interface's access stratum, to protect user plane data and RRC signalling messages between the mobile and the radio access network. As shown in Figure 12.1, LTE implements them in the non access stratum as well, to protect EPS mobility and session management messages between the mobile and the MME. This brings two main advantages. In a wide-area network, it provides two cryptographically separate levels of encryption, so that even if an intruder breaks one level of security, the information is still secured on the other. It also eases the deployment of home base stations, whose access stratum security can be more easily compromised.

12.1.2 Key Hierarchy

Network access security is based on a hierarchy of keys [7] that is illustrated in Figure 12.2. Ultimately, it relies on the shared knowledge of a user-specific key, K, which is securely stored in the home subscriber server (HSS) and securely distributed within the universal integrated circuit card (UICC). There is a one-to-one mapping between a user's IMSI and the corresponding value of K, and the authentication process relies on the fact that cloned mobiles and spoof networks will not know the correct value of K.

From K, the HSS and UICC derive two further keys, denoted CK and IK. UMTS used those keys directly for ciphering and integrity protection, but LTE uses them differently, to derive an *access security management entity* (ASME) key, denoted K_{ASME}.

From K_{ASME}, the MME and the mobile equipment derive three further keys, denoted K_{NASenc}, K_{NASint} and K_{eNB}. The first two are used for ciphering and integrity protection of non access stratum (NAS) signalling messages between the mobile and the MME,

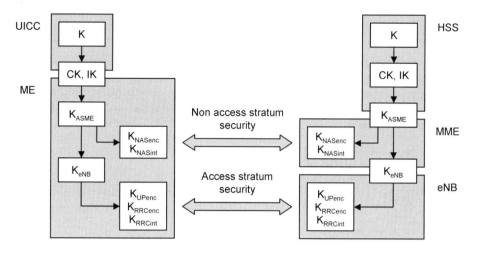

Figure 12.2 Hierarchy of the security keys used by LTE.

while the last is passed to the base station. From K_{eNB}, the base station and the mobile equipment derive three access stratum keys, denoted K_{UPenc}, K_{RRCenc} and K_{RRCint}. These are respectively used for ciphering of data, ciphering of RRC signalling messages and integrity protection of RRC signalling messages in the access stratum (AS).

This set of keys is larger than the set used by GSM or UMTS, but it brings several benefits. Firstly, the mobile stores the values of CK and IK in its UICC after it detaches from the network, while the MME stores the value of K_{ASME}. This allows the system to secure the mobile's attach request when it next switches on, in the manner described in Chapter 11. The hierarchy also ensures that the AS and NAS keys are cryptographically separate, so that knowledge of one set of keys does not help an intruder to derive the other. At the same time, the hierarchy is backwards compatible with USIMs from 3GPP Release 99.

K, CK and IK contain 128 bits each, while the other keys all contain 256 bits. The current ciphering and integrity protection algorithms use 128 bit keys, which are derived from the least significant bits of the original 256 bit keys. If LTE eventually has to upgrade its algorithms to use 256 bit keys, then it will be able to do so with ease.

12.1.3 Authentication and Key Agreement

During *authentication and key agreement* (AKA) [8], the mobile and network confirm each other's identities and agree on a value of K_{ASME}. We have already seen this procedure used as part of the larger attach procedure: Figure 12.3 shows the full message sequence.

Before the procedure begins, the MME has retrieved the mobile's IMSI from its own records or from the mobile's previous MME, or exceptionally by sending an EMM Identity Request to the mobile itself. It now wishes to confirm the mobile's identity. To start the procedure, it sends a Diameter *Authentication Information Request* to the HSS (1), in which it includes the IMSI.

The HSS looks up the corresponding secure key K and calculates an *authentication vector* that contains four elements. RAND is a random number that the MME will use

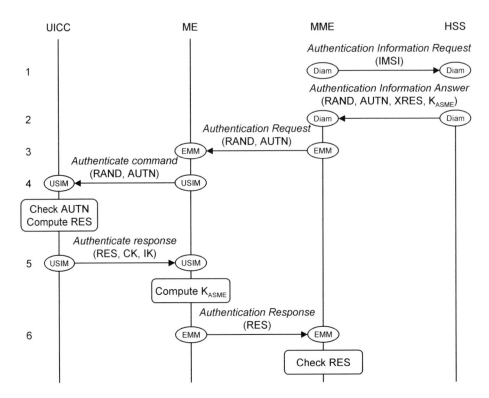

Figure 12.3 Authentication and key agreement procedure.

as an authentication challenge to the mobile. XRES is the expected response to that challenge, which can only be calculated by a mobile that knows the value of K. AUTN is an *authentication token*, which can only be calculated by a network that knows the value of K, and which includes a sequence number to prevent an intruder from recording an authentication request and replaying it. Finally, K_{ASME} is the access security management entity key, which is derived from CK and IK and ultimately from the values of K and RAND. In step 2, the HSS returns the authentication vector to the MME.

In GSM and UMTS, the HSS usually returns several authentication vectors at once, to minimize the number of separate messages that it has to handle. LTE actually discourages this technique, on the grounds that the storage of K_{ASME} has greatly reduced the number of messages that the HSS has to exchange.

The MME sends RAND and AUTN to the mobile equipment as part of an EMM *Authentication Request* (3) and the mobile equipment forwards them to the UICC (4). Inside the UICC, the USIM application examines the authentication token, to check that the network knows the value of K and that the enclosed sequence number has not been used before. If it is happy, then it calculates its response to the network's challenge, denoted RES, by combining RAND with its own copy of K. It also computes the values of CK and IK, and passes all three parameters back to the mobile equipment (5).

Using CK and IK, the mobile equipment computes the access security management entity key K_{ASME}. It then returns its response to the MME, as part of an EMM *Authentication Response* (6). In turn, the MME compares the mobile's response with the expected response that it received from the home subscriber server. If it is the same, then the MME concludes that the mobile is genuine. The system can then use the two copies of K_{ASME} to activate the subsequent security procedures, as described in the next section.

12.1.4 Security Activation

During security activation [9], the mobile and network calculate separate copies of the ciphering and integrity protection keys and start running the corresponding procedures. Security activation is carried out separately for the access and non access strata.

The MME activates non access stratum security immediately after authentication and key agreement, as shown in Figure 12.4. From K_{ASME}, the MME computes the ciphering and integrity protection keys K_{NASenc} and K_{NASint}. It then sends the mobile an EMM *Security Mode Command* (step 1), which tells the mobile to activate NAS security. The message is secured by integrity protection, but is not ciphered.

The mobile checks the integrity of the message, in the manner described below. If the message passes the integrity check, then the mobile computes its own copies of K_{NASenc} and K_{NASint} from its stored copy of K_{ASME}, and starts ciphering and integrity protection. It then acknowledges the MME's command using an EMM *Security Mode Complete* (2). On receiving the message, the MME starts downlink ciphering.

If the mobile detaches from the MME, then the two devices delete their copies of K_{NASenc} and K_{NASint}. However, the MME retains its copy of K_{ASME}, while the mobile retains its copies of CK and IK. When the mobile switches on again, it re-computes its copies of K_{NASenc} and K_{NASint} and uses the latter to apply integrity protection to the subsequent attach request. The request is not ciphered, however, because of the risk that the network will not understand it.

Access stratum security is activated later on, just before the network sets up the default radio bearer and signalling radio bearer 2. Figure 12.5 shows the message sequence. To trigger the process, the MME computes the value of K_{eNB}. It then passes K_{eNB} to the base station within the S1-AP Initial Context Setup Request, which we have already seen as step 17 of the attach procedure.

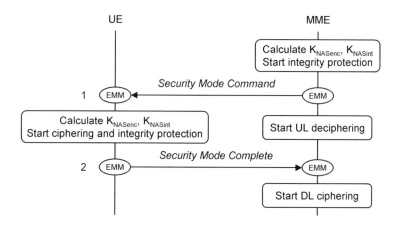

Figure 12.4 Activation of non access stratum security. Reproduced by permission of ETSI.

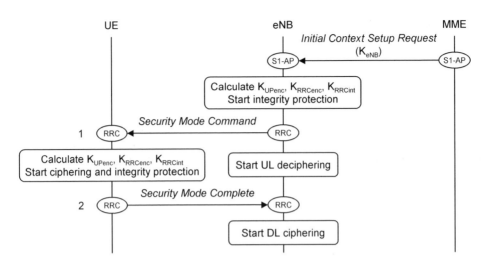

Figure 12.5 Activation of access stratum security. Reproduced by permission of ETSI.

From here, the procedure is very like the activation of NAS security. Using K_{eNB}, the base station computes the ciphering and integrity protection keys K_{UPenc}, K_{RRCenc} and K_{RRCint}. It then sends the mobile an RRC *Security Mode Command* (step 1), whose integrity is protected using K_{RRCint}. The mobile checks the integrity of the message, computes its own keys, and starts ciphering and integrity protection. It then acknowledges the base station's message using an RRC *Security Mode Complete* (step 2), at which point the base station can start downlink ciphering.

During a handover, the old base station derives a new secure key, K_{eNB}^*, either directly from K_{eNB} or from another parameter known as *next hop* (NH). It then passes K_{eNB}^* to the new base station, which uses it as the new value of K_{eNB}. If the mobile moves to RRC_IDLE, then K_{eNB}, K_{UPenc}, K_{RRCenc} and K_{RRCint} are all deleted. However, the mobile and MME both retain K_{ASME} and use it to derive a new set of access stratum keys when the mobile returns to RRC_CONNECTED.

12.1.5 Ciphering

Ciphering ensures that intruders cannot read the information that is exchanged between the mobile and the network [10–12]. The packet data convergence protocol ciphers data and signalling messages in the air interface access stratum, while the EMM protocol ciphers signalling messages in the non access stratum.

Figure 12.6 shows the ciphering process. The transmitter uses its ciphering key and other information fields to generate a pseudo-random key stream and mixes this with the outgoing data using an exclusive-OR operation. The receiver generates its own copy of the key stream and repeats the mixing process, so as to recover the original data. The algorithm is designed to be one-way, so that an intruder cannot recover the secure key from the transmitted message in a reasonable amount of computing time.

LTE currently supports three *EPS encryption algorithms* (EEAs). Two of them are SNOW 3G, which was originally used in the Release 7 standards for UMTS, and the

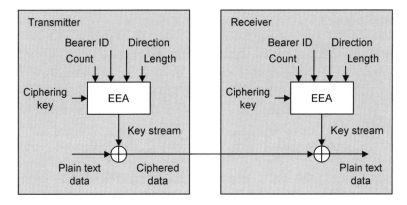

Figure 12.6 Operation of the ciphering algorithm. Reproduced by permission of ETSI.

Advanced Encryption Standard (AES). The third algorithm is a null ciphering algorithm, which means that, as in previous mobile communication systems, the air interface does not actually have to implement any ciphering at all. Unlike in previous systems, however, it is mandatory for an LTE device to let the user know whether the air interface is using ciphering or not.

12.1.6 Integrity Protection

Integrity protection [13–15] allows a device to detect modifications to the signalling messages that it receives, as a protection against problems such as man-in-the-middle attacks. The packet data convergence protocol applies integrity protection to RRC signalling messages in the air interface's access stratum, while the EMM protocol applies integrity protection to its own messages in the non access stratum.

Figure 12.7 shows the process. The transmitter passes each signalling message through an *EPS integrity algorithm* (EIA). Using the appropriate integrity protection key, the algorithm computes a 32-bit integrity field, denoted MAC-I, and appends it to the message. The receiver separates the integrity field from the signalling message and computes the expected integrity field XMAC-I. If the observed and expected integrity fields are the same, then it is happy. Otherwise, the receiver concludes that the message has been modified and discards it.

Integrity protection is mandatory for almost all of the signalling messages that the mobile and network exchange after security activation, and is based on SNOW 3G or the Advanced Encryption Standard as before. There is one exception, however: from Release 9, mobiles can use a null integrity protection algorithm for the sole purpose of making emergency voice calls without a UICC.

12.2 Network Domain Security

12.2.1 Security Protocols

Inside the fixed network, two devices often have to exchange information securely. Because the fixed network is based on IP, this can be done using standard IETF security

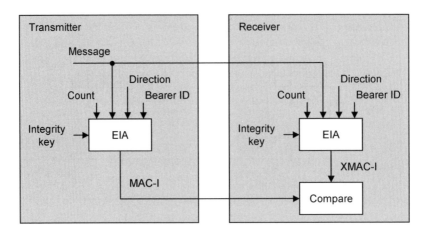

Figure 12.7 Operation of the integrity protection algorithm. Reproduced by permission of ETSI.

protocols [16]. Devices first authenticate each other and establish a security association using a protocol known as *Internet Key Exchange version 2* (IKEv2) [17, 18]. This relies either on the use of a pre-shared secret, as in the air interface's use of the secure key K, or on public key cryptography.

Encryption and integrity protection are then implemented using the *Internet Protocol Security* (IPSec) *Encapsulating Security Payload* (ESP) [19, 20]. Depending on the circumstances, the network can use ESP *transport mode*, which just protects the payload of an IP packet, or *tunnel mode*, which protects the IP header as well. These techniques are used in two parts of the LTE network, in the manner described next.

12.2.2 Security in the Evolved Packet Core

In the evolved packet core, secure communications are required between networks that are run by different network operators, so as to handle roaming mobiles. To support this, the evolved packet core is modelled using *security domains*. A security domain usually corresponds to a network operator's EPC (Figure 12.8), but the operator can divide the EPC into more than one security domain if required.

From the viewpoint of the network domain security functions, different security domains are separated by the Za interface. On this interface, it is mandatory for LTE signalling messages to be protected using ESP tunnel mode. The security functions are implemented using *secure gateways* (SEGs), although operators can include the secure gateways' functions in the network elements themselves if they wish to. There is no protection for data, which will usually end up in an insecure public network anyway. If required, the data can be protected at the application layer.

Within a security domain, the network elements are separated by the Zb interface. This interface is usually under the control of a single network operator, so protection of LTE signalling messages across this interface is optional. If the interface is secured, then support of ESP tunnel mode is mandatory, while support of ESP transport mode is optional.

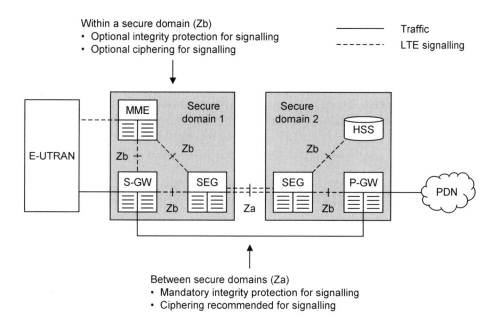

Figure 12.8 Example implementation of the network domain security architecture in the evolved packet core.

Security is the only difference between the S5 and S8 interfaces that were introduced in Chapter 2. Across the S5 interface, the serving and PDN gateways lie in the same security domain, so security functions are optional and are implemented using Zb. Across the S8 interface, the gateways lie in different security domains, so security functions are mandatory and are implemented using Za.

12.2.3 Security in the Radio Access Network

In the radio access network, network operators may also wish to secure the X2 and S1 interfaces, which connect the base stations to each other and to the evolved packet

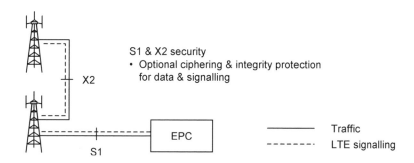

Figure 12.9 Network domain security architecture for the S1 and X2 interfaces.

core. They typically do this in two scenarios. In a wide-area network, the S1 and X2 interfaces can be implemented using microwave links rather than cables, which makes them vulnerable to intruders. In a femtocell network, a home base station communicates with the EPC across a public IP backhaul.

To handle these issues, it is optional for network operators to secure the S1 and X2 interfaces in the manner shown in Figure 12.9. If the interfaces are secured, then support of ESP tunnel mode is mandatory, while support of ESP transport mode is optional. The security functions are applied to the LTE signalling messages and to the user's data, which is a similar situation to the air interface's access stratum, but is different from the EPC.

References

1. 3GPP TS 33.401 (September 2011) *3GPP System Architecture Evolution (SAE): Security Architecture*, Release 10.
2. 3GPP TS 24.301 (September 2011) *Non-Access-Stratum (NAS) Protocol for Evolved Packet System (EPS); Stage 3*, Release 10.
3. 3GPP TS 29.272 (September 2011) *Evolved Packet System (EPS); Mobility Management Entity (MME) and Serving GPRS Support Node (SGSN) Related Interfaces Based on Diameter Protocol*, Release 10.
4. 3GPP TS 31.102 (October 2011) *Characteristics of the Universal Subscriber Identity Module (USIM) Application*, Release 10.
5. 3GPP TS 36.331 (October 2011) *Radio Resource Control (RRC); Protocol Specification*, Release 10.
6. Forsberg, D. Horn, G., Moeller, W.-D. and Niemi, V. (2010) *LTE Security*, John Wiley & Sons, Ltd, Chichester.
7. 3GPP TS 33.401 (September 2011) *3GPP System Architecture Evolution (SAE): Security Architecture*, Release 10, section 6.2.
8. 3GPP TS 33.401 (September 2011) *3GPP System Architecture Evolution (SAE): Security Architecture*, Release 10, section 6.1.
9. 3GPP TS 33.401 (September 2011) *3GPP System Architecture Evolution (SAE): Security Architecture*, Release 10, section 7.2.4.
10. 3GPP TS 33.401 (September 2011) *3GPP System Architecture Evolution (SAE): Security Architecture*, Release 10, annex B.1.
11. 3GPP TS 36.323 (March 2011) *Packet Data Convergence Protocol (PDCP) Specification*, Release 10, section 5.6.
12. 3GPP TS 24.301 (September 2011) *Non-Access-Stratum (NAS) Protocol for Evolved Packet System (EPS); Stage 3*, Release 10, section 4.4.5.
13. 3GPP TS 33.401 (September 2011) *3GPP System Architecture Evolution (SAE): Security Architecture*, Release 10, annex B.2.
14. 3GPP TS 36.323 (March 2011) *Packet Data Convergence Protocol (PDCP) Specification*, Release 10, section 5.7.
15. 3GPP TS 24.301 (September 2011) *Non-Access-Stratum (NAS) Protocol for Evolved Packet System (EPS); Stage 3*, Release 10, section 4.4.4.
16. 3GPP TS 33.401 (September 2011) *3GPP System Architecture Evolution (SAE): Security Architecture*, Release 10, sections 11, 12.
17. 3GPP TS 33.310 (September 2011) *Network Domain Security (NDS); Authentication Framework (AF), Network Domain Security (NDS); Authentication Framework (AF)*, Release 10.
18. IETF RFC 4306 (December 2005) *Internet Key Exchange (IKEv2) Protocol*.
19. 3GPP TS 33.210 (June 2011) *3G Security; Network Domain Security; IP Network Layer Security*, Release 10.
20. IETF RFC 4303 (December 2005) *IP Encapsulating Security Payload (ESP)*.

13

Quality of Service, Policy and Charging

In Chapters 2 and 11, we described how the evolved packet core transports data packets by the use of bearers and tunnels, and how it sets up a default EPS bearer for the mobile during the attach procedure. We also noted that each bearer was associated with a quality of service, which describes information such as the bearer's data rate, error rate and delay. However, we have not yet covered some important and related issues, namely how the network specifies and manages quality of service, and how it ultimately charges the user. These issues are the subject of this chapter.

We begin by defining the concept of policy and charging control, and by describing the architecture that is used for policy and charging in LTE. We continue by discussing the procedures for policy and charging control, and for session management, through which an application can request a specific quality of service from the network. We conclude with a discussion of offline and online charging.

Policy and charging control is described by two 3GPP specifications, namely TS 23.203 [1] and TS 29.213 [2]. The charging system is summarized in TS 32.240 [3] and TS 32.251 [4]. As usual, the details of the individual procedures are in the specifications for the relevant signalling protocols. Those for the evolved packet core are in [5–8], but we will introduce some more protocols that are specifically related to policy and charging in the course of the chapter.

13.1 Policy and Charging Control

13.1.1 Introduction

In Chapter 2, we explained that each EPS bearer was associated with a certain quality of service (QoS), which describes information such as the error rate, delay and total data rate of its constituent service data flows. Quality of service is itself associated with two other concepts, gating and charging, which collectively determine how the network will handle a particular bearer [9].

Firstly, *gating* determines whether or not packets are allowed to travel through a service data flow. Gating is important because the network typically configures a service data flow in three stages, by authorizing its quality of service, setting it up and finally allowing packets to flow. Together, gating and QoS make up a concept known as *policy*. Secondly, *charging* determines how the packets in a service data flow will be charged. The charging parameters include the *charging method*, which can be *offline charging*, where the user receives a regular bill, or *online charging*, for pre-paid services. They also include a *charging key*, which determines the tariff that the charging system will eventually use.

Each service data flow is associated with a *policy and charging control* (PCC) *rule*, which defines its policy and charging parameters [10]. LTE uses two types of rule. *Pre-defined rules* are permanently stored within the network and might typically be used for the standardized QoS characteristics that we will introduce below. In contrast, *dynamic rules* are composed on-the-fly.

13.1.2 Quality of Service Parameters

To add some detail to this discussion, let us look at how LTE specifies a bearer's quality of service [11, 12].

The most important QoS parameter is the *QoS class identifier* (QCI). This is an 8-bit number which acts as a pointer into a look-up table and which defines four other quantities. The *resource type* indicates whether or not the bearer has a guaranteed bit rate. If it does, then the corresponding QoS class is only applicable to dedicated bearers. The *packet error/loss rate* is an upper bound for the proportion of packets that are lost due to errors in transmission and reception. The network should apply it reliably to GBR bearers, but non GBR bearers can expect additional packet losses if the network becomes congested. The *packet delay budget* is an upper bound, with 98% confidence, for the delay that a packet receives between the mobile and the PDN gateway. Finally, the *QCI priority level* helps the scheduling process. Low numbers are associated with a high priority, and a congested network meets the packet delay budget of bearers with priority N, before moving on to bearers with priority $N + 1$.

Some of the QoS class identifiers have been standardized and are associated with the parameters listed in Table 13.1. Bearers in these classes can expect to receive a consistent quality of service, even if the mobile is roaming. Network operators can define other QoS classes for themselves, but these are only likely to work for non roaming mobiles.

The other QoS parameters are listed in Table 13.2. Each GBR bearer is associated with a guaranteed bit rate, which is a target for its long term average data rate. It is also associated with a *maximum bit rate* (MBR), which is the highest bit rate that the bearer can ever expect to receive. Despite this distinction, the specifications only support a maximum bit rate greater than the guaranteed bit rate from Release 10 onwards [13].

Non GBR bearers are collectively associated with the per UE aggregate maximum bit rate (UE-AMBR) and the per APN aggregate maximum bit rate (APN-AMBR) that we introduced in Chapter 11. These respectively limit a mobile's total data rate on all its non GBR bearers, and its total data rate on non GBR bearers that are using a particular access point name.

Finally, the *allocation and retention priority* (ARP) contains three fields. The *ARP priority level* determines the order in which a congested network should satisfy requests

Quality of Service, Policy and Charging

Table 13.1 Standardized QCI characteristics. Reproduced by permission of ETSI

QCI	Resource type	Packet error/ loss rate	Packet delay budget (ms)	QCI priority	Example services
1	GBR	10^{-2}	100	2	Conversational voice
2		10^{-3}	150	4	Real-time video
3		10^{-3}	50	3	Real-time games
4		10^{-6}	300	5	Buffered video
5	Non GBR	10^{-6}	100	1	IMS signalling
6		10^{-6}	300	6	Buffered video, TCP file transfers
7		10^{-3}	100	7	Voice, real-time video, real-time games
8		10^{-6}	300	8	Buffered video, TCP file transfers
9		10^{-6}	300	9	Buffered video, TCP file transfers

Table 13.2 Quality of service parameters

Parameter	Description	Use by GBR bearers	Use by non GBR bearers
QCI	QoS class identifier	✓	✓
ARP	Allocation & retention priority	✓	✓
GBR	Guaranteed bit rate	✓	✗
MBR	Maximum bit rate	✓	✗
APN-AMBR	Per APN aggregate maximum bit rate	✗	One field per APN
UE-AMBR	Per UE aggregate maximum bit rate	✗	One field per UE

to establish or modify a bearer, with level 1 receiving the highest priority. (Note that this parameter is different from the QCI priority level defined above.) The *pre-emption capability* field determines whether a bearer can grab resources from another bearer with a lower priority and might typically be set for emergency services. Similarly, the *pre-emption vulnerability* field determines whether a bearer can lose resources to a bearer with a higher priority.

We saw some of these parameters while discussing the medium access control protocol in Chapter 10. In the mobile's logical channel prioritization algorithm, the logical channel priority is derived from the QCI priority level in Table 13.1, while the prioritized bit rate is derived from the guaranteed bit rate in Table 13.2.

13.1.3 Policy Control Architecture

The evolved packet core contains some hardware components that we have not previously seen, to control a bearer's quality of service. These components are shown in Figure 13.1 [14].

The most important device is the *policy and charging rules function* (PCRF). This authorizes the policy and charging treatment that a service data flow will receive, either by referring to a predefined PCC rule, or by composing a dynamic PCC rule. The *policy*

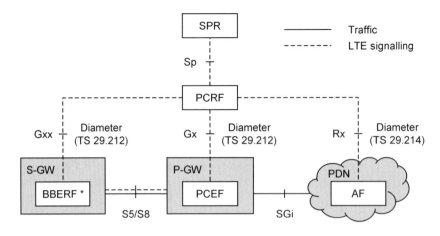

Figure 13.1 Policy control architecture for LTE. Reproduced by permission of ETSI.

and charging enforcement function (PCEF) is part of the PDN gateway. The PCEF asks the PCRF to supply a PCC rule for a service data flow and then implements its decisions; for example, by setting up new bearers, modifying the QoS of existing bearers and allowing packets to flow.

If the S5/S8 interface is based on GTP, then the PCEF maps incoming packets onto EPS bearers by the use of packet filters. If the interface is based on PMIP, then the mapping task is delegated to a *bearer binding and event reporting function* (BBERF) inside the serving gateway.

To help it specify the policy and charging rules for a service data flow, the PCRF can communicate with two other devices. The *subscription profile repository* (SPR) contains information about the users' subscriptions. It is likely to be associated with the home subscriber server, although the specifications do not define any relationship between the two devices. The *application function* (AF) runs an external application such as a voice over IP (VoIP) server and may be controlled by another party.

If the user is roaming then the PCRF's functions can be split, between a *home PCRF* (H-PCRF) in the home network and a *visited PCRF* (V-PCRF) in the visited network. The V-PCRF is used in two situations: it controls the visited network's BBERF in the case of a PMIP based S5/S8 interface, and it controls the visited network's PCEF in the case of local breakout. In both situations, the H-PCRF decides the user's quality of service, while the V-PCRF can accept or reject those decisions.

Most of the PCRF's signalling interfaces use Diameter applications, in a similar way to the interface between the MME and the home subscriber server. These include the Gx and Gxx interfaces to the PCEF and BBERF [15], the Rx interface to the application function [16], and the S9 interface between the home and visited PCRFs [17]. The Sp interface to the subscription profile repository is yet to be standardized.

Using this architecture, an application can ask for a specific quality of service in two ways. Firstly, the application function can make a QoS request by sending a signalling message directly to the PCRF. Secondly, the mobile can make a QoS request by sending

Quality of Service, Policy and Charging 205

a signalling message to the MME, which forwards the request to the PDN gateway and hence to the PCRF. We will see how the network implements these techniques shortly.

13.2 Session Management Procedures

13.2.1 IP-CAN Session Establishment

During the attach procedure from Chapter 11, we skipped over a step known as IP connectivity access network (IP-CAN) session establishment, in which the PDN gateway configured the quality of service of the default EPS bearer. Figure 13.2 shows the details [18]. In this example, we have assumed that the mobile is not roaming and that the S5/S8 interface is based on GTP.

The procedure is triggered when the PDN gateway receives a request to establish a default EPS bearer. In response, the PCEF sends a Diameter *CC-Request* to the PCRF (1). The message includes the mobile's IMSI, the access point name that the mobile is connecting to, the corresponding quality of service from the user's subscription data and any IP address that the network has allocated for the mobile. The PCRF stores the information it receives, for use in later procedures.

If the PCRF requires the user's subscription details to define the default bearer's PCC rule, but does not yet have them, then it retrieves those details from the subscription profile repository (2). The details of the interaction are yet to be specified.

The PCRF can now define the default bearer's quality of service and charging parameters, either by selecting a predefined PCC rule that the PCEF is already aware of, or by generating a dynamic rule (3). In doing so, it can either use the default quality of service that originated in the user's subscription data, or specify a new quality of service of its own. It then sends the information to the PCEF using a Diameter *CC-Answer* (4). If the PCC rule specifies the use of online charging, then the PCEF sends a credit request to the online charging system (5), in the manner described below. The PCEF can then proceed with the establishment of the default bearer.

If the user has suitable subscription details, then the PCRF can define more than one PCC rule in step 3, and can send all those PCC rules to the PCEF in step 4. This triggers

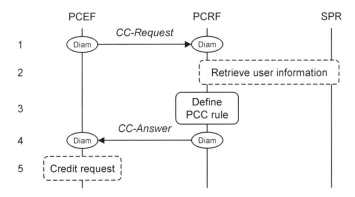

Figure 13.2 IP-CAN session establishment procedure. Reproduced by permission of ETSI.

the establishment of one or more dedicated bearers to the same APN, as well as the usual default bearer.

13.2.2 Mobile Originated QoS Request

After the attach procedure has completed, the PCRF can receive requests for a better quality of service from two sources, either the mobile itself, or an external application function. The PDN gateway can satisfy those requests in two ways, either by triggering the establishment of a dedicated bearer that has a better quality of service, or by modifying the quality of service of an existing bearer.

To illustrate this, Figure 13.3 shows how the mobile can ask for a new service data flow with an improved quality of service [19]. In this scenario, we assume that the mobile has previously made contact with an external VoIP server, using application-layer signalling messages that it has exchanged across a default EPS bearer. We also assume that the mobile's VoIP application would now like to set up a call with a better quality of service than the default EPS bearer can provide, typically with a guaranteed bit rate. Finally, we assume that the mobile is starting in ECM-CONNECTED state. When starting in ECM-IDLE, it first has to complete the service request procedure that we will describe in Chapter 14.

Inside the mobile, the VoIP application asks the LTE signalling protocols to set up a new service data flow with an improved quality of service. The protocols react by composing an ESM *Bearer Resource Allocation Request* and sending it to the MME (1). In the message, the mobile requests parameters such as the QoS class indicator, and the guaranteed and maximum bit rates for the uplink and downlink. It also specifies a traffic flow template that describes the service data flow, using parameters such as the UDP port number of the VoIP application. The MME receives the mobile's request and forwards it to the serving gateway as a GTP-C *Bearer Resource Command* (2). In turn, the serving gateway forwards the message to the appropriate PDN gateway (3).

The PDN gateway reacts by triggering the procedure for *PCEF initiated IP-CAN session modification* [20] (Figure 13.4). As before, the PCEF sends the PCRF a Diameter CC-Request (4), in which it identifies the originating mobile and states the requested quality of service. The PCRF looks up the subscription details that it previously retrieved from the subscription profile repository, or contacts the SPR if required (5).

If the mobile has previously contacted the application function, then the application may have asked the PCRF to notify it about future events involving that mobile.

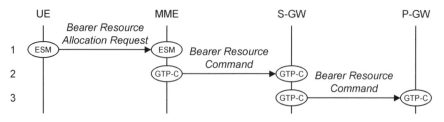

Figure 13.3 Mobile originated quality of service request. (1) Bearer resource allocation request. Reproduced by permission of ETSI.

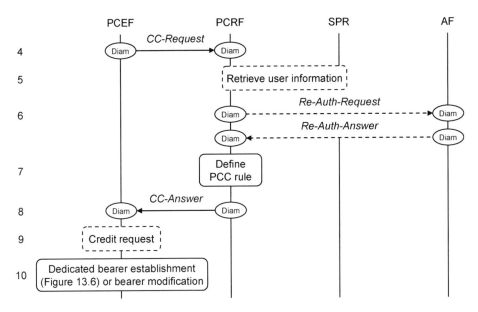

Figure 13.4 Mobile originated quality of service request. (2) PCEF initiated IP-CAN session modification. Reproduced by permission of ETSI.

If so, then the PCRF does so using a Diameter *Re-Auth-Request* (6). To meet the mobile's request, the PCRF defines a new PCC rule (7). It then returns the corresponding QoS and charging parameters to the PCEF (8), which sends a credit request to the online charging system if required (9).

Usually, the PDN gateway reacts to the new PCC rule by establishing a new dedicated bearer (10). We will show how it does this below, after discussing how the application function can make a QoS request of its own. If, however, the mobile already has an EPS bearer with the same QCI and ARP, then the PDN gateway can modify that bearer to include the new service data flow, by increasing its data rate and adding the new packet filters.

13.2.3 Server Originated QoS Request

As an alternative to the mobile originated procedure described above, an application function can ask the PCRF for a new service data flow with an improved quality of service. The message sequence is shown in Figure 13.5 [21]. In the figure, we assume as before that the mobile has made contact with an application function such as a VoIP server, using application layer signalling communications over the default bearer. The server would now like to set up a call, using a better quality of service than the default bearer can provide.

In step 1, the application function sends a Diameter *AA-Request* to the PCRF. In the message, it identifies the mobile using its IP address, and describes the requested media using parameters such as the media type, codec and port number, and the maximum uplink and downlink data rates. The PCRF looks up the mobile's subscription details, or retrieves them from the subscription profile repository if required (2) and returns an acknowledgement (3).

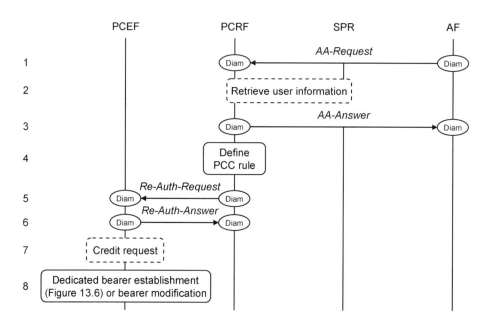

Figure 13.5 Server originated quality of service request. Reproduced by permission of ETSI.

To meet the application function's request, the PCRF defines a new PCC rule (4). It then sends the information to the PCEF in a Diameter *Re-Auth-Request* (5, 6) and the PCEF sends a credit request to the online charging system if required (7). As before, the PDN gateway reacts either by establishing a dedicated bearer in the manner described below, or by modifying the quality of service of an existing bearer (8).

13.2.4 Dedicated Bearer Establishment

In the discussions above, we showed how the mobile and the application function could both ask the PCRF for a new service data flow with an improved quality of service. The PCRF responds by defining a new PCC rule, which usually leads to the establishment of a dedicated EPS bearer in the manner shown in Figure 13.6 [22].

The procedure is triggered by one of the quality of service requests from Figures 13.3 to 13.5 (1). In response, the PDN gateway tells the serving gateway to create a new EPS bearer for the mobile, defines its quality of service, and includes an uplink tunnel endpoint identifier for use over S5/S8 and an uplink traffic flow template for the mobile (2). The serving gateway receives the message and forwards it to the MME (3).

If the application function triggered the procedure, then mobile may still be in ECM-IDLE. If it is, then the MME contacts the mobile using the paging procedure and the mobile responds with an EMM service request. We will discuss these procedures in Chapter 14.

The MME then composes an ESM *Activate Dedicated EPS Bearer Context Request*. This tells the mobile to set up a dedicated EPS bearer and includes the parameters that the MME received from the serving gateway. The MME embeds the message into an

Quality of Service, Policy and Charging

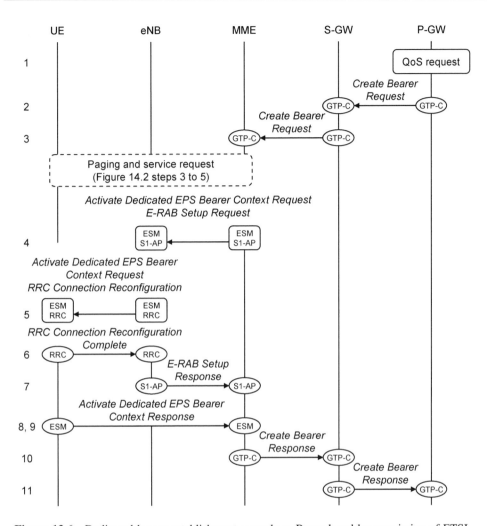

Figure 13.6 Dedicated bearer establishment procedure. Reproduced by permission of ETSI.

S1-AP *E-RAB Setup Request*, which tells the base station to set up the corresponding S1 and radio bearers and defines their qualities of service. In step 4, it sends both messages to the base station.

In response, the base station sends the mobile an RRC Connection Reconfiguration message (5), which tells the mobile how to configure the new radio bearer and which includes the ESM message that it received from the MME. The mobile configures the bearer as instructed and acknowledges its RRC message (6), which triggers a further acknowledgement from the base station back to the MME (7). The mobile then acknowledges the ESM message (8, 9), in a reply that is transported using an uplink information transfer. The two final acknowledgements (10, 11) complete the configuration of the GTP tunnels and allow data to flow.

13.2.5 Other Session Management Procedures

After the establishment of a service data flow, the mobile and application function can both ask the PCRF to modify its quality of service, for example, by increasing its data rate. In response, the PCRF modifies the corresponding PCC rule and sends the resulting parameters to the PDN gateway. In turn, the PDN gateway reacts by modifying the QoS of the corresponding bearer [23], or if necessary by extracting the modified service data flow into a new dedicated bearer.

The mobile can also ask for a connection to a second packet data network with a different access point name, using a procedure for *UE requested PDN connectivity* [24]. This triggers the procedure for IP-CAN session establishment that we discussed above and leads to the establishment of a second default EPS bearer for the new APN. The procedure is very like the ones for registration and dedicated bearer activation, so we will not discuss it in any detail. It is mandatory for the network to support connectivity to multiple access point names, but optional for the mobile.

If the S5/S8 interface is based on PMIP, then it only uses one GRE tunnel per mobile and does not distinguish the qualities of service of the corresponding data streams. Instead, the serving gateway includes a bearer binding and event reporting function, which maps incoming downlink packets onto the correct EPS bearers. During the attach procedure, the BBERF runs a process known as *Gateway control session establishment* [25], in which it establishes communications with the PCRF, forwards the subscription data that it received from the home subscriber server, and receives a PCC rule for the default EPS bearer in return. If the PCRF changes the PCC rule during IP-CAN session establishment, then it updates the BBERF using a further procedure known as *Gateway control and QoS rules provision* [26]. There are similar steps during dedicated bearer activation, bearer modification and UE requested PDN connectivity.

Finally, the detach procedure from Chapter 11 triggers a procedure known as *IP-CAN session termination* [27], in which the PCEF informs the PCRF that the mobile is detaching. As part of the procedure, the PCRF can notify the application function that the mobile is detaching, while the PCEF can send a final credit report to the online charging system and return any remaining credit.

13.3 Charging and Billing

13.3.1 High Level Architecture

LTE supports a flexible charging model, in which the cost of a session can be calculated from information such as the data volume, the data duration or the occurrence of specific events and can depend on issues such as the quality of service and the time of day. It uses the same charging architecture as the packet switched domains of UMTS and GSM [28], as shown in Figure 13.7.

There are two different charging systems: the *offline charging system* (OFCS) is suitable for post-paid services for which the subscriber receives a regular bill, while the *online charging system* (confusingly abbreviated to OCS) is suitable for pre-pay. From a technical point of view, the online charging system can affect a session in real time, for example by terminating the session if the subscriber runs out of credit, while the offline charging system cannot. We will discuss the two systems in turn in next page.

Quality of Service, Policy and Charging

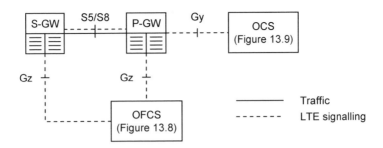

Figure 13.7 High level charging architecture.

13.3.2 Offline Charging

Figure 13.8 shows the internal architecture of the offline charging system. There are four components. *Charging trigger functions* (CTFs) monitor the subscriber's use of resources and generate *charging events*, which describe activities that the system will charge for. *Charging data functions* (CDFs) receives charging events from one or more charging trigger functions and collect them into *charging data records* (CDRs). The *charging gateway function* (CGF) post-processes the charging data records and collects them into *CDR files*, while the *billing domain* (BD) determines how much the resources have cost and sends an invoice to the user.

In LTE, the serving and PDN gateways both contain charging trigger functions. The charging data functions can be separate devices, as shown in Figure 13.8, or they can be integrated into the serving and PDN gateways. Depending on which choice is made, the Gz interface from Figure 13.7 corresponds to either the Rf or the Ga interface from Figure 13.8.

If the user is roaming, then the billing domain and charging gateway function are in the home network, while the charging data function is in the same network as the PDN gateway or serving gateway. In the case of home routed traffic, for example, the PDN gateway sends charging events to the home network's offline charging system, which processes them in the usual way. Meanwhile, the serving gateway sends charging events to the visited network's offline charging system, which uses them in two ways. Firstly, the visited network sends its charging data records to the home network, which uses them to invoice the subscriber. Secondly, the visited network uses its charging data records to invoice the home network for the subscriber's use of resources.

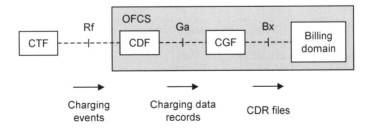

Figure 13.8 Architecture of the offline charging system. Reproduced by permission of ETSI.

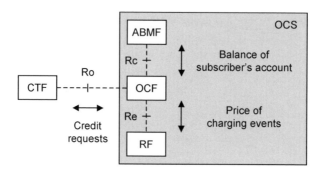

Figure 13.9 Architecture of the online charging system. Reproduced by permission of ETSI.

13.3.3 Online Charging

Figure 13.9 shows the internal architecture of the online charging system. In this architecture, the charging trigger function seeks permission to begin a session by sending a credit request to the *online charging function* (OCF). This retrieves the cost of the requested resources from the *rating function* (RF) and retrieves the balance of the subscriber's account from the *account balance management function* (ABMF). It then replies to the CTF with a credit authorization, which typically specifies how long the session can last or how much data the user can transfer. The CTF can then allow the session to proceed.

As the session continues, the charging trigger function monitors the subscriber's use of resources. If the subscriber approaches the end of the original allocation, then the CTF sends a new credit request to the online charging function, to ask for additional resources. At the end of the session, the CTF notifies the online charging function about any remaining credit and the online charging function returns the credit to the account balance management function.

In LTE, the charging trigger function lies inside the PDN gateway and is actually handled by the PCEF. The Gy interface from Figure 13.7 is identical to the Ro interface from Figure 13.9.

The online charging system is always in the subscriber's home network. If the user is roaming, then the visited network simultaneously creates charging data records using its offline charging system and uses these to invoice the home network for the subscriber's use of resources as before.

References

1. 3GPP TS 23.203 (June 2011) *Policy and Charging Control Architecture*, Release 10.
2. 3GPP TS 29.213 (September 2011) *Policy and Charging Control Signalling Flows and Quality of Service (QoS) Parameter Mapping*, Release 10.
3. 3GPP TS 32.240 (April 2011) *Telecommunication Management; Charging Management; Charging Architecture and Principles*, Release 10.
4. 3GPP TS 32.251 (September 2011) *Telecommunication Management; Charging Management; Packet Switched (PS) Domain Charging*, Release 10.
5. 3GPP TS 24.301 (September 2011) *Non-Access-Stratum (NAS) Protocol for Evolved Packet System (EPS); Stage 3*, Release 10.

6. 3GPP TS 29.274 (September 2011) *3GPP Evolved Packet System (EPS); Evolved General Packet Radio Service (GPRS) Tunnelling Protocol for Control Plane (GTPv2-C); Stage 3*, Release 10.
7. 3GPP TS 36.331 (October 2011) *Radio Resource Control (RRC); Protocol Specification*, Release 10.
8. 3GPP TS 36.413 (September 2011) *Evolved Universal Terrestrial Radio Access Network (E-UTRAN); S1 Application Protocol (S1AP)*, Release 10.
9. 3GPP TS 23.203 (June 2011) *Policy and Charging Control Architecture*, Release 10, section 4.
10. 3GPP TS 29.212 (September 2011) *Policy and Charging Control over Gx/Sd Reference Point*, Release 10, section 4.3.
11. 3GPP TS 23.203 (June 2011) *Policy and Charging Control Architecture*, Release 10, section 6.1.7.
12. 3GPP TS 23.401 (September 2011) *General Packet Radio Service (GPRS) Enhancements for Evolved Universal Terrestrial Radio Access Network (E-UTRAN) Access*, Release 10, section 4.7.3.
13. 3GPP TS 23.401 (September 2011) *General Packet Radio Service (GPRS) Enhancements for Evolved Universal Terrestrial Radio Access Network (E-UTRAN) Access*, Release 10, section 4.7.4.
14. 3GPP TS 23.203 (June 2011) *Policy and Charging Control Architecture*, Release 10, section 5.
15. 3GPP TS 29.212 (September 2011) *Policy and Charging Control over Gx/Sd Reference Point*, 3rd Generation Partnership Project, Release 10.
16. 3GPP TS 29.214 (September 2011) *Policy and Charging Control over Rx Reference Point*, Release 10.
17. 3GPP TS 29.215 (September 2011) *Policy and Charging Control (PCC) over S9 Reference Point; Stage 3*, Release 10.
18. 3GPP TS 29.213 (September 2011) *Policy and Charging Control Signalling Flows and Quality of Service (QoS) Parameter Mapping*, Release 10, section 4.1.
19. 3GPP TS 23.401 (September 2011) *General Packet Radio Service (GPRS) Enhancements for Evolved Universal Terrestrial Radio Access Network (E-UTRAN) Access*, Release 10, section 5.4.5.
20. 3GPP TS 29.213 (September 2011) *Policy and Charging Control Signalling Flows and Quality of Service (QoS) Parameter Mapping*, Release 10, section 4.3.2.1.
21. 3GPP TS 29.213 (September 2011) *Policy and Charging Control Signalling Flows and Quality of Service (QoS) Parameter Mapping*, Release 10, sections 4.3.1.1, 4.3.1.2.1.
22. 3GPP TS 23.401 (September 2011) *General Packet Radio Service (GPRS) Enhancements for Evolved Universal Terrestrial Radio Access Network (E-UTRAN) Access*, Release 10, section 5.4.1.
23. 3GPP TS 23.401 (September 2011) *General Packet Radio Service (GPRS) Enhancements for Evolved Universal Terrestrial Radio Access Network (E-UTRAN) Access*, Release 10, section 5.4.2.1.
24. 3GPP TS 23.401 (September 2011) *General Packet Radio Service (GPRS) Enhancements for Evolved Universal Terrestrial Radio Access Network (E-UTRAN) Access*, Release 10, section 5.10.2.
25. 3GPP TS 29.213 (September 2011) *Policy and Charging Control Signalling Flows and Quality of Service (QoS) Parameter Mapping*, Release 10, section 4.4.1.
26. 3GPP TS 29.213 (September 2011) *Policy and Charging Control Signalling Flows and Quality of Service (QoS) Parameter Mapping*, Release 10, section 4.4.3.
27. 3GPP TS 29.213 (September 2011) *Policy and Charging Control Signalling Flows and Quality of Service (QoS) Parameter Mapping*, Release 10, section 4.2.1.
28. 3GPP TS 32.240 (September 2011) *Telecommunication Management; Charging Management; Charging Architecture and Principles*, Release 10, section 4.

14

Mobility Management

In this chapter, we discuss the mobility management procedures that the network uses to keep track of the mobile's location.

The choice of mobility management procedures depends on the state that the mobile is in. Mobiles in RRC_IDLE use a mobile-triggered procedure known as cell reselection, whose objective is to maximize the mobile's battery life and minimize the load on the network. In contrast, mobiles in RRC_CONNECTED use the network-triggered procedures of measurements and handover, to give the base station the control it requires over mobiles that are actively transmitting and receiving. We begin the chapter by covering the procedures that switch a mobile between these states in response to changes in the user's activity, namely S1 release, paging and service requests, and continue by describing the mobility management procedures themselves.

Several specifications are relevant for this chapter. TS 36.304 [1] defines the mobility management procedures that the mobile should follow in RRC_IDLE, while TS 23.401 [2] and TS 36.300 [3] describe the signalling procedures in RRC_CONNECTED and the procedures that switch the mobile between states. As usual, the relevant stage 3 specifications [4–8] define the details of the individual signalling messages. Another specification [9] defines the measurements that the mobile has to make in both RRC states and the corresponding performance requirements.

14.1 Transitions Between Mobility Management States

14.1.1 S1 Release Procedure

After the attach procedure from Chapter 11, the mobile had completed its power-on procedures and was in the states EMM-REGISTERED, ECM-CONNECTED and RRC_CONNECTED. The user was able to communicate with the outside world, using the default EPS bearer. If the user does nothing, then the network can use a procedure known as *S1 release* [10] to take the mobile into ECM-IDLE and RRC_IDLE. As part of this procedure, the network tears down signalling radio bearers 1 and 2, and deletes all the user's data radio bearers and S1 bearers. Using the cell reselection procedure that we cover

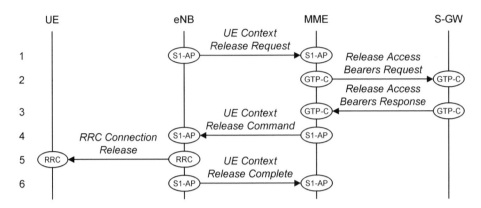

Figure 14.1 S1 release procedure, triggered by user inactivity in the base station. Reproduced by permission of ETSI.

later, the mobile can then move from one cell to another without the need to re-route the bearers, so only has a minimal impact on the network.

Figure 14.1 shows the message sequence. After a period of user inactivity, a timer expires in the base station and triggers the procedure. In response, the base station asks the MME to move the mobile into ECM-IDLE (1).

On receiving the request, the MME tells the serving gateway to tear down the S1 bearers that it was using for the mobile (2). The serving gateway does so and responds (3), but the PDN gateway is not involved, so the S5/S8 bearers remain intact. From this point, any downlink data can only travel as far as the serving gateway, where they trigger the paging procedure described below.

Using messages that we have already seen in the detach procedure, the MME tells the base station to tear down its signalling communications with the mobile (4) and the base station sends a similar message across the air interface (5). In response, the mobile tears down SRB 1, SRB 2 and all its data radio bearers, and moves into ECM-IDLE and RRC_IDLE. There is no need for it to reply. At the same time, the base station tears down its own records of the mobile's S1 and radio bearers, and sends an acknowledgement to the MME (6).

The base station can trigger the S1 release procedure for several other reasons, such as a repeated failure of the integrity check or a loss of radio communications with the mobile. The MME can also trigger the procedure starting from step 2; for example, after a failure of the authentication procedure.

14.1.2 Paging Procedure

At the end of the S1 release procedure, the mobile is in ECM-IDLE and RRC_IDLE, and its S1 and radio bearers have all been torn down. What happens if downlink data arrive at the serving gateway?

Figure 14.2 shows the answer [11]. The PDN gateway forwards an incoming data packet to the serving gateway (1), but the serving gateway is unable to send the data any further. Instead, it sends a GTP-C *Downlink Data Notification* message to the MME (2),

Mobility Management

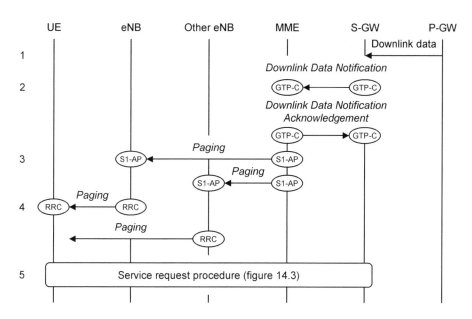

Figure 14.2 Paging procedure. Reproduced by permission of ETSI.

to tell the MME what has happened. The MME acknowledges, which stops the serving gateway from sending further notifications if more data packets arrive.

The MME now composes an S1-AP *Paging* message, and sends it to the base stations in all of the mobile's tracking areas (3). In the message, it specifies the mobile's S-TMSI, and all the information that the base stations will need to calculate the paging frame and the paging occasion. In response, each base station sends a single RRC *Paging* message to the mobile (4), in accordance with the discontinuous reception procedure from Chapter 8. The mobile receives the message from one of the base stations and responds using the service request procedure described below (5).

14.1.3 Service Request Procedure

The mobile runs the *service request* procedure [12] if it is in ECM-IDLE state but wishes to communicate with the network. The procedure can be triggered in two ways, either by the paging procedure described above, or internally, for example if the user tries to contact a server while the mobile is idle. During the procedure, the network moves the mobile into ECM-CONNECTED, re-establishes SRB 1 and SRB 2, and re-establishes the mobile's data radio bearers and S1 bearers. The mobile can then exchange data with the outside world.

The message sequence is shown in Figure 14.3. There is actually little new material here, as most of the messages have already appeared within the procedures for registration and session management.

The mobile starts by establishing a signalling connection with a serving eNB, using the procedures for contention based random access and RRC connection establishment. It then composes an EMM *Service Request*, which asks the serving MME to move it into

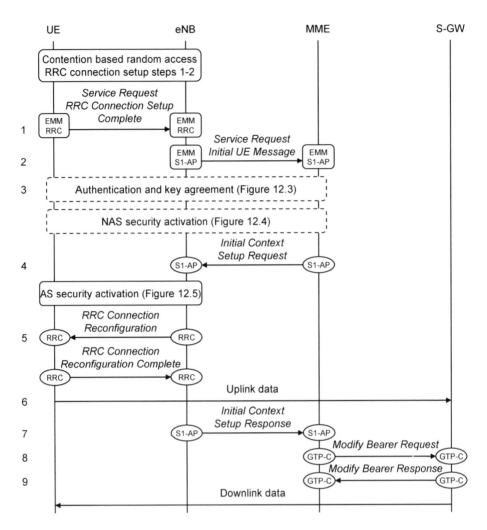

Figure 14.3 Service request procedure. Reproduced by permission of ETSI.

ECM-CONNECTED. It embeds the request into its RRC Connection Setup Complete, and sends both messages to the base station (1).

The base station extracts the mobile's request and forwards it to the MME (2). Optionally, the MME can now authenticate the mobile and can update non access stratum security using the resulting keys (3). These steps are mandatory if the EMM message failed the integrity check and are optional otherwise.

The MME now tells the base station to set up the mobile's S1 and radio bearers (4). The message includes several parameters that the MME stored after the attach procedure, such as the identity of the serving gateway, the tunnel endpoint identifiers for the uplink S1 bearers, the mobile's radio access capabilities and the security key K_{eNB}. The base station reacts to the last of these by activating access stratum security.

Mobility Management

The base station can now send an RRC message to the mobile (5), which configures SRB 2 and the mobile's data radio bearers. The mobile treats the message as an implicit acceptance of its EMM service request, returns an acknowledgement and can now send any uplink data to the outside world (6). In turn, the base station sends an acknowledgement to the MME (7) and includes tunnel endpoint identifiers (TEIDs) for the serving gateway to use on the downlink.

The MME now forwards the downlink TEIDs to the serving gateway and identifies the target base station (8). The serving gateway responds (9) and can now send downlink data to the mobile, notably any data that triggered the paging procedure in Figure 14.2.

14.2 Cell Reselection in RRC_IDLE

14.2.1 Objectives

For mobiles in RRC_IDLE and ECM-IDLE, the mobility management procedures have two main aims. The first is to maximize the mobile's battery life. To achieve this, the mobile usually wakes up only once in every discontinuous reception cycle, to monitor the network for paging messages and to make the measurements described below. Using this technique, the mobile can spend most of its time in a low power state.

The second aim is to minimize the signalling load on the network. To achieve this, the mobile decides by itself whether to stay with the previous cell or move to a new cell, by following a procedure known as *cell reselection*. It only informs the network about its location if it moves into a tracking area in which it was not previously registered.

In this section, we discuss the mobility management procedures that these mobiles use. We will follow the steps used in Release 8, but will also note some new features that have been introduced in Release 9.

14.2.2 Cell Reselection on the Same LTE Frequency

We start the cell reselection procedure [13, 14] by assuming that the network is only using a single LTE carrier frequency. If this is the case, then the mobile wakes up once every discontinuous reception cycle, in the same subframes that it is already monitoring for paging messages. In those subframes, the mobile measures the reference signal received power (RSRP) from the serving cell. If the RSRP is high enough, then the mobile can continue camping on that cell and does not have to measure any neighbouring cells at all. This technique minimizes the number of measurements that the mobile performs and the time for which it is awake, so maximizes its battery life.

This situation continues until the RSRP falls below the following threshold:

$$S_{\text{rxlev}} \leq S_{\text{IntraSearchP}} \quad (14.1)$$

In this equation, $S_{\text{IntraSearchP}}$ is a threshold that the serving cell advertises as part of SIB 3. S_{rxlev} depends on the RSRP of the serving cell and is calculated using Equation (11.2). If the above condition is met, then the mobile starts to measure neighbouring cells that are on the same LTE carrier frequency as the serving cell. To do this, it runs the acquisition procedure from Chapter 7, identifies each neighbouring cell, checks the identity of the corresponding network and any closed subscriber group, and measures the RSRP.

Unlike in earlier systems, the mobile can find neighbouring LTE cells by itself: the base station does not have to advertise an LTE neighbour list as part of its system information. This brings three benefits. Firstly, the network operator can configure the radio access network more easily in LTE than before. Secondly, there is no risk of a mobile missing nearby cells due to errors in the neighbour list. Thirdly, it is easier for an operator to introduce home base stations, which the user can install in locations that are unknown to the surrounding macrocell network. However, the base station can still identify individual neighbouring cells in SIB 4 using their physical cell identities and can describe them using optional cell-specific parameters that we will see next.

After finding and measuring the neighbouring cells, the mobile computes the following ranking scores:

$$R_s = Q_{meas,s} + Q_{hyst}$$
$$R_n = Q_{meas,n} - Q_{offset,s,n} \tag{14.2}$$

Here, R_s and R_n are the ranking scores of the serving cell and one of its neighbours, while $Q_{meas,s}$ and $Q_{meas,n}$ are the corresponding reference signal received powers. Q_{hyst} is a hysteresis parameter that the base station advertises in SIB 3, which discourages the mobile from bouncing back and forth between cells as the signal levels fluctuate. $Q_{offset,s,n}$ is an optional cell-specific offset, which the serving cell can advertise in SIB 4 to encourage or discourage the mobile to or from individual neighbours.

The mobile then switches to the best ranked cell, provided that three conditions are met. Firstly, the mobile must have been camped on the serving cell for at least one second. Secondly, the new cell must be suitable, according to the criteria laid out in Chapter 11. Finally, the new cell must be better ranked than the serving cell for a time of $T_{reselection, EUTRA}$, which is advertised in SIB 3 and has a value of 0 to 7 seconds. The mobile uses the same procedure if any of the neighbouring cells belongs to a closed subscriber group, except that the mobile must also belong to the group in order to camp on a CSG cell.

There is one new feature in 3GPP Release 9. From this release, a mobile can also start measurements of neighbouring cells if the reference signal received quality (RSRQ) falls below the following threshold:

$$S_{qual} \leq S_{IntraSearchQ} \tag{14.3}$$

Here, $S_{IntraSearchQ}$ is a threshold that the base station advertises as part of SIB 3. S_{qual} depends on the serving cell's RSRQ and is calculated using Equation (11.5).

14.2.3 Cell Reselection to a Different LTE Frequency

If the network is using more than one LTE carrier frequency, then the serving cell advertises the other carriers as part of SIB 5. As before, it may include offsets to the individual neighbouring cells, but is not obliged to do so. The serving cell does, however, associate each carrier frequency with a priority from 0 to 7, where 7 is the highest priority. It also specifies a priority for the serving frequency as part of SIB 3. The network can use these priorities to encourage or discourage the mobile to or from individual carriers, a feature that is particularly useful in layered networks, as microcells are usually on a different carrier frequency from macrocells and usually require a higher priority.

If another carrier is involved, then the processes of measurement triggering and cell reselection depend on whether its priority is higher than that of the current carrier, or the same, or lower. First, let us consider carriers with a higher priority than the current one. The mobile always measures cells on higher priority frequencies, no matter how strong the signal from the serving cell. It makes the measurements separately from the discontinuous reception cycle, as the mobile cannot look for paging messages on one carrier and measure cells on another at the same time. However, the mobile only has to measure one carrier frequency every minute, so the load on the mobile is small.

The mobile moves to a new cell on a higher priority carrier if three conditions are met. We have already seen the first two: the mobile must have been camped on the serving cell for at least one second, and the new cell must be suitable according to the criteria from Chapter 11. Finally, the new cell's RSRP must meet the following condition, for a time of at least $T_{\text{reselection, EUTRA}}$:

$$S_{\text{rxlev, x, n}} > \text{Thresh}_{\text{x, HighP}} \tag{14.4}$$

Here, $\text{Thresh}_{\text{x, HighP}}$ is a threshold for frequency x that the serving cell advertises in SIB 5. $S_{\text{rxlev, x, n}}$ depends on the new cell's RSRP, and is calculated using Equation (11.2). The mobile does not measure the RSRP from the serving cell in making this decision, so it moves to a higher priority frequency whenever it finds a cell that is good enough.

Now consider carriers with the same priority as the current one. The mobile starts measuring these if the following condition is satisfied:

$$S_{\text{rxlev}} \leq S_{\text{NonIntraSearchP}} \tag{14.5}$$

Here, $S_{\text{NonIntraSearchP}}$ is a threshold that the base station advertises in SIB 3, while S_{rxlev} depends on the serving cell's RSRP as before. The mobile moves to a new cell on that frequency using nearly the same criteria as it did for a cell on the same frequency. The only difference is in Equation (14.2), where the serving cell can optionally add a frequency-specific offset $Q_{\text{offset, frequency}}$ to the cell-specific offset $Q_{\text{offset, s, n}}$.

Finally, consider carriers with a lower priority than the current one. The mobile starts measuring these using the same criterion as for an equal priority frequency. However, it only moves to a new cell on that frequency if several conditions are met. The first two are as before: the mobile must have been camped on the serving cell for at least one second and the new cell must be suitable according to the criteria from Chapter 11. The mobile must be unable to find a cell on the original frequency, or on a frequency with an equal or higher priority, which it can move to. Finally, the RSRPs of the serving and neighbouring cells must satisfy the following conditions, for a time of at least $T_{\text{reselection, EUTRA}}$:

$$\begin{aligned} S_{\text{rxlev}} &< \text{Thresh}_{\text{Serving, LowP}} \\ S_{\text{rxlev, x, n}} &> \text{Thresh}_{\text{x, LowP}} \end{aligned} \tag{14.6}$$

As before, $\text{Thresh}_{\text{Serving, LowP}}$ and $\text{Thresh}_{\text{x, LowP}}$ are thresholds that the base station advertises in SIB 5, while S_{rxlev} and $S_{\text{rxlev, x, n}}$ depend on the RSRP of the serving and neighbouring cells.

There is one adjustment for mobiles that belong to a closed subscriber group. If a mobile is camped on a non CSG cell and detects a suitable CSG cell that is the highest ranked on another carrier, then it moves to that CSG cell, irrespective of the priority of the new carrier frequency. This does not apply to CSG cells on the same carrier frequency, however.

There are two new features in Release 9. Firstly, the mobile can also start measurements of neighbouring cells on an equal or lower priority frequency if the RSRQ falls below the following threshold:

$$S_{qual} \leq S_{NonIntraSearchQ} \tag{14.7}$$

Secondly, the base station can optionally replace the criteria for moving to a new cell on a higher or lower priority frequency (Equations 14.4 and 14.6), with similar criteria that are based on the RSRQ instead of the RSRP.

14.2.4 Fast Moving Mobiles

In the above algorithms, the mobile can only move to a neighbouring cell whose received signal power has been above a suitable threshold for a time of at least $T_{reselection, EUTRA}$. Usually, the value of $T_{reselection, EUTRA}$ is fixed. For fast moving mobiles, however, the use of a fixed value can introduce unwanted delays into the procedure, and can even prevent a mobile from moving to a new cell altogether.

To deal with this problem, the mobile measures the rate at which it is making cell reselections, ignoring any reselections that cause it to bounce back and forth between neighbouring cells. Depending on the result, it places itself either in a normal mobility state, or in a state of medium or high mobility.

In the medium and high mobility states, the mobile makes two adjustments. Firstly, it reduces the reselection time, $T_{reselection, EUTRA}$, using a state-dependent scaling factor that lies between 0.25 and 1. This reduces the delays in the cell reselection procedure, and allows the mobile to move to a neighbouring cell more quickly. Secondly, the mobile reduces the hysteresis parameter Q_{hyst} from Equation (14.2). This makes the mobile less likely to stick in the current cell, so also eases the process of cell reselection. The network specifies all the necessary thresholds and adjustments as part of SIB 3.

14.2.5 Tracking Area Update Procedure

After the mobile reselects to a new cell, it reads the cell's system information and examines the tracking area code. If the mobile has moved into a tracking area in which it was not previously registered, then it tells the evolved packet core using a procedure known as a *tracking area update* [15].

Figure 14.4 shows a basic version of the procedure. The diagram assumes that the mobile is starting in ECM-IDLE and RRC_IDLE and also that the mobile has stayed in the same MME pool area and S-GW service area, so that the MME and serving gateway can remain unchanged.

The procedure begins in a similar way to the service request that we saw earlier. The mobile runs the random access procedure and steps 1 and 2 of RRC connection establishment, so as to move temporarily into RRC_CONNECTED. It then composes an EMM *Tracking Area Update Request* and sends it to the base station by embedding it into its RRC Connection Setup Complete (1). In turn, the base station forwards the message to the serving MME (2). As before, the MME can authenticate the mobile and update non access stratum security (3).

The MME examines the new tracking area and decides in this example that the serving gateway can remain unchanged. In response, it sends the mobile an EMM *Tracking Area*

Mobility Management

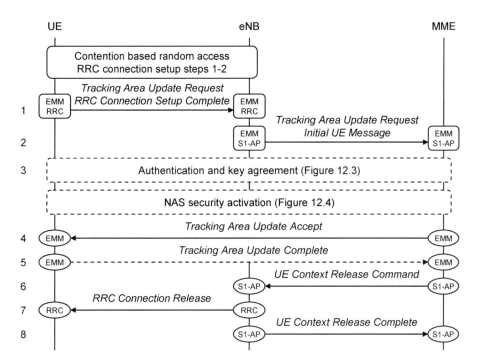

Figure 14.4 Tracking update procedure, starting from RRC_IDLE, with retention of the MME and serving gateway. Reproduced by permission of ETSI.

Update Accept (4), in which it lists the tracking areas in which the mobile is now registered and optionally gives the mobile a new globally unique temporary identity. If the GUTI does change, then the mobile sends an acknowledgement (5). The MME can now tell the base station to move the mobile back into RRC_IDLE, in messages that we saw as part of S1 release (steps 6 to 8).

The mobile also runs the tracking area update procedure periodically, even if it stays in the same tracking area, to tell the MME that it is still switched on and in an area of LTE coverage. The timer has a default value of 54 minutes, but the MME can choose a different value in its Attach Accept or Tracking Area Update Accept messages.

There are two complications. If the mobile has moved into a new MME pool area, then the new base station will not be connected to the old MME. Instead, the base station chooses a new MME after step 1, using the same technique that we saw during the attach procedure. The new MME retrieves the mobile's details from the old MME and contacts the home subscriber server to update its record of the mobile's location. It also tells the serving gateway that it is now looking after the mobile.

Independently, the mobile may have moved into a new S-GW service area. In place of the last interaction from the paragraph above, the MME tells the new serving gateway to set up a new set of EPS bearers for the mobile and to redirect the S5/S8 bearers by contacting the PDN gateway. The MME also tells the old serving gateway to tear down its bearers. If the MME has changed as well, then these steps are performed by the new and old MME respectively.

14.2.6 Network Reselection

If a mobile is configured for automatic network selection and is roaming in a visited network, then it periodically runs a procedure for *network reselection* whenever it is in RRC_IDLE, to search for networks that have the same country code but a higher priority [16, 17]. The search period is stored in the USIM and has a default value of 60 minutes [18].

The procedure is the same as the earlier one for network and cell selection, except that the conditions for a suitable cell (Equations 11.2 and 11.5) are modified as follows:

$$S_{\text{rxlev}} = Q_{\text{rxlevmeas}} - Q_{\text{rxlevmin}} - Q_{\text{rxlevminoffset}} - P_{\text{compensation}} \qquad (14.8)$$

$$S_{\text{qual}} = Q_{\text{qualmeas}} - Q_{\text{qualmin}} - Q_{\text{qualminoffset}} \qquad (14.9)$$

In the first equation, the original serving cell specifies the parameter $Q_{\text{rxlevminoffset}}$ as part of SIB 1. This parameter increases the minimum RSRP that is required in the destination cell and prevents a mobile from selecting a high priority network that only contains poor cells. The same applies to the parameter $Q_{\text{qualminoffset}}$ from Release 9.

14.3 Measurements in RRC_CONNECTED

14.3.1 Objectives

If a mobile is in ECM-CONNECTED and RRC_CONNECTED states, then the mobility management procedures are completely different. In these states, the mobile can be transmitting and receiving at a high data rate, so it is important for the network to decide which cell a mobile is communicating with.

The system achieves this by a two-step procedure. In the first step, the mobile measures the signal levels from the serving cell and its nearest neighbours, and sends measurement reports to the serving eNB. In the second step, the serving eNB can use these measurements to request a handover to a neighbouring cell.

14.3.2 Measurement Procedure

The measurement procedure [19, 20] is shown in Figure 14.5. To start the procedure, the serving eNB sends the mobile an RRC Connection Reconfiguration message (step 1). We have already seen this message being used to reconfigure a mobile's radio bearers in places such as the attach procedure, but it can also be used to specify the measurements that a mobile should be making.

In the RRC message, the measurements are specified using an information element known as a *measurement configuration*. This contains a list of *measurement objects*, each of which describes an LTE carrier frequency to measure and the corresponding downlink bandwidth. It also contains a list of *reporting configurations*, which tell the mobile when to report the results and which are described next. Finally, it defines each individual measurement using a *measurement identity*, which simply pairs up a measurement object with a reporting configuration.

As before, the mobile can identify the neighbouring cells by itself, so the base station does not have to list them. The measurement object does, however, contain several

Mobility Management

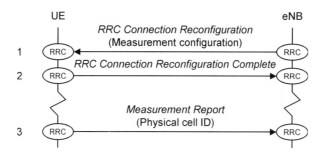

Figure 14.5 Measurement reporting procedure.

optional fields, such as a list of cells that the mobile should ignore when making measurements, and frequency- and cell-specific offsets for use in measurement reporting. In these fields, the base station identifies each cell using its physical cell identity.

The mobile acknowledges (step 2) and makes the measurements. Eventually, the mobile sends an RRC *Measurement Report* to the serving eNB (step 3) and identifies the cell that triggered the report using its physical cell identity. In response, the serving eNB can run the handover procedure that we will cover shortly.

14.3.3 Measurement Reporting

The base station tells the mobile when to return a measurement report using the reporting configuration that we introduced above. Measurement reports can be *periodic*, with a period from 120 milliseconds to 60 minutes. More often, however, measurement reports are triggered by various *measurement events*, which happen when signal levels cross over thresholds. Table 14.1 lists the events that are used for measurements of other LTE cells. (We will see some other measurement events when we discuss inter-system operation in Chapter 15.)

As an example, let us consider event A3, which might typically be used to trigger a handover within LTE. The mobile reports measurement event A3 to the network when the following condition is met:

$$M_n + \mathrm{Of}_n + \mathrm{Oc}_n > M_s + \mathrm{Of}_s + \mathrm{Oc}_s + \mathrm{Off} + \mathrm{Hys} \qquad (14.10)$$

Table 14.1 Measurement events used to report neighbouring LTE cells

Event	Release	Description
A1	R8	Serving cell rises above threshold
A2	R8	Serving cell falls below threshold
A3	R8	Neighbouring LTE cell rises above serving cell + offset
A4	R8	Neighbouring LTE cell rises above threshold
A5	R8	Serving cell falls below threshold 1
		Neighbouring LTE cell rises above threshold 2
A6	R10	Neighbouring LTE cell rises above secondary cell + offset

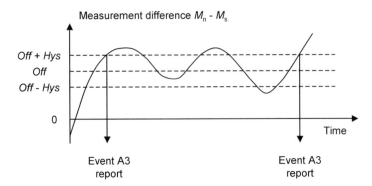

Figure 14.6 Operation of measurement event A3, in the case where the frequency- and cell-specific offsets are zero.

It then stops any further reporting of measurement event A3 until the following condition is met:

$$M_n + Of_n + Oc_n < M_s + Of_s + Oc_s + Off - Hys \qquad (14.11)$$

In these equations, M_s and M_n are the mobile's measurements of the serving and neighbouring cells respectively. As part of the reporting configuration, the base station can tell the mobile to measure either the cells' reference signal received power, or their reference signal received quality [21].

Hys is a hysteresis parameter for measurement reporting. If the mobile sends a measurement report to its serving eNB, then Hys prevents any more reports until the signal levels have changed by $2 \times$ Hys. Similarly, Off is a hysteresis parameter for handovers. If the measurement report triggers a handover, then Off prevents the mobile from moving back to the original cell until the signal levels have changed by $2 \times$ Off. Of_s and Of_n are the optional frequency-specific offsets noted earlier, while Oc_s and Oc_n are the cell-specific offsets.

Figure 14.6 shows the effect. In this diagram, the frequency- and cell-specific offsets are zero, while the vertical axis shows the difference between the measurements of the neighbouring and serving cells, $M_n - M_s$. The mobile sends a measurement report once this quantity exceeds Off + Hys, but does not send another until it has fallen below Off − Hys.

14.3.4 Measurement Gaps

If a neighbouring cell is on the same carrier frequency as the serving eNB, then the mobile can measure it at any time. If it is on a different frequency, then things are more difficult. Unless the mobile has an expensive dual frequency receiver, it cannot transmit and receive on one frequency and make measurements on another at the same time.

To deal with this problem, the base station can define *measurement gaps* (Figure 14.7) as part of the measurement configuration. Measurements gaps are subframes in which the base station promises not to schedule any transmissions to or from the mobile. During these subframes, the mobile can move to another carrier frequency and make a measurement, confident that it will not miss any downlink data or uplink transmission

Mobility Management

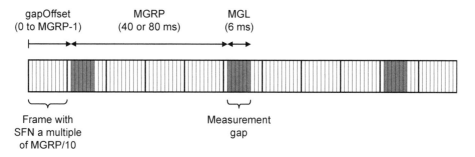

Figure 14.7 Measurement gaps.

opportunities. Each gap has a *measurement gap length* (MGL) of six subframes and a *measurement gap repetition period* (MGRP) of either 40 or 80 subframes. A mobile receives an individual offset within this period as part of its measurement configuration.

14.4 Handover in RRC_CONNECTED

14.4.1 X2 Based Handover Procedure

After receiving a measurement report, the serving eNB may decide to hand the mobile over to another cell. There are a few scenarios, but the most common is the basic *X2-based handover procedure*. This involves a change of base station using signalling messages over the X2 interface, but no change of serving gateway or MME.

The procedure begins in Figure 14.8, which follows the numbering scheme from TS 36.300 [22]. The mobile identifies a neighbouring cell in a measurement report (1, 2) and the old base station decides to hand the mobile over (3). Using an X2-AP *Handover Request* (4), it asks the new base station to take control of the mobile and includes the new cell's global ID, the identity of the mobile's serving MME, the security key $K_{eNB}{}^*$ and the mobile's radio access capabilities. It also identifies the bearers that it would like to transfer and describes their qualities of service.

The new base station examines the list of bearers and identifies the bearers that it is willing to accept (5). It might reject some if the new cell is overloaded, for example, or if the new cell has a smaller bandwidth than the old one. It then composes an RRC Connection Reconfiguration message, which tells the mobile how to communicate with the new cell. In the message, it gives the mobile a new C-RNTI and includes the configurations of SRB 1, SRB 2 and the data radio bearers that it is willing to accept. As described in Chapter 9, it can optionally include a preamble index for the non contention based random access procedure. The new base station embeds its RRC message into an X2-AP *Handover Request Acknowledge*, which acknowledges the old base station's request and lists the bearers that it will accept. It then sends both messages to the old base station (6).

The old base station extracts the RRC message and sends it to the mobile (7). At the same time, it sends an X2-AP *SN Status Transfer* to the new base station (8), which identifies the PDCP service data units that it has successfully received on the uplink, on bearers that are using RLC acknowledged mode. It also forwards any uplink packets that it has received out of sequence, any downlink packets that the mobile has not yet acknowledged and any more downlink packets that arrive from the serving gateway.

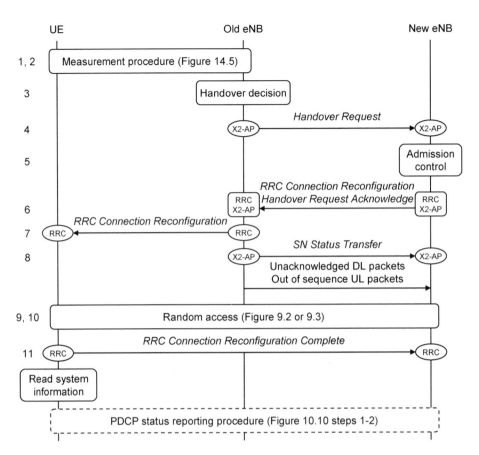

Figure 14.8 X2 based handover procedure. (1) Handover decision, preparation and execution. Reproduced by permission of ETSI.

On receiving the RRC message, the mobile reconfigures itself for the new cell and runs either the non contention based or the contention based random access procedure (9, 10), depending on whether or not it received a preamble index. It then acknowledges the RRC message (11) and reads the new cell's system information. Optionally, the base station and mobile can also run the PDCP status reporting procedure from Chapter 10, to minimize the amount of duplicate packet re-transmission.

There is one task remaining, shown in Figure 14.9. The serving gateway is still sending downlink packets to the old base station, so we need to contact it and change the downlink path. To do this, the new base station sends an S1-AP *Path Switch Request* to the MME (12), in which it lists the bearers that it has accepted and includes tunnel endpoint identifiers for the serving gateway to use on the downlink. The MME forwards the TEIDs to the serving gateway (13), along with the IP address of the new base station.

On receiving the message, the serving gateway redirects the GTP-U tunnels for the bearers that the base station accepted (14) and deletes the ones that were rejected. To indicate the end of the data stream, it also sends a GTP-U *End marker* packet to the old base station, which forwards it to the new one. Once the new base station has sent all the

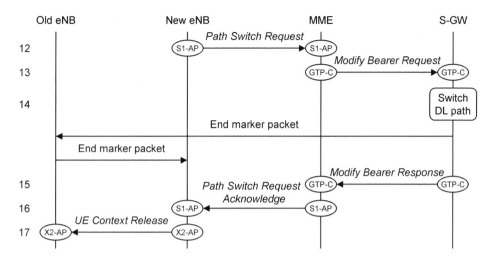

Figure 14.9 X2 based handover procedure. (2) Handover completion. Reproduced by permission of ETSI.

forwarded packets to the mobile and has reached the end marker, it can be confident that no more will arrive and can start transmitting any new packets that arrive direct from the serving gateway.

To conclude the procedure, the serving gateway sends an acknowledgement to the MME (15), in which it includes TEIDs for the base station to use on the uplink. The MME forwards the TEIDs to the new base station (16), which tells the old one that the handover has completed successfully (17). Once the old base station has received the message and forwarded the end marker packet, it can delete all the resources that were associated with the mobile.

As a result of the handover procedure, the mobile can move into a tracking area in which it was not previously registered. If so, then it runs the tracking area update procedure after the handover has completed. There are a few simplifications, as the mobile is starting and finishing in RRC_CONNECTED state and there is no need to change the MME or the serving gateway.

14.4.2 Handover Variations

There are several variations to the basic procedure described above [23]. The first leads to a simplification. If the same base station controls both cells, then we can leave out all the X2 messages from the sequence above and can keep the downlink path unchanged.

If the mobile changes base station and moves into a new S-GW serving area, then the MME has to change the serving gateway in place of steps 13 to 15. To do this, the MME tells the old serving gateway to tear down the bearers that it was using for the mobile and tells the new serving gateway to set up a new set of bearers. In turn, the new serving gateway contacts the PDN gateway, to change the traffic path on S5/S8.

As we noted in Chapter 2, the X2 interface is optional. If there is no X2 interface between the two base stations, then the procedure described above is unsuitable, so the handover is instead carried out using the *S1-based handover procedure*. In that procedure, the base stations communicate by way of the MME, using messages that are exchanged on the S1 interface instead of on X2.

If the mobile moves into a new MME pool area, then the S1-based handover procedure is mandatory. As before, the old base station requests a handover by contacting the old MME. The old MME hands control of the mobile over to a new MME and the new MME forwards the handover request to the new base station.

References

1. 3GPP TS 36.304 (October 2011) *User Equipment (UE) Procedures in Idle Mode*, Release 10.
2. 3GPP TS 23.401 (September 2011) *General Packet Radio Service (GPRS) Enhancements for Evolved Universal Terrestrial Radio Access Network (E-UTRAN) Access*, Release 10.
3. 3GPP TS 36.300 (October 2011) *Evolved Universal Terrestrial Radio Access (E-UTRA) and Evolved Universal Terrestrial Radio Access Network (E-UTRAN); Overall Description; Stage 2*, Release 10.
4. 3GPP TS 24.301 (September 2011) *Non-Access-Stratum (NAS) Protocol for Evolved Packet System (EPS); Stage 3*, Release 10.
5. 3GPP TS 29.274 (September 2011) *3GPP Evolved Packet System (EPS); Evolved General Packet Radio Service (GPRS) Tunnelling Protocol for Control Plane (GTPv2-C); Stage 3*, Release 10.
6. 3GPP TS 36.331 (October 2011) *Radio Resource Control (RRC); Protocol Specification*, Release 10.
7. 3GPP TS 36.413 (September 2011) *Evolved Universal Terrestrial Radio Access Network (E-UTRAN); S1 Application Protocol (S1AP)*, Release 10.
8. 3GPP TS 36.423 (September 2011) *Evolved Universal Terrestrial Radio Access Network (E-UTRAN); X2 Application Protocol (X2AP)*, Release 10.
9. 3GPP TS 36.133 (October 2011) *Requirements for Support of Radio Resource Management*, Release 10.
10. 3GPP TS 23.401 (September 2011) *General Packet Radio Service (GPRS) Enhancements for Evolved Universal Terrestrial Radio Access Network (E-UTRAN) Access*, Release 10, section 5.3.5.
11. 3GPP TS 23.401 (September 2011) *General Packet Radio Service (GPRS) Enhancements for Evolved Universal Terrestrial Radio Access Network (E-UTRAN) Access*, Release 10, section 5.3.4.3.
12. 3GPP TS 23.401 (September 2011) *General Packet Radio Service (GPRS) Enhancements for Evolved Universal Terrestrial Radio Access Network (E-UTRAN) Access*, Release 10, section 5.3.4.1.
13. 3GPP TS 36.304 (October 2011) *User Equipment (UE) Procedures in Idle Mode*, Release 10, section 5.2.4.
14. 3GPP TS 36.133 (October 2011) *Requirements for Support of Radio Resource Management*, Release 10, section 4.2.
15. 3GPP TS 23.401 (September 2011) *General Packet Radio Service (GPRS) Enhancements for Evolved Universal Terrestrial Radio Access Network (E-UTRAN) Access*, Release 10, section 5.3.3.
16. 3GPP TS 23.122 (September 2011) *Non-Access-Stratum (NAS) Functions Related to Mobile Station (MS) in Idle Mode*, Release 10, section 4.4.3.3.
17. 3GPP TS 36.304 (October 2011) *User Equipment (UE) Procedures in Idle Mode*, Release 10, section 5.2.3.
18. 3GPP TS 31.102 (October 2011) *Characteristics of the Universal Subscriber Identity Module (USIM) Application*, Release 10, section 4.2.6.
19. 3GPP TS 36.331 (October 2011) *Radio Resource Control (RRC); Protocol Specification*, Release 10, sections 5.5, 6.3.5.
20. 3GPP TS 36.133 (October 2011) *Requirements for Support of Radio Resource Management*, Release 10, section 8.
21. 3GPP TS 36.214 (March 2011) *Physical Layer; Measurements*, Release 10, sections 5.1.1, 5.1.3.
22. 3GPP TS 36.300 (October 2011) *Evolved Universal Terrestrial Radio Access (E-UTRA) and Evolved Universal Terrestrial Radio Access Network (E-UTRAN); Overall Description; Stage 2*, Release 10, section 10.1.2.
23. 3GPP TS 23.401 (September 2011) *General Packet Radio Service (GPRS) Enhancements for Evolved Universal Terrestrial Radio Access Network (E-UTRAN) Access*, Release 10, section 5.5.1.

15

Inter-System Operation

In the early stages of rolling out the technology, LTE will only be available in large cities and in isolated hotspots. In other areas, network operators will continue to use older technologies such as GSM, UMTS and cdma2000. Similarly, most LTE mobiles will actually be multiple mode devices that also support some or all of those technologies. To handle this situation, LTE has been designed so that it can inter-operate with other mobile communication systems, particularly by handing mobiles over if they move outside the coverage area of LTE. Such inter-operation is the subject of this chapter.

We begin by discussing the most important aspect, namely inter-operation with the earlier 3GPP technologies of UMTS and GSM. These systems cooperate closely with LTE, using a technique known as optimized handover that can transfer mobiles with no packet loss and with a minimal break in communications. Inter-operation with non 3GPP technologies such as WiMAX is not usually optimized, and can result in handover delays and packet loss. There is, however, one exception to this rule: the specifications define optimized handovers between LTE and cdma2000, in support of cdma2000 network operators who are upgrading to LTE.

This chapter uses similar specifications to the previous one. The most important ones cover the procedures that a mobile should follow in RRC_IDLE [1], inter-operation with GSM and UMTS in RRC_CONNECTED [2] and inter-operation with non 3GPP systems [3]. Other specifications define the measurements that the mobile makes in both RRC states [4] and the individual signalling messages [5–8]. Two 3GPP2 specifications [9, 10] define inter-operation between LTE and cdma2000, from the viewpoint of the latter system. In addition, there are several detailed accounts of the other systems that we cover in this chapter, including titles on GSM and UMTS [11–14], cdma2000 [15–17] and WiMAX [18, 19].

15.1 Inter-Operation with UMTS and GSM

15.1.1 S3-Based Architecture

LTE can inter-operate with 2G/3G core networks using two possible architectures. The usual architecture [20] is shown in Figure 15.1. The diagram only shows the components

---

An Introduction to LTE: LTE, LTE-Advanced, SAE and 4G Mobile Communications, First Edition. Christopher Cox.
© 2012 John Wiley & Sons, Ltd. Published 2012 by John Wiley & Sons, Ltd.

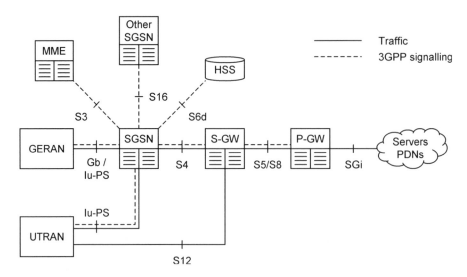

Figure 15.1 Architecture for inter-operation with UMTS and GSM, using an enhanced S3/S4 based SGSN.

and interfaces that are actually relevant to inter-operation: it omits, for example, the signalling interface between the MME and the serving gateway, as well as the policy and charging control architecture from Chapter 13. It also omits the circuit switched domain of the 2G/3G core network, which we will cover in Chapter 16.

When a mobile hands over from LTE to UMTS or GSM, the MME transfers control of the mobile to a serving GPRS support node (SGSN). In contrast, the PDN gateway and serving gateway both stay in the data path. The use of the PDN gateway allows the system to retain all the mobile's EPS bearers after the handover and to maintain its communications with the outside world. The use of the serving gateway allows this device to be a common point of contact for roaming mobiles, whichever technology they are using, and eases the implementation of certain procedures such as paging. The architecture also retains the home subscriber server (HSS), which acts as a common database for the 2G, 3G and 4G networks.

Data packets usually flow through the SGSN as well. However, the optional S12 interface also allows packets to flow directly between the serving gateway and the radio access network of UMTS, to replicate a UMTS technique known as *direct tunnelling*. If the network is using direct tunnelling, the SGSN only handles the mobile's signalling messages, so it bears a close resemblance to the MME.

In the original 2G/3G packet switched domain, the network elements communicated using an older version of the GPRS tunnelling protocol, denoted GTPv1-C [21]. To use the architecture in Figure 15.1, the SGSN must be enhanced so that it also supports GTPv2-C signalling messages on the S3, S4 and S16 interfaces, and Diameter communications on S6d. If the system is to support cell reselection and handover back to LTE, then the UMTS and GSM radio access networks have to be enhanced as well, so that they can tell their mobiles about the neighbouring LTE carrier frequencies and cells.

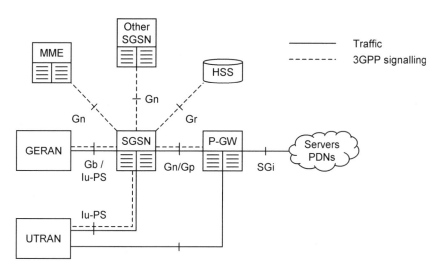

Figure 15.2 Architecture for inter-operation with UMTS and GSM, using a legacy Gn/Gp based SGSN.

15.1.2 Gn/Gp-Based Architecture

Figure 15.2 is an alternative architecture [22]. Here, the SGSN does not have to be enhanced at all: it only has to handle GTPv1-C signalling messages on the Gn and Gp interfaces and messages written using the *mobile application part* (MAP) on Gr. Instead, the architecture requires more powerful versions of the MME and PDN gateway, which can handle messages written using GTPv1-C as well as GTPv2-C.

This Gn/Gp-based architecture has a few limitations. Firstly, it only supports handovers to and from LTE networks whose S5/S8 interfaces are based on GTP, not PMIP. Secondly, it does not support a technique known as idle mode signalling reduction, which we will cover below. In the sections that follow, we will mainly focus on the S3-based inter-operation architecture from Figure 15.1.

15.1.3 Bearer Management

There are a few points to make about bearer management when using these architectures. In the 2G/3G packet switched domain, the equivalent of an EPS bearer is a *packet data protocol* (PDP) *context*. EPS bearers and PDP contexts look exactly the same in the user plane, but they are managed by different signalling messages in the control plane. The S3 based architecture from Figure 15.1 uses both types of data structure: the evolved packet core refers to EPS bearers using signalling messages across the S4 and S5/S8 interfaces, while the mobile and radio access networks refer to PDP contexts. The SGSN handles the conversion between the two, using QoS mapping rules that are defined in the specifications [23]. On the other hand, the Gn/Gp based architecture from Figure 15.2 uses PDP contexts alone, with the MME converting EPS bearers to PDP contexts during the handover procedure.

In LTE, a mobile requests a default EPS bearer at the time that it registers with the evolved packet core. In UMTS and GSM, by contrast, a mobile only requests a PDP context later on, when it wishes to communicate with the outside world. Both architectures handle this in a manner consistent with the older technologies: a mobile can register with the evolved packet core through a UMTS or GSM radio access network in the manner described below, but it does not receive a PDP context right away.

We also noted in Chapter 13 that an LTE mobile does not request a dedicated EPS bearer explicitly: instead, the mobile requests an improved quality of service, and the network decides whether to respond by creating a new dedicated bearer or by improving the QoS of an existing bearer. In UMTS and GSM, by contrast, a mobile can explicitly request the creation of an additional PDP context, using a procedure known as secondary PDP context activation. To handle this distinction, an S3 based network will always create a new dedicated EPS bearer in response to a successful secondary PDP context activation request by the mobile [24]. In a final distinction, a PDP context is associated with either an IPv4 address or an IPv6 address, whereas an EPS bearer can be associated with both. If an EPS bearer has an IPv4 address and an IPv6 address, then the network maps it onto two separate PDP contexts.

15.1.4 Power-On Procedures

As we noted in Chapter 11, the USIM contains prioritized lists of home networks and of any networks that the user or network operator have specified. Each network can be associated with a prioritized list of radio access technologies, including not just LTE, but also UMTS, GSM and cdma2000. During the procedures for network and cell selection, the mobile can use these lists to look for cells that belong to those other radio access technologies [25]. It does this using the procedures that are appropriate for each technology: those for UMTS and GSM are in [26] and [27] respectively.

If the mobile selects a UMTS or GSM cell, then it runs the appropriate attach procedure and registers with an SGSN [28]. It does not activate a PDP context right away: instead, it only does so later on, when it needs to communicate with the outside world. If the network is using the S3 based architecture from Figure 15.1, then the mobile always reaches the outside world through a PDN gateway, even if it has no LTE capability. In the case of the Gn/Gp based architecture from Figure 15.2, an LTE capable mobile uses a PDN gateway, while other mobiles can continue using a gateway GPRS support node (GGSN) as before.

15.1.5 Cell Reselection in RRC_IDLE

In RRC_IDLE state, a mobile can switch to a 2G or 3G cell using the cell reselection procedure from Chapter 14 [29, 30]. To support this procedure, the base station lists the carrier frequencies of neighbouring UMTS and GSM cells using information in SIB 6 and SIB 7 respectively. Each carrier frequency is associated with a priority level, which has to be different from the priority of the serving LTE cell. As in the case of LTE, the network does not provide the mobile with a full neighbour list: instead, the mobile is responsible for identifying the individual neighbours. The actual algorithm is the same one that we saw in Chapter 14 for reselection to another LTE frequency with a higher or lower priority. The only difference is that some of the quantities are re-interpreted to ones that are suitable for the target radio access technology.

After the reselection, the mobile runs a procedure known as a *Routing area update* [31], which is the equivalent of a tracking area update from LTE. During that procedure, the mobile registers with a suitable SGSN, which takes control of the mobile from the MME. The mobile can then move around its new radio access network, using the normal cell reselection procedures for UMTS or GSM.

Later on, the mobile can move back into a region of LTE coverage. If the 2G/3G network has been enhanced to support LTE and is broadcasting information about neighbouring LTE frequencies in its system information, then this can trigger a cell reselection back to LTE. After the reselection, the mobile requests a tracking area update and the MME takes control back from the SGSN.

15.1.6 Idle Mode Signalling Reduction

If the mobile is near the edge of the LTE coverage area, then it can easily bounce back and forth between cells that are using LTE and cells that are using UMTS or GSM. There is then a risk that the mobile will execute a large number of routing and tracking area updates, leading to excessive signalling.

To avoid this problem, S3-based networks can optionally implement a technique known as *idle mode signalling reduction* (ISR) [32]. This technique behaves very like the registration of a mobile in multiple tracking areas. When using idle mode signalling reduction, the network can simultaneously register the mobile in a routing area that is served by an S3-based SGSN and in one or more tracking areas that are served by an MME. The mobile can then freely reselect between cells that are using the three radio access technologies and only has to inform the network if it moves into a routing or tracking area in which it is not currently registered.

If downlink data arrive from the PDN gateway, then the paging procedure is modified so that the serving gateway contacts both the MME and the SGSN. The MME pages the mobile in all the tracking areas in which it is registered, as before, while the SGSN pages the mobile throughout its routing area. After the mobile's response, the network moves it back into connected mode using the appropriate radio access technology.

15.1.7 Measurements in RRC_CONNECTED

In RRC_CONNECTED state, the base station can tell the mobile to make measurements of neighbouring UMTS and GSM cells using the measurement procedure from Chapter 14 [33]. This time, however, there is some expectation that the network will provide the mobile with a neighbour list, as the corresponding requirements on the mobile only apply in the presence of such a neighbour list [34]. We will see how the network might populate its neighbour list as part of Chapter 19.

Measurement reporting can be triggered by two measurement events, B1 and B2, which are listed in Table 15.1. To illustrate their behaviour, the mobile reports event B2 on completion of two separate threshold-crossing events. Firstly, the signal from the LTE serving cell must lie below one threshold:

$$M_s < \text{Thresh}_1 - \text{Hys} \qquad (15.1)$$

Table 15.1 Measurement events used to report neighbouring UMTS, GSM and cdma2000 cells

Event	Description
B1	Neighbouring cell from other technology rises above threshold
B2	Serving cell falls below threshold 1
	Neighbouring cell from other technology rises above threshold 2

Secondly, the signal from the non-LTE neighbour must lie above another threshold:

$$M_n + Of_n > Thresh_2 + Hys \quad (15.2)$$

The mobile then stops any further reporting of event B2 until either of the following conditions is met:

$$M_s > Thresh_1 + Hys \quad (15.3)$$

$$M_n + Of_n < Thresh_2 - Hys \quad (15.4)$$

In these equations, M_s is the reference signal received power or reference signal received quality from the serving LTE cell. M_n is the measurement of the neighbouring cell, namely the *received signal strength indicator* (RSSI) in the case of GSM and either the *received signal code power* (RSCP) or SINR per chip (Ec/No) in the case of UMTS [35]. $Thresh_1$ and $Thresh_2$ are thresholds, Hys is a hysteresis parameter for measurement reporting and Of_n is an optional frequency-specific offset.

Using this procedure, the mobile can inform the serving eNB if the signal from the serving LTE cell is sufficiently weak, while the signal from a neighbouring UMTS or GSM cell is sufficiently strong. On the basis of this measurement report, the base station can hand the mobile over to UMTS or GSM.

15.1.8 Handover in RRC_CONNECTED

Handovers between LTE and UMTS or GSM operate in a similar way to handovers within LTE. In particular, the MME and SGSN exchange signalling messages before the actual handover takes place, to prepare the 3G network, minimize the gap in communications, and minimize the risk of packet loss. Such handovers are known as *optimized handovers*.

To illustrate the procedure, Figures 15.3, 15.4 and 15.5 show the procedure that is typically used for a handover from LTE to UMTS [36]. From the viewpoint of LTE, the procedure is similar to a normal S1 based handover, while from the viewpoint of UMTS, it looks like a combined hard handover and radio network controller (RNC) relocation. In these figures, we have assumed that the network is using a GTP based S5/S8 interface and an S3 based SGSN, and is not using the S12 direct tunnelling interface. We also assume that the handover uses a technique known as indirect data forwarding, in which the eNB forwards unacknowledged downlink packets to the target RNC by way of the serving gateway. Finally, we assume that the serving gateway remains unchanged.

The procedure begins in Figure 15.3. On the basis of a measurement report, the eNB decides to hand the mobile over to a UMTS cell. It tells the MME that a handover

Inter-System Operation

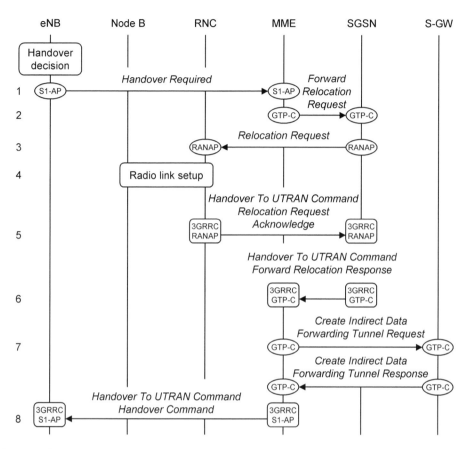

Figure 15.3 Handover from LTE to UMTS. (1) Handover preparation. Reproduced by permission of ETSI.

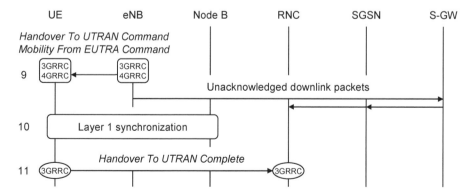

Figure 15.4 Handover from LTE to UMTS. (2) Handover execution. Reproduced by permission of ETSI.

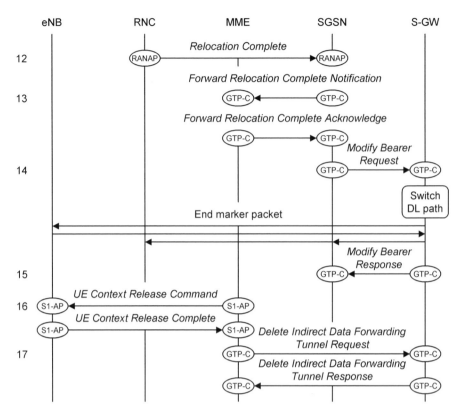

Figure 15.5 Handover from LTE to UMTS. (3) Handover completion. Reproduced by permission of ETSI.

is required (1) and identifies the target Node B, together with the corresponding radio network controller and routing area. The MME identifies a suitable SGSN, asks it to relocate the mobile and describes the EPS bearers that the mobile is currently using (2).

The SGSN reacts by mapping the EPS bearers' QoS parameters onto the parameters for the corresponding PDP contexts. It then asks the RNC to establish resources for the mobile and describes the bearers that it would like to set up (3), using a *Relocation Request* that is written using the UMTS *Radio access network application part* (RANAP). The RNC configures the Node B (4) and returns an acknowledgement to the SGSN (5). In its acknowledgement, the RNC lists the bearers that it is willing to accept and includes an embedded *Handover to UTRAN Command*, which tells the mobile how to communicate with the target cell. This last message is written using the UMTS RRC protocol, which the diagram denotes as 3GRRC.

The SGSN acknowledges the MME's request, tells it which bearers have been accepted and attaches the UMTS RRC message that it received from the RNC (6). It also includes a tunnel endpoint identifier (TEID), which the serving gateway can use to forward any downlink packets that the mobile has not yet acknowledged. The MME sends the TEID to the serving gateway (7), which creates the tunnel as requested. The MME can then tell the eNB to hand the mobile over to UMTS (8). In the message, the MME states which bearers should be retained and which should be released, and includes the UMTS RRC message that it received from the SGSN.

To trigger the handover (Figure 15.4), the eNB sends the mobile a *Mobility from EUTRA Command* (9). This message is written using the LTE RRC protocol, which the diagram denotes as 4GRRC. Embedded in the message is the UMTS handover command that was originally written by the target RNC. The mobile reads both RRC messages, switches to UMTS, synchronizes with the new cell (10) and sends an acknowledgement to the RNC (11). At the same time, the eNB starts to return any unacknowledged downlink packets to the serving gateway, as well as any new packets that continue to arrive. The serving gateway forwards the packets to the SGSN using the tunnel that it has just created, and the SGSN forwards the packets to the target RNC.

The network still has to release the old resources and redirect the data path from the serving gateway (Figure 15.5). To achieve this, the RNC tells the SGSN that the relocation procedure is complete (12) and the SGSN forwards this information to the MME (13). The SGSN also tells the serving gateway to redirect its downlink path (14) so as to send future downlink packets to the SGSN. The serving gateway does so, indicates the end of the data stream by sending an end marker packet on the old downlink path to the eNB and returns an acknowledgement to the SGSN (15).

On the expiry of a timer, the MME tells the eNB to release the mobile's resources (16), and tells the serving gateway to tear down the indirect forwarding tunnel that it created earlier (17). At the same time, the mobile notices that it has moved into a new 2G/3G routing area. It responds by running a routing area update, unless it is using idle mode signalling reduction and is already registered there.

The handover procedure from LTE to GSM is very similar. The specifications also support handovers from UMTS and GSM back to LTE, but these are less important, as the network will have few regions where there is coverage from LTE but not from other technologies.

15.2 Inter-Operation with Generic Non 3GPP Technologies

15.2.1 Network Based Mobility Architecture

LTE is also designed to inter-operate with non 3GPP technologies, such as wireless local area networks and WiMAX. There are two main architectures [37], which handle issues such as mobility and roaming using functions in the network and mobile respectively.

Figure 15.6 shows the architecture when using network based mobility. The architecture retains the HSS and the PDN gateway, which respectively act as a common subscriber database and a common point of contact with the outside world. The mobile is controlled using devices inside the non 3GPP access network, while the *3GPP authentication, authorization and accounting* (AAA) *server* authenticates the mobile and grants it access to services through the PDN gateway. The signalling interfaces to the AAA server use protocols that are based on Diameter [38].

The architecture makes a distinction between *trusted* and *untrusted* access networks. A trusted network provides adequate security for the mobile's radio communications using techniques such as ciphering and integrity protection, while an untrusted network does not. It is the AAA server's responsibility to decide whether a particular network is trusted or untrusted.

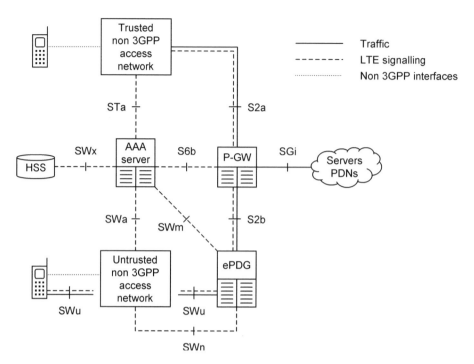

Figure 15.6 Architecture for inter-operation with non 3GPP access technologies, using network based mobility.

When using the trusted network architecture, the mobile communicates directly with the access network using a non 3GPP radio access technology such as WiMAX. Packets are forwarded between the access network and the PDN gateway using generic routing encapsulation (GRE), which we have already seen as one of the options for the S5/S8 interface of the evolved packet core. The GRE tunnels are managed using signalling messages between the access network and the PDN gateway, which are written using either proxy mobile IP version 6 (PMIPv6) [39], or an older protocol known as *mobile IP version 4* (MIPv4) [40].

The untrusted network architecture is suitable for scenarios such as insecure wireless local area networks and includes an extra component known as the *evolved packet data gateway* (ePDG). Together, the mobile and gateway secure the data that are exchanged across the SWu interface by means of ciphering and integrity protection, using IPSec in tunnel mode. The access network is forced to send uplink traffic through the gateway by means of signalling messages over SWn.

The architecture also retains the policy and charging rules function (PCRF), which acts as a common source of policy and charging control with the evolved packet core. The PDN gateway contains a policy and charging enforcement function (PCEF), as before, while the trusted access network and the evolved packet data gateway both contain a bearer binding and event reporting function (BBERF).

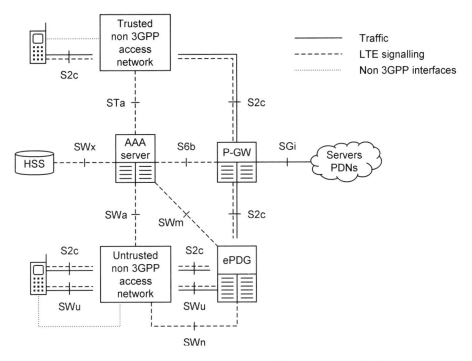

Figure 15.7 Architecture for inter-operation with non 3GPP access technologies, using host based mobility.

15.2.2 Host Based Mobility Architecture

Figure 15.7 shows an alternative architecture, in which the mobility management functions lie in the mobile. When using this architecture, the mobile communicates with the non 3GPP access network in the same way as before. However, the packets are routed directly between the mobile and the PDN gateway, using a tunnel that runs across the S2c interface. The tunnel is managed using a signalling protocol known as *dual stack mobile IP version 6* (DSMIPv6) [41], which is implemented in the mobile and the PDN gateway. There is no mobility management functionality in the access network, which simply acts as a router.

15.2.3 Attach Procedure

To illustrate the procedures for inter-operation with non 3GPP networks, let us consider the case of network-based mobility across a trusted access network (Figure 15.6), with the signalling messages written using PMIPv6. This option might be suitable for mobile WiMAX and is also a useful prerequisite for the discussion of cdma2000 below.

If the mobile switches on in a region of non 3GPP coverage, then it accesses the system using the attach procedure that is summarized in Figure 15.8 [42]. The mobile starts by establishing communications with the access network (1), using techniques that

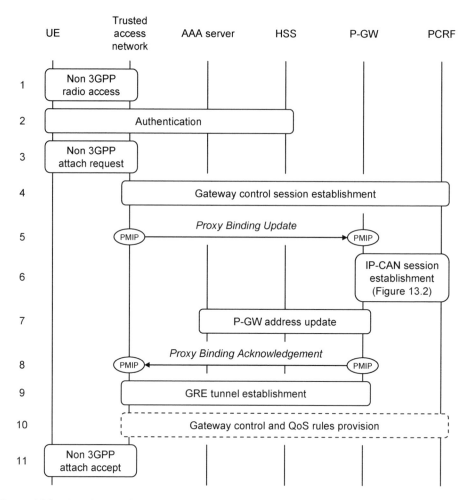

Figure 15.8 Attach procedure to a trusted non 3GPP access network, with network based mobility using PMIPv6. Reproduced by permission of ETSI.

correspond to the LTE procedures of network selection, cell selection and RRC connection establishment. The mobile and AAA server can then authenticate each other (2). During this process, the AAA server retrieves the mobile's identity from the home subscriber server, along with subscription data such as the default access point name and the corresponding quality of service parameters. It then forwards this information to the access network.

The mobile can then send an attach request to the access network (3), using messages that correspond to steps 1 and 2 from the LTE attach procedure. In its request, the mobile states whether it supports IP version 4, IP version 6 or both and can request a preferred access point name. The access network reacts by selecting a suitable access point name and a corresponding PDN gateway. In step 4, the access network's bearer binding and event reporting function runs the procedure of gateway control session establishment that we described in Chapter 13. In this procedure, it sets up signalling communications with the PCRF, forwards the subscription data that it received from the home subscriber server and receives a policy and charging control rule in return.

The access network then sends a PMIPv6 *Proxy Binding Update* message to the chosen PDN gateway (5), which identifies the mobile and requests the establishment of a GRE tunnel across the S2a interface. The PDN gateway allocates an IP address for the mobile and runs the usual procedure of IP-CAN session establishment (6). It also tells the AAA server about the access point name that the mobile is using (7) so that the AAA server can store this information in the HSS for use in future handovers. The PDN gateway can then reply to the access network with an acknowledgement that includes the mobile's IP address (8), after which the PDN gateway and access network complete the configuration of their GRE tunnel (9).

If the PCRF changed the policy control and charging rule during IP-CAN session establishment, then it updates the access network using the procedure of gateway control and QoS rules provision (10). Finally, the access network accepts the mobile's attach request, and includes its IP address and the selected access point name (11). The mobile can now communicate through the PDN gateway, using the functions of the non 3GPP access network.

The attach procedures for the other network architectures are similar. In the case of host based mobility over S2c (Figure 15.7), the procedure sets up a tunnel for the exchange of data and signalling messages between the mobile and the PDN gateway. In the case of an untrusted network, the procedure also sets up a tunnel between the mobile and the evolved packet data gateway, for the secure exchange of information using IPSec.

15.2.4 Cell Reselection and Handover

A mobile can switch from LTE to a non 3GPP access network and back again. However the generic non 3GPP architecture does not include any signalling interfaces to the MME, or any interfaces that would allow it to exchange data packets with the serving gateway. As a result, it does not support the optimized handovers that we have previously seen.

This has two implications. Firstly, the MME cannot send the mobile any advance information about the non 3GPP network, such as the parameters that the mobile requires to access the target cell. Instead, the mobile has to carry out cell reselection and handover by detaching from LTE and attaching to the non 3GPP network. The resulting procedure [43] is similar to the one from Figure 15.8, and can lead to a significant gap in communications. Secondly, there is no mechanism for forwarding unacknowledged downlink packets from LTE to the non 3GPP network. This implies that packets are likely to be lost during the communication gap.

The handover procedure does, however, offer more functions than the basic attach procedure. During the authentication and attach request procedures (steps 2 and 3 of Figure 15.8), the mobile indicates that it is handing over to the target access network, and that it would like to retain its original session and IP address. In response, the AAA server retrieves the identity of the PDN gateway and the APN from the home subscriber server, which stored them at the end of the LTE attach procedure from Chapter 11, and forwards these to the access network. The access network can then contact the original PDN gateway in steps 4 and 5, and the PDN gateway can continue using the mobile's old IP address instead of allocating a new one. As a result, the mobile retains its session and IP address during the handover, so it can maintain its communications with any external servers.

15.3 Inter-Operation with cdma2000 HRPD

15.3.1 System Architecture

As we saw in Chapter 1, most cdma2000 operators are planning to migrate their systems to LTE. In the early stages of this process, network operators will only have LTE cells in hotspots and large cities, leading to frequent handovers between the two systems. It is highly desirable that these handovers should be as efficient as possible. To support this, the LTE specifications define an optimized variant of the non 3GPP architecture for use with cdma2000 high rate packet data (HRPD), also known as cdma2000 evolution data optimized (EV-DO). An HRPD network that supports this architecture is known as *evolved HRPD* (eHRPD).

Figure 15.9 shows the architecture that is used [44–46]. The architecture is based on the one for network-based mobility across a trusted non 3GPP access network, but with two new interfaces. Firstly, the MME can exchange signalling messages with an HRPD *evolved access network/evolved packet control function* (eAN/ePCF) across the S101 interface. The interface uses the *S101 application protocol* (S101-AP) [47], which simply transports HRPD signalling messages between the two devices. Secondly, the serving gateway can exchange data packets with an *HRPD serving gateway* (HSGW) across the S103 interface, so as to minimize the risk of packet loss during a handover.

15.3.2 Preregistration with cdma2000

If a mobile switches on in a region of LTE coverage, then it attaches to LTE in the usual way. However, the network can then tell it to run a procedure known as *HRPD*

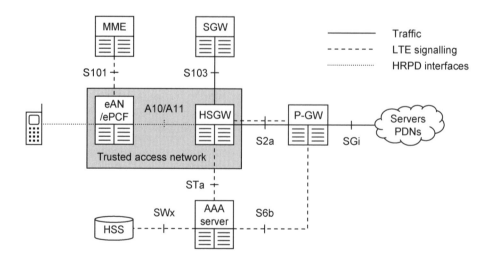

Figure 15.9 Architecture for inter-operation with cdma2000 HRPD.

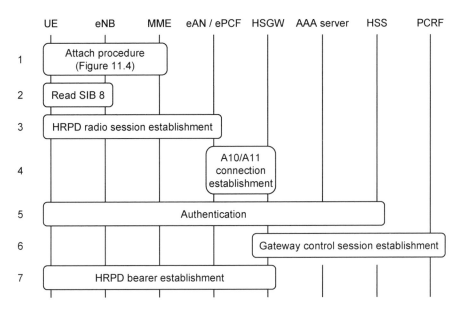

Figure 15.10 Procedure for preregistration with cdma2000 HRPD. Reproduced by permission of ETSI.

preregistration [48–50]. Using this procedure, the mobile establishes a dormant session in the HRPD network, which can be used later on to speed up any subsequent reselection or handover. The procedure is summarized in Figure 15.10.

After running the usual LTE attach procedure (1), the mobile reads SIB 8, which contains the parameters it will need for reselection to cdma2000 (2). One of these parameters is a preregistration trigger, which tells the mobile whether it should preregister with an HRPD network. By setting this trigger on a cell-by-cell basis, the network operator can tell a mobile to preregister with HRPD if it is near the edge of the LTE coverage area.

If the trigger is set, then the mobile registers with the HRPD access network using HRPD messages that are tunnelled over the air interface and over S1 and S101 (3). The network establishes a signalling connection for the mobile with the HSGW (4) and the mobile authenticates itself to the 3GPP AAA server (5). During the authentication process, the AAA server retrieves the mobile's identity from the home subscriber server, along with information such as the mobile's access point name and PDN gateway, and sends these to the eAN/ePCF.

Inside the HSGW, the bearer binding and event reporting function runs the procedure of gateway control session establishment (6), to establish communications with the PCRF. Finally, the HSGW exchanges a set of HRPD messages with the mobile (7), to establish a dormant set of HRPD bearers that mirror the ones in the evolved packet core. The existence of these bearers will speed up any subsequent reselection or handover to HRPD.

Later on, the evolved packet core may set up new EPS bearers for the mobile, or may modify the quality of service of existing bearers. If this happens, then the mobile exchanges additional messages with the HSGW, so as to keep the HRPD bearers in step.

15.3.3 Cell Reselection in RRC_IDLE

In RRC_IDLE state, a mobile can reselect to a cdma2000 HRPD cell using the same procedure that it used for UMTS and GSM. The only difference is that SIB 8 contains a list of neighbouring cdma2000 HRPD cells, so the mobile is not expected to find these by itself.

Once reselection is complete, there are two possibilities. If the mobile has preregistered with HRPD, then it contacts the HRPD access network to indicate that it has arrived and the network sets up a new set of resources for the mobile. The effect is similar to the cdma2000 handover procedure described below, although with fewer steps. If the mobile has not preregistered, then it has to run the generic non 3GPP handover procedure that we covered earlier.

Note that there is no analogue to idle mode signalling reduction for cdma2000. Network operators may therefore wish to configure the cell reselection parameters so as to minimize the risk of mobiles bouncing back and forth between cdma2000 and LTE.

15.3.4 Measurements and Handover in RRC_CONNECTED

In RRC_CONNECTED state, the base station can tell the mobile to measure the signal-to-interference ratios of neighbouring cdma2000 cells, in the same way that it did for UMTS and GSM. The measurement reports use measurement events B1 and B2, as before.

Based on such a measurement report, the base station can decide to hand the mobile over to a cdma2000 HRPD network. To do this, it uses the procedure shown in the figures that follow [51–53]. In this procedure, we assume that the mobile has already preregistered with the HRPD network. If it has not done so, then it instead has to run the generic non 3GPP handover procedure that we covered earlier.

To begin the procedure (Figure 15.11), the mobile identifies a neighbouring HRPD cell in a measurement report (1) and the base station decides to hand the mobile over (2). In step 3, the base station tells the mobile to contact the HRPD network, using the signalling path that it established during preregistration. In response, the mobile composes an HRPD *Connection Request*, in which it requests the parameters of a physical *traffic channel* that it can use for communication across the HRPD air interface. In steps 4, 5 and 6, the message is forwarded across the Uu, S1 and S101 interfaces to the eAN/ePCF.

On receiving the message, the eAN/ePCF retrieves the mobile's access point name and PDN gateway identity, which it stored during the preregistration procedure, and sends these to the HSGW (7). In response, the HSGW returns an IP address to which the serving gateway can forward any unacknowledged downlink data packets. The eAN/ePCF can then allocate a set of radio resources for the mobile and can compose an HRPD *Traffic Channel Assignment*, which provides the mobile with the details that it requested in step 2. It sends the message to the MME by embedding it in an S101-AP direct transfer (8), which also contains the forwarding address that the HSGW sent earlier. The MME sends the forwarding address to the serving gateway (9), which responds by creating the requested tunnel.

The MME can now send the traffic channel assignment to the base station (10). In turn, the base station sends the message to the mobile (11) and also starts to return any unacknowledged downlink packets to the serving gateway. Using the forwarding address that it received earlier, the serving gateway can send these packets across the S103 interface to the HSGW, along with any other packets that arrive on the downlink later

Inter-System Operation

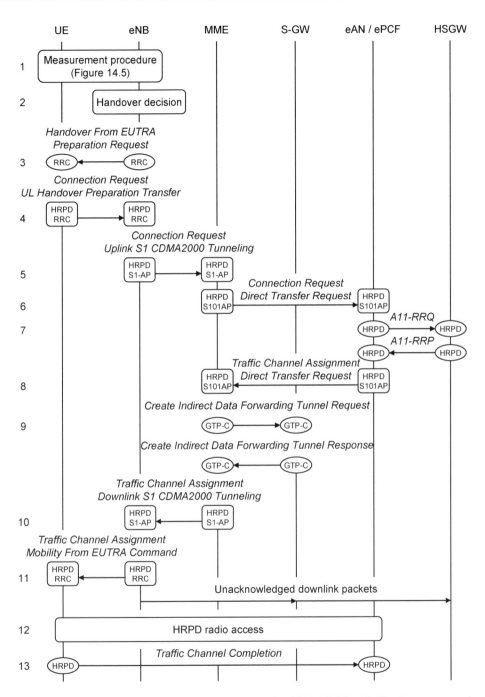

Figure 15.11 Inter-system handover from LTE to cdma2000 HRPD. (1) Handover preparation and execution. Reproduced by permission of ETSI.

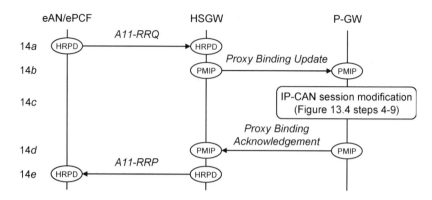

Figure 15.12 Inter-system handover from LTE to cdma2000 HRPD. (2) Bearer activation. Reproduced by permission of ETSI.

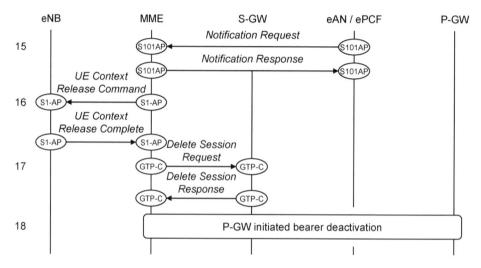

Figure 15.13 Inter-system handover from LTE to cdma2000 HRPD. (3) Handover completion. Reproduced by permission of ETSI.

on. On receiving the base station's message, the mobile switches to cdma2000, acquires the specified traffic channel (12) and acknowledges the HRPD network's message using an HRPD *Traffic Channel Completion* (13).

In step 14 (Figure 15.12), the HRPD access network tells the HSGW to activate the bearers that it set up during preregistration. The HSGW asks the PDN gateway to establish a GRE tunnel to carry the data, which triggers the procedure of IP-CAN session modification in which the PDN gateway retrieves the corresponding policy and charging control rule. After the acknowledgements, traffic can flow on the uplink and downlink between the HRPD access network, the HSGW and the PDN gateway. If necessary, the network repeats these steps for every access point name that the mobile is using.

We still need to tear down the resources that the mobile was using in LTE (Figure 15.13). To achieve this, the HRPD access network tells the MME that the handover has completed (15). In response, the MME tells the base station to tear down the mobile's resources (16) and sends a similar message to the serving gateway (17). At about the same time, the PDN gateway runs a procedure known as *PDN GW initiated bearer deactivation* (18) [54], which releases the remaining resources in the evolved packet core.

References

1. 3GPP TS 36.304 (October 2011) *User Equipment (UE) Procedures in Idle Mode*, Release 10.
2. 3GPP TS 23.401 (September 2011) *General Packet Radio Service (GPRS) Enhancements for Evolved Universal Terrestrial Radio Access Network (E-UTRAN) Access*, Release 10.
3. 3GPP TS 23.402 (September 2011) *Architecture Enhancements for Non-3GPP Accesses*, Release 10.
4. 3GPP TS 36.133 (October 2011) *Requirements for Support of Radio Resource Management*, Release 10.
5. 3GPP TS 24.301 (September 2011) *Non-Access-Stratum (NAS) Protocol for Evolved Packet System (EPS); Stage 3*, Release 10.
6. 3GPP TS 29.274 (September 2011) *3GPP Evolved Packet System (EPS); Evolved General Packet Radio Service (GPRS) Tunnelling Protocol for Control Plane (GTPv2-C); Stage 3*, Release 10.
7. 3GPP TS 36.331 (October 2011) *Radio Resource Control (RRC); Protocol Specification*, Release 10.
8. 3GPP TS 36.413 (September 2011) *Evolved Universal Terrestrial Radio Access Network (E-UTRAN); S1 Application Protocol (S1AP)*, Release 10.
9. 3GPP2 A.S0022-0 (April 2011) *Interoperability Specification (IOS) for Evolved High Rate Packet Data (eHRPD) Radio Access Network Interfaces and Interworking with Enhanced Universal Terrestrial Radio Access Network (EUTRAN)*, version 2.0.
10. 3GPP2 X.S0057-0 (September 2011) *E-UTRAN - eHRPD Connectivity and Interworking: Core Network Aspects*, version 2.0.
11. Sauter, M. (2010) *From GSM to LTE: An Introduction to Mobile Networks and Mobile Broadband*, John Wiley & Sons, Ltd, Chichester.
12. Eberspächer, J., Vögel, H.-J., Bettstetter, C. and Hartmann, C. (2008) *GSM: Architecture, Protocols and Services*, 3rd edn, John Wiley & Sons, Ltd, Chichester.
13. Johnson, C. (20080 *Radio Access Networks for UMTS: Principles and Practice*, John Wiley & Sons, Ltd, Chichester.
14. Kreher, R. and Ruedebusch, T. (2007) *UMTS Signaling: UMTS Interfaces, Protocols, Message Flows and Procedures Analyzed and Explained*, 2nd edn, John Wiley & Sons, Ltd, Chichester.
15. Etemad, K. (2004) *CDMA2000 Evolution: System Concepts and Design Principles*, John Wiley & Sons, Ltd, Chichester.
16. Vanghi, V., Damnjanovic, A. and Vojcic, B. (2004) *The cdma2000 System for Mobile Communications: 3G Wireless Evolution*, Prentice Hall.
17. Yang, S. (2004) *3G CDMA2000 Wireless System Engineering*, Artech.
18. Ahmadi, S (2010) *Mobile WiMAX: A Systems Approach to Understanding IEEE 802.16m Radio Access Technology*, Academic Press.
19. Andrews, J. G., Ghosh, A. and Muhamed, R. (2007) *Fundamentals of WiMAX: Understanding Broadband Wireless Networking*, Prentice Hall.
20. 3GPP TS 23.401 (September 2011) *General Packet Radio Service (GPRS) Enhancements for Evolved Universal Terrestrial Radio Access Network (E-UTRAN) Access*, Release 10, sections 4.2, 4.4.
21. 3GPP TS 29.060 (September 2011) *General Packet Radio Service (GPRS); GPRS Tunnelling Protocol (GTP) Across the Gn and Gp Interface*, Release 10.
22. 3GPP TS 23.401 (September 2011) *General Packet Radio Service (GPRS) Enhancements for Evolved Universal Terrestrial Radio Access Network (E-UTRAN) Access*, Release 10, annex D.
23. 3GPP TS 23.401 (September 2011) *General Packet Radio Service (GPRS) Enhancements for Evolved Universal Terrestrial Radio Access Network (E-UTRAN) Access*, Release 10, annex E.
24. 3GPP TS 23.060 (September 2011) *General Packet Radio Service (GPRS); Service Description; Stage 2*, Release 10, section 9.2.2.1.1A.
25. 3GPP TS 23.122 (September 2011) *Non-Access-Stratum (NAS) Functions Related to Mobile Station (MS) in Idle Mode*, Release 10, section 4.4.
26. 3GPP TS 25.304 (October 2011) *User Equipment (UE) Procedures in Idle Mode and Procedures for Cell Reselection in Connected Mode*, Release 10, section 5.2.6.
27. 3GPP TS 43.022 (April 2011) *Functions Related to Mobile Station (MS) in Idle Mode and Group Receive Mode*, Release 10, section 4.5.

28. 3GPP TS 23.401 (September 2011) *General Packet Radio Service (GPRS) Enhancements for Evolved Universal Terrestrial Radio Access Network (E-UTRAN) Access*, Release 10, section 5.3.2.2.
29. 3GPP TS 36.304 (October 2011) *User Equipment (UE) Procedures in Idle Mode*, Release 10, section 5.2.4.5.
30. 3GPP TS 36.133 (October 2011) *Requirements for Support of Radio Resource Management*, Release 10, section 4.2.2.5.
31. 3GPP TS 23.401 (September 2011) *General Packet Radio Service (GPRS) Enhancements for Evolved Universal Terrestrial Radio Access Network (E-UTRAN) Access*, Release 10, sections 5.3.3.3, 5.3.3.6.
32. 3GPP TS 23.401 (September 2011) *General Packet Radio Service (GPRS) Enhancements for Evolved Universal Terrestrial Radio Access Network (E-UTRAN) Access*, Release 10, annex J.
33. 3GPP TS 36.331 (October 2011) *Radio Resource Control (RRC); Protocol Specification*, Release 10, sections 5.5, 6.3.5.
34. 3GPP TS 36.133 (October 2011) *Requirements for Support of Radio Resource Management*, Release 10, section 8.1.2.4.
35. 3GPP TS 36.214 (March 2011) *Physical Layer; Measurements*, Release 10, section 5.1.
36. 3GPP TS 23.401 (September 2011) *General Packet Radio Service (GPRS) Enhancements for Evolved Universal Terrestrial Radio Access Network (E-UTRAN) Access*, Release 10, section 5.5.2.1.
37. 3GPP TS 23.402 (September 2011) *Architecture Enhancements for Non-3GPP Accesses*, Release 10, sections 4.2.2, 4.2.3.
38. 3GPP TS 29.273 (June 2011) *Evolved Packet System (EPS); 3GPP EPS AAA Interfaces*, Release 10.
39. 3GPP TS 29.275 (September 2011) *Proxy Mobile IPv6 (PMIPv6) Based Mobility and Tunnelling Protocols; Stage 3*, Release 10.
40. 3GPP TS 29.279 (April 2011) *Mobile IPv4 (MIPv4) Based Mobility Protocols; Stage 3*, Release 10.
41. IETF RFC 5555 (June 2009) *Mobile IPv6 Support for Dual Stack Hosts and Routers (DSMIPv6)*.
42. 3GPP TS 23.402 (September 2011) *Architecture Enhancements for Non-3GPP Accesses*, Release 10, section 6.2.1.
43. 3GPP TS 23.402 (September 2011) *Architecture Enhancements for Non-3GPP Accesses*, Release 10, section 8.2.2.
44. 3GPP TS 23.402 (September 2011) *Architecture Enhancements for Non-3GPP Accesses*, Release 10, section 9.1.
45. 3GPP2 A.S0022-0 (April 2010) *Interoperability Specification (IOS) for Evolved High Rate Packet Data (eHRPD) Radio Access Network Interfaces and Interworking with Enhanced Universal Terrestrial Radio Access Network (EUTRAN)*, version 2.0, section 1.4.
46. 3GPP2 X.S0057-0 (September 2010) *E-UTRAN - eHRPD Connectivity and Interworking: Core Network Aspects*, version 3.0, section 4.
47. 3GPP TS 29.276 (September 2011) *Optimized Handover Procedures and Protocols between E-UTRAN Access and cdma2000 HRPD Access*, Release 10.
48. 3GPP TS 23.402 (September 2011) *Architecture Enhancements for non-3GPP Accesses*, Release 10, section 9.3.1.
49. 3GPP2 A.S0022-0 (April 2010) *Interoperability Specification (IOS) for Evolved High Rate Packet Data (eHRPD) Radio Access Network Interfaces and Interworking with Enhanced Universal Terrestrial Radio Access Network (EUTRAN)*, version 2.0, section 3.2.1.
50. 3GPP2 X.S0057-0 (September 2010) *E-UTRAN - eHRPD Connectivity and Interworking: Core Network Aspects*, version 3.0, section 13.1.1.
51. 3GPP TS 23.402 (September 2011) *Architecture Enhancements for non-3GPP Accesses*, Release 10, section 9.3.2.
52. 3GPP2 A.S0022-0 (April 2010) *Interoperability Specification (IOS) for Evolved High Rate Packet Data (eHRPD) Radio Access Network Interfaces and Interworking with Enhanced Universal Terrestrial Radio Access Network (EUTRAN)*, version 2.0, section 3.2.2.2.
53. 3GPP2 X.S0057-0 (September 2010) *E-UTRAN - eHRPD Connectivity and Interworking: Core Network Aspects*, version 3.0, section 13.1.2.
54. 3GPP TS 23.401 (September 2011) *General Packet Radio Service (GPRS) Enhancements for Evolved Universal Terrestrial Radio Access Network (E-UTRAN) Access*, Release 10, section 5.4.4.1.

16

Delivery of Voice and Text Messages over LTE

As we explained in Chapter 1, LTE was designed as a data pipe: a system that would deliver information to and from the user, but would not concern itself with the overlying application. For most data services, such as web browsing and emails, the applications are separate from the delivery system and are supplied by third parties, so this approach works well. For voice and text messages, however, the applications have previously been supplied by the network operator and have been tightly integrated into the delivery system. This is a very different principle from the one adopted for LTE.

There are two main approaches to the delivery of voice over LTE, both of which can be implemented in two ways. The first approach is to treat voice like any other data service and to deliver it using a voice over IP server that lies outside the LTE network. This approach can be implemented either by a third party service provider, or by a separate 3GPP network known as the IP multimedia subsystem. The second approach is to connect LTE to the circuit switched domains of 2G and 3G, and to use their existing capabilities for placing voice calls. This approach can be implemented using two other techniques, which are known as circuit switched fallback and voice over LTE via generic access. In turn, each of these four techniques can be adapted for the delivery of text messages, using either SMS or a proprietary messaging application.

Before considering any of these techniques, however, let us take a brief look at the market for voice and SMS.

16.1 The Market for Voice and SMS

To illustrate the importance of voice and SMS to mobile network operators, Figure 16.1 shows the revenue that Western European operators have been earning from voice, messaging services such as SMS and other data applications. The information is from market research by Analysys Mason and uses operators' data up to 2011 and forecasts thereafter.

There is a striking contrast between the information in Figure 16.1 and the information about worldwide network traffic that we presented in Figures 1.5 and 1.6: data applications

---

An Introduction to LTE: LTE, LTE-Advanced, SAE and 4G Mobile Communications, First Edition. Christopher Cox.
© 2012 John Wiley & Sons, Ltd. Published 2012 by John Wiley & Sons, Ltd.

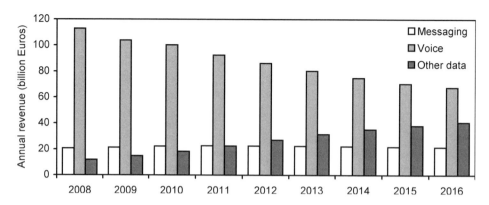

Figure 16.1 Revenue earned by network operators in Western Europe from voice, messaging and other data services, with operators' data up to 2011 and forecasts thereafter. Data supplied by Analysys Mason.

supply most of the operators' traffic, but voice supplies most of their revenue. Note that the imbalance between voice and data traffic is even more extreme in Western Europe than elsewhere: for example, data comprised about 90% of Western European traffic in 2011, compared with about 70% worldwide.

Voice data rates are low (usually 64 kbps in the circuit switched domains of 2G and 3G) and now only make up a small percentage of the total network traffic. However, voice applications provide many valuable features to the user, notably supplementary services such as voicemail and call forwarding, communication with fixed phones on the public switched telephone network and the ability to place emergency calls. Because of this, operators can still charge a large premium for voice services. This premium is falling, but, despite this, voice still makes a disproportionate contribution to operator revenue. A similar but more extreme situation applies in the case of messaging services, which make a negligible contribution to network traffic. In contrast, mobile data services often require a high data rate, yet do not provide the extra value that could justify correspondingly high charges.

16.2 Third Party Voice over IP

The simplest technique is to provide a voice over IP (VoIP) service through a third party supplier such as Skype, using the same principles as any other data application. Figure 16.2 shows the architecture that might typically be used for a voice service, although the details will differ from one provider to another. A similar approach can be used for text messages.

In this architecture, the user sets up a call by exchanging VoIP signalling messages with an external VoIP server, and ultimately with another fixed or mobile phone. From LTE's point of view, these signalling messages look just like any other kind of data and are transported in exactly the same way.

During call setup, the VoIP server can send LTE signalling messages to the policy and charging rules function (PCRF), so as to request a dedicated EPS bearer to transport the call. By doing this, the system can ensure a good quality of service for the user, at least over the path between the mobile and the PDN gateway. Using media gateways,

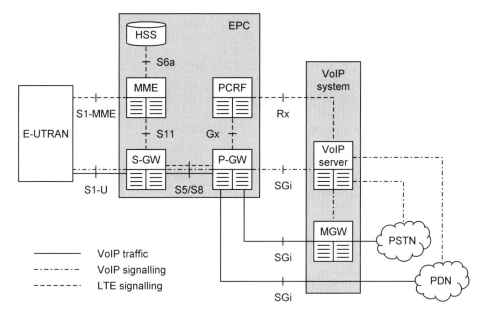

Figure 16.2 Architecture of a generic third party VoIP system.

the VoIP system can also convert IP packets to and from the data streams that are used by traditional circuit switched networks. This allows the user to make a call from a VoIP device to a 2G or 3G phone, or to a land line.

This approach requires little investment by the network operator, and still generates revenue through a partnership agreement with the service provider. Nevertheless, the revenue will undoubtedly be less than the operator could have achieved by going it alone.

There are also some technical issues to consider. If the user moves outside the region of LTE coverage, then the system may be able to handle the call using the 2G or 3G packet switched domains, or may be able to convert it to a traditional circuit switched call. If it does not support either of these approaches, then the call will have to be dropped. Furthermore, Skype does not support emergency calls at the time of writing, so cannot yet be used to replace a traditional voice service.

16.3 The IP Multimedia Subsystem

16.3.1 IMS Architecture

The *IP multimedia subsystem* (IMS) is a separate 3GPP network, which communicates with the evolved packet core and with the packet switched domains of UMTS and GSM so as to control real time IP multimedia services such as voice over IP. A full description of the IMS is outside the scope of this book, so we will limit ourselves to a summary. There is more information in books on the IMS [1, 2] and the relevant 3GPP specifications [3, 4].

The IMS was originally specified in 3GPP Release 5, which was frozen in 2002. Despite this, few 3G operators ever introduced it. There were two reasons for this: it is a complex system that requires significant investment from the network operator and it delivered few services that 3G networks could not provide in other ways. The IMS is, however, very suitable for the delivery of voice and SMS over LTE and can be viewed as a sophisticated 3GPP version of the third party VoIP system that we discussed above. This use of the IMS is being promoted by the *voice over LTE* (VoLTE) initiative of the *GSM Association* (GSMA) [5], based on earlier work by an industrial collaboration known as *One Voice*. It is likely to be the main long term solution for the delivery of voice over LTE.

Figure 16.3 shows a simplified version of the IMS and its relationship with LTE. The most important components are *call session control functions* (CSCFs), of which there are three types. The *serving CSCF* (S-CSCF) controls the mobile, in a similar way to an MME. Every mobile is registered with a serving CSCF, which sets up calls with other devices on the mobile's behalf and which contacts the mobile if an incoming call arrives. The *proxy CSCF* (P-CSCF) is the mobile's first point of contact with the IMS. It compresses the signalling messages that the mobile exchanges with the IMS so as to reduce their load on the LTE transport network, and secures those messages by encryption and integrity protection. It also communicates with the PCRF over the Rx interface, so as to guarantee the quality of service of the IP multimedia streams. Finally, the *interrogating CSCF* (I-CSCF) is the first point of contact for signalling messages that arrive from another IMS.

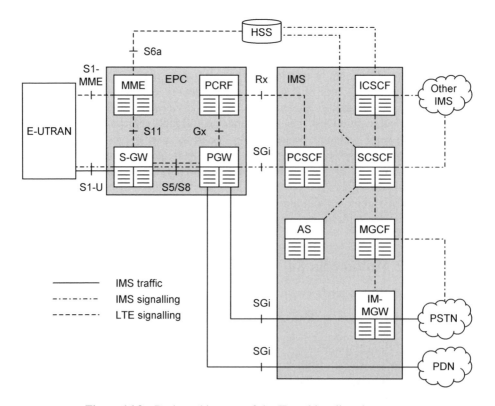

Figure 16.3 Basic architecture of the IP multimedia subsystem.

The *IMS media gateway* (IM-MGW) and *media gateway control function* (MGCF) allow the IMS to communicate with traditional circuit switched networks such as the public switched telephone network. The IMS media gateway converts VoIP streams to circuit switched streams and back again, while the media gateway control function carries out the same role for signalling messages, controls the IMS media gateway and is itself controlled by the serving CSCF. *Application servers* (ASs) provide the user with supplementary services such as voicemail, and the home subscriber server (HSS) is a central database that is shared with the evolved packet core. There are other components and interfaces that the figure does not show, but these are beyond our scope.

The main IMS signalling protocol is the *session initiation protocol* (SIP) [6], an IETF protocol that is widely used by other VoIP systems. Using this protocol, the IMS network elements communicate with each other and with the mobile, to carry out tasks such as registering the mobile and setting up a call. SIP messages can contain embedded information written using the *session description protocol* (SDP) [7], which defines the properties of the media such as the requested data rate and the supported codecs.

The mobile identifies itself to the IMS using a *private identity*, which acts in a similar way to the IMSI. It also identifies itself to the outside world using one or more *public identities*, which act like phone numbers or email addresses. These identities are stored in the *IP multimedia services identity module* (ISIM), which is a UICC application similar to the USIM.

16.3.2 IMS Procedures

The two most important IMS procedures are registration and call setup.

If a mobile supports the IP multimedia subsystem, then it registers with the IMS [8] after it has completed the attach procedure from Chapter 11. Network operators typically use a separate access point name for the IMS, so the mobile begins by connecting to that APN using the UE requested PDN connectivity procedure from Chapter 13. The network responds by setting up a default EPS bearer for signalling communications with the IMS, most likely using QoS class identifier 5 (Table 13.1). Using this bearer, the mobile sends a SIP registration request to the proxy CSCF, which forwards it to a suitable serving CSCF. In the registration procedure that follows, the mobile and serving CSCF authenticate each other, and the mobile registers its public identities with the serving CSCF.

If an incoming call arrives, then the serving CSCF contacts the mobile by sending it a SIP signalling message over the EPS bearer. If the mobile is in ECM-IDLE state, then the message triggers the LTE paging procedure from Chapter 14. The mobile moves back to ECM-CONNECTED by means of a service request and responds.

During call setup [9], the two mobiles exchange SIP signalling messages with embedded SDP session descriptions, so as to set up the call and negotiate the voice codec that they will use. At the same time, the two IP multimedia subsystems negotiate the quality of service. In each IMS, the proxy CSCF asks the PCRF to supply a quality of service that is appropriate for the selected codec, using the procedure from Chapter 13 for a server originated QoS request. In response, the PDN gateway sets up a dedicated EPS bearer that will carry the call. The IMS can also set up a call to a traditional circuit switched network, using the media and signalling conversion functions provided by the IMS media gateway and the media gateway control function.

A related technique is known as *single radio voice call continuity* (SRVCC) [10]. If a mobile moves outside the coverage area of LTE, then the network can use this technique to transfer the mobile from VoIP communications over the IMS, to traditional circuit switched communications over GSM, UMTS or cdma2000 1xRTT. The technique is based on the inter-system handover procedures described in Chapter 15, but with extra steps to transfer the call without loss of service.

The IMS offers full support for emergency calls over LTE from Release 9, using features such as the null integrity protection algorithm that we noted in Chapter 12.

16.3.3 SMS over the IMS

The IMS can also deliver SMS messages, using a technique known as *SMS over generic IP access* [11]. Figure 16.4 shows the architecture. The only new component is the *IP short message gateway* (IP-SM-GW), which acts as an interface between the IMS and the network elements that handle SMS messages.

To transmit an SMS message, the mobile embeds it into a SIP signalling message and sends it to the IP multimedia subsystem. Inside the IMS, the IM-SM-GW extracts the embedded SMS message and sends it to a standard SMS device known as the *SMS interworking MSC* (SMS-IWMSC). This device then forwards the message to the SMS *service centre* (SC), which stores the message for delivery to its destination. The same sequence is used in reverse to deliver a message, except that the SMS interworking MSC is replaced by another device known as the *SMS gateway MSC* (SMS-GMSC).

16.4 Circuit Switched Fallback

16.4.1 Architecture

Many network operators are rolling out LTE before they roll out the IP multimedia subsystem. Because of this, users may wish to place voice calls from LTE devices in the absence of the IMS. To deal with this possibility, 3GPP has standardized a technique known as *circuit switched* (CS) *fallback* [12] as a possible interim approach.

Using circuit switched fallback, a user can make voice calls by reverting to traditional circuit switched communications over UMTS or GSM. The architecture (Figure 16.5) builds on the architecture for inter-operation with the 2G/3G packet switched domain, by supporting the circuit switched domain as well.

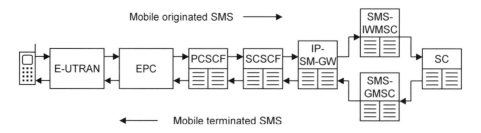

Figure 16.4 Architecture used to deliver SMS messages over the IP multimedia subsystem.

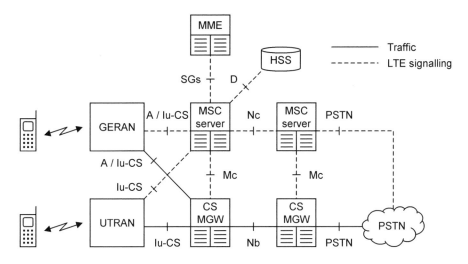

Figure 16.5 Circuit switched fallback architecture. Reproduced by permission of ETSI.

In this architecture, an MME can communicate with a mobile switching centre (MSC) server that supports circuit switched fallback, across a new interface that is denoted SGs. The signalling messages are written using the *SGs application protocol* (SGsAP) [13]. Using these messages, a mobile can register with an MSC server and set up a circuit switched voice call and can also send or receive an SMS message using a related technique called *SMS over SGs*. The details are described next.

16.4.2 Combined EPS/IMSI Attach Procedure

When a mobile switches on, it registers with a serving MME using the attach procedure from Chapter 11. This procedure is modified for mobiles that support circuit switched fallback, so as to register the mobile with an MSC server as well. Figure 16.6 shows the resulting message sequence [14].

In its attach request (1), the mobile uses an information element known as the EPS attach type to request a *combined EPS/IMSI attach*. This indicates that it is configured for circuit switched fallback and/or SMS over SGs and would like to register with an MSC server.

The MME runs steps 3 to 16 of the attach procedure (2), which cover the steps required for identification and security, location updates and default bearer creation. The MME then identifies a suitable MSC server and sends it an SGsAP *Location Update Request* (3). In response, the MSC server creates an association between the mobile and the MME and runs a standard procedure known as a *location update* (4), in which it registers the mobile and updates the circuit switched domain's copy of the mobile's location. The MSC server then acknowledges the original request from the MME (5) and the MME runs the remaining steps of the attach procedure (6). In its attach accept, the MME indicates whether the result was a combined EPS/IMSI attach, or an EPS only attach.

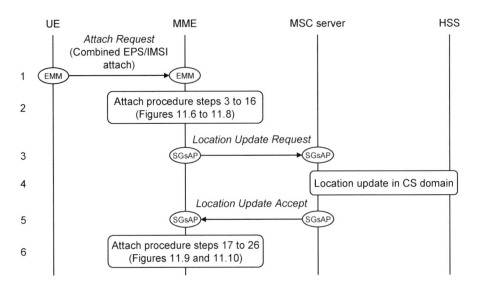

Figure 16.6 Combined EPS/IMSI attach procedure. Reproduced by permission of ETSI.

16.4.3 Voice Call Setup

Figure 16.7 shows the procedure for a mobile originated voice call using circuit switched fallback [15]. It assumes that the mobile has previously registered with the circuit switched domain using the combined EPS/IMSI attach described above and that the mobile is starting in RRC_IDLE state. It also assumes that the network can hand over the mobile's packet switched services to the 2G/3G packet switched domain. The procedure is based on the service request procedure from Chapter 14 and on the procedure for inter-system handover from Chapter 15.

The mobile starts with the usual procedures for random access and RRC connection establishment, and then writes an EMM message known as an *Extended Service Request*. This indicates that the mobile would like to place a voice call using circuit switched fallback and includes the reason for the request, namely a mobile originated call, a mobile terminated call or an emergency call. The mobile sends the message to the base station as part of its RRC Connection Setup Complete (1) and the base station forwards the message to the MME (2).

In response, the MME tells the base station to set up a signalling connection for the mobile in the usual way (3), but also sets a CS fallback indicator that requests a handover to the circuit switched domain. The base station activates access stratum security (4) and tells the mobile to move into ECM_CONNECTED state (5). The mobile's response (6) triggers an acknowledgement from the base station to the MME (7).

As part of step 5, the base station also tells the mobile to measure the signals that it can receive from nearby UMTS and GSM cells. This step is optional in theory but important in practice, as the base station does not yet know which cell it should hand the mobile over to. The mobile makes the measurements and transmits a measurement report (8).

Using the measurement report, the base station starts the procedure for handover from LTE to UMTS or GSM (9). Once the mobile is communicating with the target radio access network, it can contact the MSC server using a *Connection Management* (CM) *Service*

Voice and Text Messages over LTE

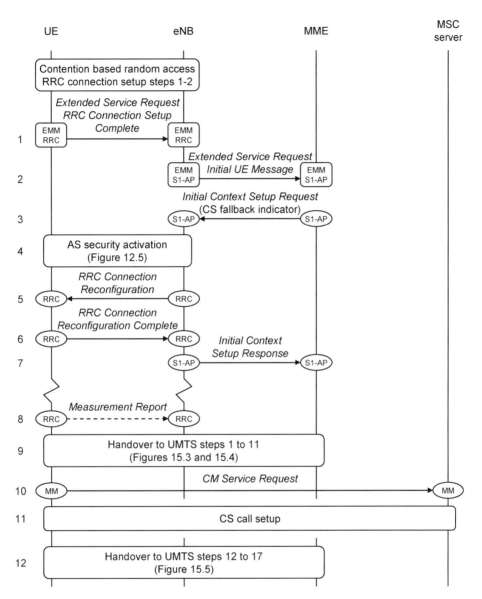

Figure 16.7 Procedure for mobile originated call setup using circuit switched fallback, with the packet switched services handed over. Reproduced by permission of ETSI.

Request that is written using the circuit switched domain's *mobility management* (MM) protocol (10). The mobile can then initiate the 2G/3G procedure for establishing a circuit switched voice call (11). At the same time, the network can complete the handover by transferring the mobile's EPS bearers to the 2G/3G packet switched domain (12), so that the user can continue any previous data sessions. On completion of the call, the network can hand the mobile back to LTE.

If the network or mobile does not support packet switched handovers, then circuit switched fallback is implemented using a different procedure, in which the base station releases the mobile's S1 connection and tells the mobile to carry out a cell reselection to 2G or 3G. As a result, the mobile's packet switched services are suspended for the duration of the call.

If a mobile terminated call arrives for the mobile, then the incoming signalling message arrives at the MSC server. This notices that the mobile is registered with an MME and sends the MME an SGsAP *Paging Request*. The MME runs the paging procedure from Chapter 14 and the mobile responds with an extended service request as before. The call setup procedure then continues in a similar way to the procedure described above. If a mobile terminated call arrives while idle mode signalling reduction is active, then the MSC server sends an SGsAP paging request to the MME in the same manner as before. The MME contacts the SGSN, both devices page the mobile and the mobile responds in the usual way. There are no direct communications between the MSC server and the SGSN, so the MSC server remains unaware of the use of ISR.

Circuit switched fallback only requires a few enhancements to the system and is likely to be the main interim solution for the delivery of voice over LTE. However, it has several drawbacks [16]. Firstly, the mobile can only use the technique if it is simultaneously in the coverage area of LTE and a 2G or 3G cell. Secondly, the procedure involves delays of a few seconds while the mobile makes its measurements and the network hands it over. Thirdly, inter-system handovers have traditionally been one of the least reliable aspects of a mobile telecommunication system. If this unreliability is repeated in the case of LTE, then circuit switched fallback will result in many dropped calls.

Overall, the resulting service degradation makes it doubtful whether circuit switched fallback is actually an acceptable solution. To compound the problem further, all the above drawbacks will also apply to emergency calls.

16.4.4 SMS over SGs

We can use similar ideas for the delivery of SMS messages, using the architecture shown in Figure 16.8. Thankfully, the mobile does not have to hand over to a 2G or 3G network before sending or receiving a message: instead, the network delivers SMS messages by embedding them into signalling messages between the MME and the MSC server. The technique is therefore known as *SMS over SGs* [17].

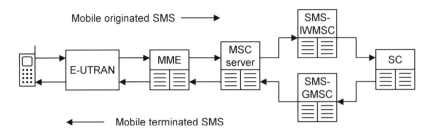

Figure 16.8 Architecture for SMS over SGs.

To use this technique, the mobile registers with the circuit switched domain using the combined EPS/IMSI attach procedure described in the previous page. It can then send an SMS message to the MME, by embedding it into an EMM message known as *Uplink NAS transport*. The MME can then forward the SMS message to the MSC server, by embedding it into an SGsAP *Uplink Unitdata* message. From there, the SMS message is passed to the SMS interworking MSC and the service centre in the usual way.

If a mobile terminated SMS reaches the MSC server while the mobile is in ECM-IDLE state, then the MSC server alerts the MME using an SGsAP Paging Request. This triggers the paging procedure from Chapter 14 and the mobile replies by initiating a service request in the usual way. The MME can then retrieve the SMS message from the MSC server and can deliver it to the mobile.

16.4.5 Circuit Switched Fallback to cdma2000 1xRTT

The 3GPP specifications also support circuit switched fallback to cdma2000 1xRTT networks. The architecture is shown in Figure 16.9 [18].

The technique uses similar principles to the ones we saw in Chapter 15 for inter-operation with packet switched cdma2000 HRPD networks. After attaching to LTE, the mobile preregisters with a cdma2000 1xRTT MSC, using 1xRTT signalling messages that are transported across the S102 interface [19]. If the user wishes to place a voice call later on, then the mobile sends an extended service request to the MME in the manner described above and the MME tells the base station to initiate a handover to cdma2000 1xRTT. Note, however, that the mobile can only transfer its packet switched sessions if it supports voice communications on a 1xRTT carrier frequency at the same time as data communications on a different HRPD carrier frequency.

If an incoming call arrives for the mobile, then the 1xRTT MSC sends the mobile a paging message across the S102 interface. This triggers an extended service request from the mobile and the call setup procedure continues in a similar way. The mobile can

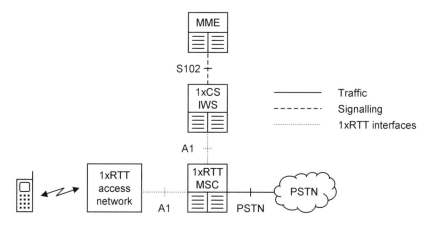

Figure 16.9 Architecture for circuit switched fallback to cdma2000 1xRTT. Reproduced by permission of ETSI.

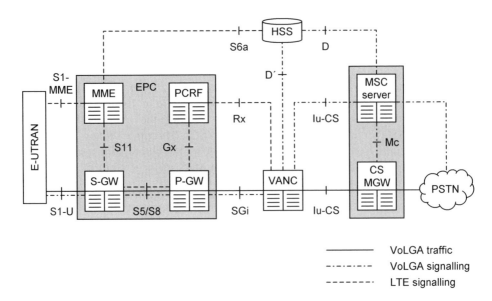

Figure 16.10 VoLGA architecture.

also send and receive SMS messages, by tunnelling them to and from the 1xRTT MSC over S102.

16.5 VoLGA

Voice over LTE via generic access (VoLGA) is another technique for delivering voice calls to LTE devices using the existing capabilities of the 2G/3G circuit switched domain. It is promoted by an industry initiative known as the *VoLGA forum* [20], which has defined the technique in three main specifications [21–23].

The technique is based on earlier 3GPP specifications [24], through which a mobile can reach the 2G/3G core network through a generic access network such as a wireless local area network. As shown in Figure 16.10, the VoLGA architecture exploits this by connecting the PDN gateway to the 2G/3G circuit switched domain through a device called a *VoLGA access network controller* (VANC). From the viewpoint of the PDN gateway, this device looks like any other server in the outside world, but from the viewpoint of the circuit switched domain, it looks like a generic radio access network. As a result, the LTE system can impersonate the behaviour of a generic radio access network and can give the mobile access to the services of the 2G/3G circuit switched domain.

VoLGA has found less support from the industry than the other approaches described earlier, and appears unlikely to be widely adopted. However, this situation might be re-assessed if deployment of the IMS is delayed.

References

1. Camarillo, G. and Garcia-Martin, M.-A. (2008) *The 3G IP Multimedia Subsystem (IMS): Merging the Internet and the Cellular Worlds*, 3rd edn, John Wiley & Sons, Ltd, Chichester.

2. Poikselkä, M., Holma, H., Hongisto, J., Kallio, J. and Toskala, A. (2012) *Voice over LTE (VoLTE)*, John Wiley & Sons, Ltd, Chichester.
3. 3GPP TS 23.228 (September 2011) *IP Multimedia Subsystem (IMS); Stage 2*, Release 10.
4. 3GPP TS 24.229 (September 2011) *IP Multimedia Call Control Protocol Based on Session Initiation Protocol (SIP) and Session Description Protocol (SDP); Stage 3*, Release 10.
5. GSM Association (2011) *VoLTE~GSM World*. Available at: http://www.gsmworld.com/our-work/mobile_broadband/VoLTE.htm (accessed 12 December, 2011).
6. IETF RFC 3261 (June 2002) *SIP: Session Initiation Protocol*.
7. IETF RFC 4566 (July 2006) *SDP: Session Description Protocol*.
8. 3GPP TS 23.228 (September 2011) *IP Multimedia Subsystem (IMS); Stage 2*, Release 10, section 5.2.
9. 3GPP TS 23.228 (September 2011) *IP Multimedia Subsystem (IMS); Stage 2*, Release 10, sections 5.4, 5.6, 5.7.
10. 3GPP TS 23.216 (June 2011) *Single Radio Voice Call Continuity (SRVCC); Stage 2*, Release 10.
11. 3GPP TS 23.204 (September 2011) *Support of Short Message Service (SMS) Over Generic 3GPP Internet Protocol (IP) Access; Stage 2*, Release 10.
12. 3GPP TS 23.272 (September 2011) *Circuit Switched (CS) Fallback in Evolved Packet System (EPS); Stage 2*, Release 10.
13. 3GPP TS 29.118 (September 2011) *Mobility Management Entity (MME) – Visitor Location Register (VLR) SGs Interface Specification*, Release 10.
14. 3GPP TS 23.272 (September 2011) *Circuit Switched (CS) Fallback in Evolved Packet System (EPS); Stage 2*, Release 10, section 5.2.
15. 3GPP TS 23.272 (September 2011) *Circuit Switched (CS) Fallback in Evolved Packet System (EPS); Stage 2*, Release 10, sections 6.2, 6.4.
16. Disruptive Analysis (2009) *A Critical Examination of CSFB as a Solution for Providing Voice-on-LTE*.
17. 3GPP TS 23.272 (September 2011) *Circuit Switched (CS) Fallback in Evolved Packet System (EPS); Stage 2*, Release 10, section 8.2.
18. 3GPP TS 23.272 (September 2011) *Circuit Switched (CS) Fallback in Evolved Packet System (EPS); Stage 2*, Release 10, annex B.
19. 3GPP TS 29.277 (September 2011) *Optimised Handover Procedures and Protocol between EUTRAN Access and Non-3GPP Accesses (S102); Stage 3*, Release 10.
20. VoLGA Forum (2010) *VoLGA Forum - Start*. Available at: http://www.volga-forum.com (accessed 12 December, 2011).
21. VoLGA Forum (April 2010) *Voice over LTE via Generic Access; Requirements Specification; Phase 2*, version 2.0.0.
22. VoLGA Forum (June 2010) *Voice over LTE via Generic Access; Stage 2 Specification; Phase 2*, version 2.0.0.
23. VoLGA Forum (June 2010) *Voice over LTE via Generic Access; Stage 3 Specification; Phase 2*, version 2.0.0.
24. 3GPP TS 43.318 (April 2011) *Generic Access Network (GAN); Stage 2*, Release 10.

17

Enhancements in Release 9

At the time when the specifications for LTE Release 8 were being written, there was competition between LTE and WiMAX for support from network operators and equipment vendors. In view of this competition, 3GPP understandably wanted to finalize the specifications for Release 8 as soon as they could. To help achieve this, they delayed some of the peripheral features of the system until Release 9, which was frozen in December 2009.

This chapter covers the capabilities that were introduced in Release 9, which include the multimedia broadcast/multicast service, location services and dual layer beamforming. Release 9 also enhances the self optimization capabilities of LTE, a feature that we will defer until Chapter 19. The contents of Release 9 are described in the relevant 3GPP release summary [1] and there is a useful review in a regularly updated white paper by 4G Americas [2].

17.1 Multimedia Broadcast/Multicast Service

17.1.1 Introduction

Mobile cellular networks are normally used for one-to-one communication services such as phone calls and web browsing, but they can also be used for one-to-many services such as mobile television. There are two types of one-to-many service: *broadcast services* are available to anyone, while *multicast services* are only available to users who have subscribed to a multicast group. Broadcast and multicast services require several different techniques from traditional unicasting. For example, the network has to distribute the data using IP multicast, while the encryption techniques have to be modified to ensure that all subscribing users can receive the information stream.

UMTS implements these techniques using the *multimedia broadcast/multicast service* (MBMS) [3], which was introduced in 3GPP Release 6. Although it is not widely used, it has proved popular in a few markets such as Japan and South Korea. The LTE multimedia broadcast/multicast service was introduced in Release 9. Despite the service's name, LTE currently only supports broadcast services, which do not require the user to subscribe to a multicast group.

To transmit MBMS data streams, LTE uses an air interface technique known as *multicast/broadcast over a single frequency network* (MBSFN). MBSFN was fully specified in Release 8, but we will cover both MBMS and MBSFN in this chapter so as to keep the two related issues together.

17.1.2 Multicast/Broadcast over a Single Frequency Network

When delivering a broadcast or multicast service, the radio access network transmits the same information stream from several nearby cells. This is different from the usual situation in a mobile telecommunication system, in which nearby cells are transmitting completely different information. LTE exploits this feature to improve the transmission of broadcast and multicast services, using the technique of MBSFN (Figure 17.1).

When using MBSFN, nearby base stations are synchronized so that they broadcast the same content at the same time and on the same sub-carriers. The mobile receives multiple copies of the information, which are identical except for their different arrival times, amplitudes and phases. The mobile can then process the information streams using exactly the same techniques that we introduced in Chapter 4 for handling multipath. It is not even aware that the information is coming from multiple cells.

Because the extra cells are transmitting the same information stream, they do not cause any interference to the mobile: instead, they contribute to the received signal power. This increases the mobile's SINR and maximum data rate, particularly at the edge of the cell where interference is usually high. In turn, this allows LTE to reach a target spectral efficiency of 1 bit s^{-1} Hz^{-1} for the delivery of MBMS [4], equivalent to 16 mobile TV channels in a 5 MHz bandwidth at a rate of 300 kbps each.

17.1.3 Implementation of MBSFN in LTE

In LTE, the synchronized base stations that we introduced above lie in a geographical region known as an *MBSFN area*. MBSFN areas can overlap, so that one base station can transmit multiple sets of content from multiple MBSFN areas.

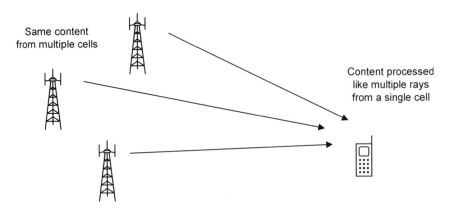

Figure 17.1 Multicast/broadcast over a single frequency network.

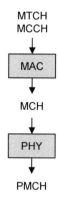

Figure 17.2 Channels used for MBMS and MBSFN.

In each MBSFN area, the LTE air interface delivers MBMS using the channels shown in Figure 17.2. There are two logical channels. The multicast traffic channel (MTCH) carries broadcast traffic such as a television station, while the multicast control channel (MCCH) carries RRC signalling messages that describe how the traffic channels are being transmitted. Each MBSFN area contains one multicast control channel and multiple instances of the multicast traffic channel.

The multicast traffic and control channels are transported using the multicast channel (MCH) and the physical multicast channel (PMCH). Each MBSFN area contains multiple instances of the PMCH, each of which carries either the multicast control channel, or one or more multicast traffic channels.

There are a few differences between the transmission techniques used for the PMCH and the PDSCH [5]. The PMCH implements the MBSFN techniques described above, and always uses the extended cyclic prefix to handle the long delay spreads that result from the use of multiple base stations. It is transmitted on antenna port 4, to keep it separate from the base station's other transmissions, and does not use transmit diversity, spatial multiplexing or hybrid ARQ. Each instance of the channel uses a fixed modulation scheme and coding rate, which are configured by means of RRC signalling.

The PMCH uses a different set of reference signals from usual, known as *MBSFN reference signals*. These are tagged with the MBSFN area identity instead of the physical cell identity, to ensure that the mobile can successfully combine the reference signals that it receives from different cells.

To date, MBMS is only delivered on carriers that are shared with unicast traffic. This is achieved using time division multiplexing. In any one cell, each downlink subframe is allocated either to unicast traffic on the PDSCH, or to broadcast traffic on the PMCH, according to a mapping that is defined in SIB 2. Thus a particular subframe contains either the PDSCH, or the PMCH, but not both. Furthermore, each broadcast subframe is allocated to a single MBSFN area. Within an MBSFN area, each broadcast subframe carries either signalling messages on the multicast control channel, or broadcast traffic on a single instance of the multicast traffic channel.

A broadcast subframe still starts with a PDCCH control region, but this is only used for uplink scheduling and uplink power control commands, and is only one or two symbols

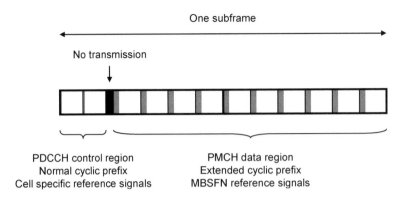

Figure 17.3 Example structure of an MBSFN subframe, for a 2-symbol control region using the normal cyclic prefix.

long. The control region uses the cell's usual cyclic prefix duration (normal or extended) and the cell specific reference signals. The rest of the subframe is occupied by the PMCH, which uses the extended cyclic prefix and the MBSFN reference signals. Figure 17.3 shows the resulting slot structure, for the case where the control region uses the normal cyclic prefix and contains two symbols. Note the gap in the downlink transmission, as the cell changes from one cyclic prefix duration to the other.

Eventually, it is envisaged that LTE will also support cells that are dedicated to MBMS. These cells will only use the downlink reference and synchronization signals, the physical broadcast channel and the physical multicast channel and will not support any uplink transmissions. They have not yet been fully specified, even in Release 10, but are worth mentioning because they can optionally use a special slot structure in support of very large cells. In this slot structure, the subcarrier spacing is reduced to 7.5 kHz, which increases the symbol duration to 133.3 µs and the cyclic prefix duration to 33.3 µs. This option appears in the air interface specifications from Release 8 [6], but is not yet usable and can generally be ignored.

17.1.4 Architecture of MBMS

Figure 17.4 shows the architecture that is used for the delivery of MBMS over LTE [7, 8]. The *broadcast/multicast service centre* (BM-SC) receives MBMS content from a content provider. The *MBMS gateway* (MBMS-GW) distributes the content to the appropriate base stations, while the *multicell/multicast coordination entity* (MCE) schedules the transmissions from all the base stations in a single MBSFN area.

The BM-SC indicates the start of each MBMS session by sending a signalling message across the SGmb interface. The interface is not defined by the 3GPP specifications, but the message describes the session's quality of service and tells the MBMS gateway to reserve resources for it. The message is propagated across the Sm [9], M3 [10] and M2 [11] interfaces, the last of which also defines the modulation scheme, coding rate and subframe allocation that the base stations in the MBSFN area should use.

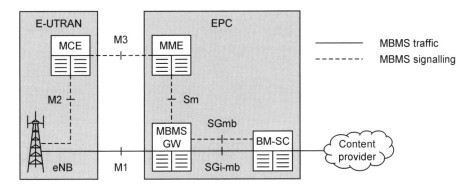

Figure 17.4 Architecture for MBMS in LTE.

The BM-SC then broadcasts the data across the SGi-mb interface using IP multicast. Each MBMS gateway forwards the data to the appropriate base stations across the M1 interface, along with a header that indicates each packet's transmission time with an accuracy of 10 milliseconds [12]. By combining this with the scheduling information that it receives from the multicell/multicast coordination entity, the base station can establish the exact transmission time for each packet.

17.1.5 Operation of MBMS

The best way to understand the operation of MBMS is to look at the procedures within the mobile [13–16]. After it switches on, the mobile reads SIB 2. By doing this, it discovers which subframes have been reserved for MBSFN transmissions on the PMCH and which have been reserved for unicast transmissions on the PDSCH. The MBSFN subframes repeat with a period of 1 to 32 frames. They do not clash with the subframes used by the synchronization signals or the physical broadcast channel, or with paging subframes.

If the user wishes to receive broadcast services, then the mobile continues by reading a new system information block, SIB 13. This lists the MBSFN areas that the cell belongs to. For each MBSFN area, it also defines the subframes that carry the multicast control channel, and the modulation scheme and coding rate that those transmissions will use.

The mobile can now receive the multicast control channel, which carries a single RRC signalling message, *MBSFN Area Configuration*. This message lists the physical multicast channels that the MBSFN area is using. For each PMCH, the message lists the corresponding multicast traffic channels and defines the subframes that will be used, the modulation scheme, the coding rate and a parameter known as the *MCH scheduling period* that lies between 8 and 1024 frames.

The mobile can now receive each instance of the PMCH. It still has to discover one more piece of information, namely the way in which the PMCH subframes are shared amongst the various multicast traffic channels that it carries. The base station signals this information in a new MAC control element known as *MCH scheduling information*, which it transmits at the beginning of each MCH scheduling period. Once it has read this control element, the mobile can receive each instance of the MTCH.

If the contents of the multicast control channel change, then the base station alerts the mobiles by writing a PDCCH scheduling command using a variant of DCI format 1C and addressing it to the *MBMS radio network temporary identifier* (M-RNTI).

17.2 Location Services

17.2.1 Introduction

Location services (LCS) [17, 18], also known as *location based services* (LBS), allow an application to find out the geographical location of a mobile. UMTS supported location services from Release 99, but they were only introduced into LTE from Release 9.

The biggest single motivation is emergency calls. This issue is especially important in the USA, where the Federal Communications Commission requires network operators to localize an emergency call to an accuracy between 50 and 300 metres, depending on the type of positioning technology used [19]. Location services are also of increasing importance to the user, for applications such as navigation and interactive games. Other applications include lawful interception by the police or security services and the use of a mobile's location to support network-based functions such as handover.

17.2.2 Positioning Techniques

In common with other mobile communication systems, LTE can calculate a mobile's position using three different techniques. The most accurate and increasingly common technique is the use of a *global navigation satellite system* (GNSS), a collective term for satellite navigation systems such as the *Global Positioning System* (GPS). There are two variants. With *UE based positioning*, the mobile has a complete satellite receiver and calculates its own position. The network can send it information to assist this calculation, such as an initial position estimate and a list of visible satellites. With *UE assisted positioning*, the mobile has a more basic satellite receiver, so it sends a basic set of measurements to the network and the network calculates its position. Whichever method is adopted, the measurement accuracy is typically around 10 metres.

The second technique is known as *downlink positioning* or *observed time difference of arrival* (OTDOA). Here, the mobile measures the times at which signals arrive from its serving cell and the nearest neighbours, and reports the time differences to the network. The network can then calculate the mobile's position by triangulation. The timing measurements are made on a new set of reference signals, known as *positioning reference signals* [20], which are transmitted on a new antenna port, number 6. The positioning accuracy is limited by multipath, typically to around 100 metres [21], so the technique has difficulty in meeting the requirements of the US FCC. It is often used as a backup to satellite positioning, as a mobile may not be able to receive a satisfactory satellite signal if it is surrounded by tall buildings or is indoors.

The last technique is known as *enhanced cell ID positioning*. Here, the network estimates the mobile's position from its knowledge of the serving cell identity and additional information such as the mobile's timing advance. The positioning accuracy depends on the cell size, being excellent in femtocells (provided that the base station's position is actually known), but very poor in macrocells.

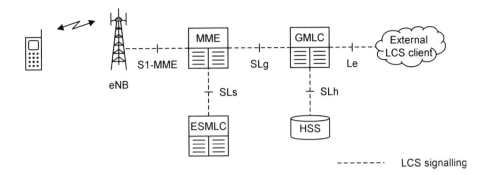

Figure 17.5 Architecture for location services in LTE.

17.2.3 Location Service Architecture

Figure 17.5 shows the main hardware components that LTE uses for location services [22, 23]. The *gateway mobile location centre* (GMLC) receives location requests from external clients across the Le interface. It retrieves the identity of the mobile's serving MME from the home subscriber server and forwards the location request to the MME. In turn, the MME delegates responsibility for calculating a mobile's position to the *evolved serving mobile location centre* (E-SMLC).

Two other components can be separate devices or can be integrated into the GMLC. The *privacy profile register* (PPR) contains the users' privacy details, which determine whether a location request from an external client will actually be accepted. The *pseudonym mediation device* (PMD) retrieves a mobile's IMSI using the identity supplied by the external client.

The architecture uses several signalling protocols. The GMLC communicates with the home subscriber server and the MME using Diameter applications [24, 25], while the MME communicates with the E-SMLC using the *LCS application protocol* (LCS-AP) [26]. The MME can also send positioning-related information to the mobile using *supplementary service* (SS) messages that are embedded into EMM messages on the air interface [27]. The E-SMLC communicates with the mobile and the base station using the *LTE positioning protocol* (LPP) [28], the messages being transported by embedding them into lower-level LCS-AP and EMM messages. The GMLC can communicate with the external client using a few different techniques, such as the *open service architecture* (OSA).

17.2.4 Location Service Procedures

To illustrate the operation of location services, Figure 17.6 shows how the network might respond to a location request from an external client [29]. The diagram assumes that the mobile is not roaming and that the GMLC can communicate directly with the mobile's serving MME. It also assumes the use of a mobile-assisted or mobile-based positioning technique.

The procedure begins when the external client asks the GMLC for the mobile's position and optionally velocity (1). The client typically identifies the mobile using its IP address.

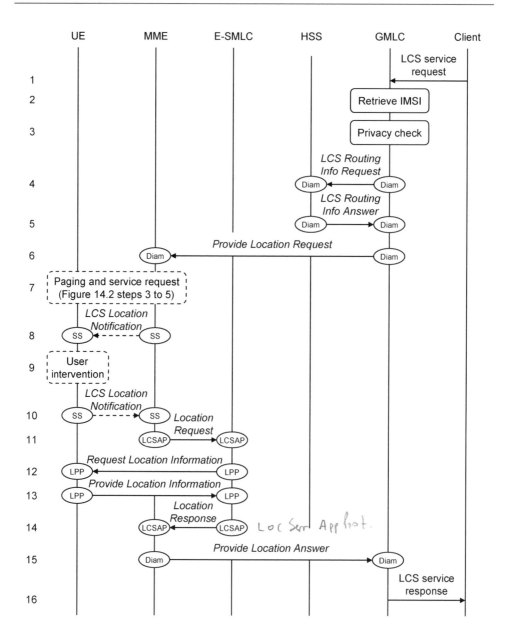

Figure 17.6 Mobile terminated location request procedure. Reproduced by permission of ETSI.

The GMLC retrieves the mobile's IMSI from the pseudonym mediation device (2) and interrogates the privacy profile register to establish whether the location request can be accepted (3). It then retrieves the identity of the mobile's serving MME from the home subscriber server (4, 5) and forwards the location request there (6). As part of that message, it can specify the location estimate's quality of service in terms of the positional accuracy and response time, the information being obtained either from the client or from the mobile's subscription data.

If the mobile is in ECM-IDLE state, then the MME wakes it up using the paging procedure, to which the mobile responds by initiating a service request (7). If the privacy information indicates that the user should be notified of an incoming location request, then the MME sends a notification message to the mobile (8). The mobile asks the user whether the request can be accepted (9) and indicates the response to the MME (10). Unless the user withholds permission, the MME selects an E-SMLC and forwards the location request there (11). The message includes the mobile's location capabilities, which are supplied as part of its non access stratum capabilities during the attach procedure.

Using the mobile's capabilities and the requested quality of service, the E-SMLC decides the positioning technique that it will use. Assuming the use of a mobile-assisted or mobile-based technique, it sends the mobile a location request (12). In its message, the E-SMLC specifies the selected positioning technique and supplies supporting information such as the satellites that should be visible or the positioning reference signals that nearby base stations are transmitting. The message is transported to the mobile by embedding it into lower-level LCS-AP and EMM messages.

The mobile makes the measurements that have been requested and sends a response to the E-SMLC (13). In turn, the E-SMLC returns its position estimate to the client (14, 15, 16).

17.3 Other Enhancements in Release 9

17.3.1 Dual Layer Beamforming

In dual layer beamforming (Chapter 5), the base station transmits two simultaneous data streams using the same set of resource blocks, by processing the data using two parallel sets of antenna weights. It can then direct the data either to two different mobiles, or to two antennas on the same mobile.

Dual layer beamforming is first supported in LTE Release 9 [30–32]. To use it, the base station configures the mobile into a new transmission mode, mode 8, and schedules it using a new DCI format, 2B. It then transmits to the mobile using either or both of two new antenna ports, numbers 7 and 8.

Ports 7 and 8 use a new set of UE-specific reference signals, which behave in the same way as the reference signals used for single layer beamforming on port 5. To achieve this, the base station processes the reference signals using the same antenna weights that it applies to the PDSCH, so that the weights are transparent to the mobile, and are removed as a side-effect of channel estimation and equalization. In addition, the base station tags the reference signals using the RNTI of the target mobile and only transmits them in the physical resource blocks that the mobile is actually using.

17.3.2 Commercial Mobile Alert System

The US Federal Communications Commission established the *commercial mobile alert system* (CMAS) in response to the US Warning Alert and Response Network act of 2006. Using this system, participating network operators can transmit three types of emergency message: Presidential alerts about local, regional or national emergencies, imminent threat alerts about natural disasters such as hurricanes, and child abduction emergency alerts.

In Release 9, LTE supports CMAS by generalizing its earthquake and tsunami warning system to a *public warning system* (PWS) that covers both types of information [33]. The base station continues to send earthquake and tsunami warnings on SIBs 10 and 11, and transmits commercial mobile alerts on a new system information block, SIB 12.

17.3.3 Enhancements to Earlier Features of LTE

In earlier chapters, we saw two other enhancements that Release 9 makes to earlier features of the system. Firstly, Release 9 improves the algorithms for cell selection and reselection, to include measurements of the reference signal received quality as well as the reference signal received power. Secondly, Release 9 supports emergency calls over the IP multimedia subsystem, by adding features such as a null integrity protection algorithm.

References

1. 3rd Generation Partnership Project (2011) *FTP directory*. Available at: ftp://ftp.3gpp.org/Information/WORK_PLAN/Description_Releases/ (accessed 12 December, 2011).
2. 4G Americas (February 2011) *4G Mobile Broadband Evolution: 3GPP Release 10 and Beyond*.
3. 3GPP TS 23.246 (June 2011) *Multimedia Broadcast/Multicast Service (MBMS); Architecture and Functional Description*, Release 10.
4. 3GPP TR 25.913 (January 2009) *Requirements for Evolved UTRA (E-UTRA) and Evolved UTRAN (E-UTRAN)*, Release 10, section 7.5.
5. 3GPP TS 36.211 (September 2011) *Physical Channels and Modulation*, Release 10, sections 6.5, 6.10.2.
6. 3GPP TS 36.211 (September 2011) *Physical Channels and Modulation*, Release 10, sections 6.2.3, 6.10.2.2, 6.12.
7. 3GPP TS 23.246 (June 2011) *Multimedia Broadcast/Multicast Service (MBMS); Architecture and Functional Description*, Release 10, sections 4.2.2, 4.3.2, 5.
8. 3GPP TS 36.440 (September 2011) *Evolved Universal Terrestrial Radio Access Network (E-UTRAN); General Aspects and Principles for Interfaces Supporting Multimedia Broadcast Multicast Service (MBMS) within E-UTRAN*, Release 10, section 4.
9. 3GPP TS 29.274 (September 2011) *3GPP Evolved Packet System (EPS); Evolved General Packet Radio Service (GPRS) Tunnelling Protocol for Control Plane (GTPv2-C); Stage 3*, Release 10, section 7.13.
10. 3GPP TS 36.444 (June 2011) *Evolved Universal Terrestrial Radio Access Network (E-UTRAN); M3 Application Protocol (M3AP), M3 Application Protocol (M3AP)*, Release 10.
11. 3GPP TS 36.443 (September 2011) *Evolved Universal Terrestrial Radio Access Network (E-UTRAN); M2 Application Protocol (M2AP), M2 Application Protocol (M2AP)*, Release 10.
12. 3GPP TS 25.446 (June 2011) *MBMS Synchronisation Protocol (SYNC)*, Release 10.
13. 3GPP TS 36.331 (October 2011) *Radio Resource Control (RRC); Protocol Specification*, Release 10, sections 5.8, 6.2.2 (*MBSFNAreaConfiguration*), 6.3.1, 6.3.7.
14. 3GPP TS 36.321 (October 2011) *Medium Access Control (MAC) Protocol Specification*, Release 10, sections 5.12, 6.1.3.7.
15. 3GPP TS 36.213 (September 2011) *Physical Layer Procedures*, Release 10, section 11.
16. 3GPP TS 36.212 (September 2011) *Multiplexing and Channel Coding*, Release 10, section 5.3.3.1.4.
17. 3GPP TS 23.271 (March 2011) *Functional Stage 2 Description of Location Services (LCS)*, Release 10.
18. 3GPP TS 36.305 (October 2011) *Evolved Universal Terrestrial Radio Access Network (E-UTRAN); Stage 2 Functional Specification of User Equipment (UE) Positioning in E-UTRAN*, Release 10.
19. Federal Communications Commission (2011) *Wireless 911 Services*. Available at: http://www.fcc.gov/guides/wireless-911-services (accessed 12 December, 2011).
20. 3GPP TS 36.211 (September 2011) *Physical Channels and Modulation*, Release 10, section 6.10.4.
21. 3GPP R1-090768 (February 2009) *Performance of DL OTDOA with Dedicated LCS-RS*, Release 10.
22. 3GPP TS 23.271 (March 2011) *Functional Stage 2 Description of Location Services (LCS)*, Release 10, section 5.

23. 3GPP TS 36.305 (October 2011) *Evolved Universal Terrestrial Radio Access Network (E-UTRAN); Stage 2 Functional Specification of User Equipment (UE) Positioning in E-UTRAN*, Release 10, section 6.
24. 3GPP TS 29.173 (April 2011) *Location Services (LCS); Diameter-Based SLh Interface for Control Plane LCS*, Release 10.
25. 3GPP TS 29.172 (September 2011) *Location Services (LCS); Evolved Packet Core (EPC) LCS Protocol (ELP) between the Gateway Mobile Location Centre (GMLC) and the Mobile Management Entity (MME); SLg Interface*, Release 10.
26. 3GPP TS 29.171 (June 2011) *Location Services (LCS); LCS Application Protocol (LCS-AP) between the Mobile Management Entity (MME) and Evolved Serving Mobile Location Centre (E-SMLC); SLs Interface*, Release 10.
27. 3GPP TS 24.171 (April 2011) *Control Plane Location Services (LCS) Procedures in the Evolved Packet System (EPS)*, Release 10.
28. 3GPP TS 36.355 (October 2011) *Evolved Universal Terrestrial Radio Access (E-UTRA); LTE Positioning Protocol (LPP)*, Release 10.
29. 3GPP TS 23.271 (March 2011) *Functional Stage 2 Description of Location Services (LCS)*, Release 10, sections 9.1.1, 9.1.15, 9.3a.
30. 3GPP TS 36.211 (September 2011) *Physical Channels and Modulation*, Release 10, section 6.10.3.
31. 3GPP TS 36.212 (September 2011) *Multiplexing and Channel Coding*, Release 10, section 5.3.3.1.5B.
32. 3GPP TS 36.213 (September 2011) *Physical Layer Procedures*, Release 10, section 7.1.5A.
33. 3GPP TS 22.268 (October 2011) *Public Warning System (PWS) Requirements*, Release 10.

18

LTE-Advanced and Release 10

Release 10 enhances the capabilities of LTE, to make the technology compliant with the International Telecommunication Union's requirements for IMT-Advanced. The resulting system is known as LTE-Advanced. This chapter covers the new features of LTE-Advanced, by focussing on carrier aggregation, relaying and enhancements to multiple antenna transmission on the uplink and downlink. We will also look ahead to the new features that are being considered for Release 11 and beyond.

For the most part, the Release 10 enhancements are designed to be backwards compatible with Release 8. Thus a Release 10 base station can control a Release 8 mobile, normally with no loss of performance, while a Release 8 base station can control a Release 10 mobile. In the few cases where there is a loss of performance, the degradation has been kept to a minimum.

TR 36.912 [1] is 3GPP's submission to the ITU for LTE-Advanced and is a useful summary of the new features of the system. The contents of Releases 10 and 11 are also described in the relevant 3GPP release summaries [2] and are reviewed in a regularly updated white paper by 4G Americas [3].

18.1 Carrier Aggregation

18.1.1 Principles of Operation

The ultimate goal of LTE-Advanced is to support a maximum bandwidth of 100 MHz. This is an extremely large bandwidth, which is most unlikely to be available as a contiguous allocation in the foreseeable future. To deal with this problem, LTE-Advanced allows a mobile to transmit and receive on up to five *component carriers* (CCs), each of which has a maximum bandwidth of 20 MHz. This technique is known as *carrier aggregation* (CA) [4].

There are three scenarios, shown in Figure 18.1. In *inter-band aggregation*, the component carriers are located in different frequency bands. This is the most challenging scenario, because the mobile may require different radio components to support each band and because the cell's coverage area in each band may be very different. The component carriers are separated by a multiple of 100 kHz, which is the usual LTE carrier spacing.

An Introduction to LTE: LTE, LTE-Advanced, SAE and 4G Mobile Communications, First Edition. Christopher Cox.
© 2012 John Wiley & Sons, Ltd. Published 2012 by John Wiley & Sons, Ltd.

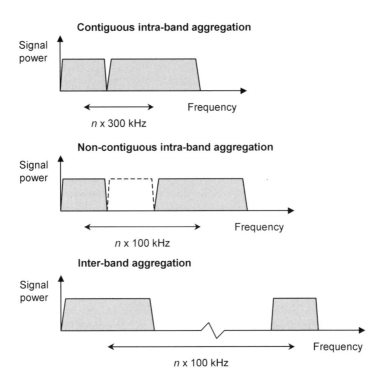

Figure 18.1 Carrier aggregation scenarios.

In *non contiguous intra-band aggregation*, the component carriers are in the same band. This deals with some of the difficulties noted above, and simplifies the design of the mobile and the network. In *contiguous intra-band aggregation*, the component carriers are in the same band and are adjacent to each other. They are separated by a multiple of 300 kHz, which is consistent with the orthogonality requirement from Chapter 4, so that the different sets of sub-carriers are orthogonal to each other and do not interfere.

There are two more restrictions. In FDD mode, the allocations on the uplink and downlink can be different, but the number of downlink component carriers is always greater than or equal to the number used on the uplink. In TDD mode, each component carrier uses the same TDD configuration, so it allocates its subframes to the uplink and downlink in the same way.

The component carriers are organized into one *primary cell* (PCell) and up to four *secondary cells* (SCells). The primary cell contains one component carrier in TDD mode, or one downlink CC and one uplink CC in FDD mode. It is used in exactly the same way as a cell in Release 8. Secondary cells are only used by mobiles in RRC_CONNECTED and are added or removed by means of mobile-specific signalling messages. Each secondary cell contains one component carrier in TDD mode, or one downlink CC and optionally one uplink CC in FDD mode.

Carrier aggregation only affects the physical layer and the medium access control protocol on the air interface, and the RRC, S1-AP and X2-AP signalling protocols. There is no impact on the radio link control or packet data convergence protocols and no impact on data transport in the fixed network.

18.1.2 UE Capabilities

Ultimately, carrier aggregation will allow a mobile to transmit and receive using five component carriers in a variety of frequency bands. It is handled by a single UE category, category 8, which supports a peak data rate of 3000 Mbps in the downlink and 1500 Mbps in the uplink. Despite this, a category 8 mobile does not have to support every feature of category 8 and does not have to support such a high peak data rate: instead, the mobile declares its support for individual features as part of its UE capabilities. Furthermore, 3GPP has not introduced full support for category 8 right away: instead, Release 10 only supports some of its features.

There are two aspects to this [5–7]. Firstly, the specifications only support carrier aggregation in a limited number of frequency bands. This limits the complexity of the specifications, because some of the radio frequency requirements have to be defined individually for each band or band combination. Release 10 supports intra-band contiguous aggregation in FDD band 1 (1920–1980 MHz uplink and 2110–2170 MHz downlink) and TDD band 40 (2300–2400 MHz) and also supports inter-band aggregation in FDD bands 1 and 5 (824–849 MHz uplink and 869–894 MHz downlink). As part of its capabilities, a mobile declares which bands and band combinations it supports.

Secondly, a mobile declares a capability known as the *CA bandwidth class* for each of its supported bands or band combinations. The CA bandwidth class states the number of component carriers that the mobile supports and the total number of resource blocks that it can handle. Table 18.1 lists the classes that are used by LTE-Advanced. Release 10 supports a maximum of 200 resource blocks across two component carriers by the use of class C, while classes D, E and F are reserved for future releases.

18.1.3 Scheduling

In Chapter 8, we described how a mobile looks for PDCCH scheduling messages in common and UE-specific search spaces, which lie within the downlink control region at the start of every subframe. Release 10 continues to use this process, but with a few modifications [8, 9].

Table 18.1 Carrier aggregation bandwidth classes. Reproduced by permission of ETSI

Carrier aggregation bandwidth class	Release	Maximum number of component carriers	Maximum number of resource blocks
A	R10	1	100
B	R10	2	100
C	R10	2	200
D	R11 or beyond	Not yet determined	Not yet determined
E	R11 or beyond	Not yet determined	Not yet determined
F	R11 or beyond	Not yet determined	Not yet determined

Each component carrier is independently scheduled and generates an independent set of hybrid ARQ feedback bits. The system does, however, support cross carrier scheduling: the base station can trigger an uplink or downlink transmission on one component carrier using a scheduling message on another. Release 10 implements cross carrier scheduling by adding a *carrier indicator field* (CIF) to each DCI format, which indicates the carrier to be used for the subsequent transmission.

Using cross carrier scheduling, a base station can transmit its scheduling messages on the component carrier that has the greatest coverage, so as to maximize the reliability of successful reception. It can also use the technique to balance the loads from traffic and scheduling across the different component carriers. The common search space is always on the primary cell, but the UE-specific search spaces can be on the primary cell or on any of the secondary cells.

18.1.4 Data Transmission and Reception

Carrier aggregation does not affect data transmission in the downlink, but it does lead to some changes in the uplink. In Release 8, a mobile uses SC-FDMA, which assumes that the mobile is transmitting on a single contiguous block of sub-carriers. In Release 10, this assumption is no longer valid: instead, the mobile uses a more general technique known as *discrete Fourier transform spread orthogonal frequency division multiple access* (DFT-S-OFDMA). This multiple access technique is the same as SC-FDMA, except that it supports transmission on a non contiguous allocation of sub-carriers.

To exploit the new multiple access technique, the specifications are relaxed in two other ways. Firstly, a mobile can transmit on each component carrier using sub-carriers that are grouped into two blocks, rather than one. These transmissions are scheduled using a new uplink resource allocation scheme, known as type 1. Secondly, a mobile can transmit on the PUCCH and PUSCH at the same time. Both features are optional for the mobile, which declares support for them as part of its capabilities [10].

A mobile's peak output power is higher when using DFT-S-OFDMA than when it is using SC-FDMA. This puts greater demands on the mobile's power amplifier, which increases the cost of the amplifier and the uplink power consumption.

18.1.5 Uplink and Downlink Feedback

Carrier aggregation leads to a few changes in the transmission of uplink control information [11–13]. The most important is that the mobile only transmits the PUCCH on the primary cell. However, it can send uplink control information using the PUSCH on the primary cell or on any of the secondary cells.

If the mobile needs to send hybrid ARQ acknowledgements to the base station, then it groups them together onto a single component carrier. When using the PUCCH, it can send the acknowledgements in two ways. The first is to transmit on multiple PUCCH resources using PUCCH format 1b, in a similar way to the use of ACK/NACK multiplexing in TDD mode. The second way is to use a new PUCCH format, number 3. This format handles the simultaneous transmission of up to 10 hybrid ARQ bits in FDD mode and 20 in TDD mode, together with an optional scheduling bit, using resource block pairs that are shared amongst five mobiles.

There are no significant changes to the procedure for uplink transmission and reception. In particular, a base station sends its PHICH acknowledgements on the same cell (primary or secondary) that the mobile used for its uplink data transmission.

18.1.6 Other Physical Layer and MAC Procedures

Carrier aggregation introduces a few other changes to the physical layer and MAC procedures from Chapters 8 to 10. As noted earlier, the base station adds and removes secondary cells using mobile-specific RRC Connection Reconfiguration messages. In addition, it can quickly activate and deactivate a secondary cell by sending a MAC *Activation/Deactivation* control element to the target mobile [14].

As before, the base station can control the power of a mobile's PUSCH transmissions using DCI formats 0, 3 and 3A. In LTE-Advanced, each component carrier has a separate power control loop. When using DCI format 0, the base station identifies the component carrier using the carrier indicator field from earlier. In the case of formats 3 and 3A, it assigns a different value of TPC-PUSCH-RNTI to each component carrier and uses that value as the target of the power control command.

18.1.7 RRC Procedures

Carrier aggregation introduces a few changes to the RRC procedures from Chapters 11 to 15, but not many. In RRC_IDLE state, the mobile carries out cell selection and reselection using one cell at a time, as before. The RRC connection setup procedure is unchanged too: at the end of the procedure, the mobile is only communicating with a primary cell. Once the mobile is in RRC_CONNECTED state, the base station can add or remove secondary cells using mobile-specific RRC Connection Reconfiguration messages.

In RRC_CONNECTED state, the mobile measures individual neighbouring cells in much the same way as before. The serving cell corresponds to the primary cell in measurement events A3, A5 and B2, and to either the primary or a secondary cell in measurement events A1 and A2. There is also a new measurement event, A6, which the mobile reports if the power from a neighbour cell rises sufficiently far above the power from a secondary cell. The base station might use this measurement report to trigger a change of secondary cell.

During a handover, the new base station tells the mobile about the new secondary cells using its RRC Connection Reconfiguration command, in the same way that it conveyed the random access preamble index in Release 8. This allows the network to change all the secondary cells as part of the handover procedure and also to hand a mobile over between base stations with differing support for Releases 8 and 10.

18.2 Enhanced Downlink MIMO

18.2.1 Objectives

Release 8 included full support for downlink single user MIMO, using a maximum of four antenna ports and four transmission layers. However, it only included rudimentary support for downlink multiple user MIMO.

Release 9 introduced support for dual layer beamforming, in which the base station transmits to two receive antennas that are located on one or two mobiles. As we noted in Chapter 5, it is best to implement downlink multiple user MIMO using the same basic principles. The dual layer beamforming technique is therefore extended as part of release 10, by providing full support for downlink multiple user MIMO and by increasing the maximum number of base station antenna ports to eight. We will cover the details below.

The same technique can also be used to support single user MIMO, with a maximum of eight antenna ports and eight transmission layers. A network might typically prefer single user MIMO in uncorrelated channel conditions or to maximize the peak data rate to a single mobile, while using multiple user MIMO in correlated channel conditions or to maximize the cell capacity. In addition, the network might select any intermediate point between the two extremes, such as the transmission of two layers to each of four mobiles.

The peak downlink data rate in Release 10 is 1200 Mbps. This is four times greater than in Release 8 and results from the use of two component carriers, each of which carries eight transmission layers rather than four. Eventually, LTE should support a peak downlink data rate of 3000 Mbps, through the use of five component carriers.

18.2.2 Downlink Reference Signals

Reference signals have two functions: they provide an amplitude and phase reference in support of channel estimation and demodulation, and they provide a power reference in support of channel quality measurements and frequency-dependent scheduling. In the Release 8 downlink, the cell specific reference signals support both of these functions, at least in transmission modes 1 to 6.

In principle, the designers of LTE-Advanced could have supported eight antenna MIMO in the same way as four antenna MIMO, by adding four new antenna ports that each carried the cell specific reference signals. However, this approach would have led to a few difficulties. The reference signals would occupy more resource elements, which would increase the overhead for Release 10 mobiles that recognized them and increase the interference for Release 8 mobiles that did not. They would also do nothing to improve the performance of multiple user MIMO.

Instead, Release 10 introduces some new downlink reference signals [15], in which the two functions are split. UE specific reference signals support channel estimation and demodulation, in a similar way to the demodulation reference signals on the uplink, and are transmitted on antenna ports 7 to 14. The signals on ports 7 and 8 are the same ones used by dual layer beamforming, while those on ports 9 to 14 support eight antenna single and multiple user MIMO.

The base station precodes the UE specific reference signals using the same precoding matrix that it applies to the PDSCH. This makes the precoding operation transparent to the mobile, so the base station can apply any precoding matrix it likes. This improves the performance of multiple user MIMO, which requires a free choice of precoding matrix to ensure that the signals reach the mobiles with the correct constructive or destructive interference. Furthermore, the base station only transmits the UE-specific reference signals in the physical resource blocks that the target mobile is actually using. As a result, the reference signals do not cause any overhead or interference for the other mobiles in the cell.

The base station also sends *CSI reference signals* on eight more antenna ports, numbered from 15 to 22. These signals support channel quality measurements and frequency dependent scheduling, in a similar way to the sounding reference signals on the uplink. The signals do not have to be sent so often, so the base station can configure their transmission interval to a value between 5 and 80 ms. They cause some overheads for Release 10 mobiles, but the long transmission interval implies that the overheads are acceptably small. They can also cause CRC failures for Release 8 mobiles that do not recognize them, but the base station can avoid these by scheduling Release 8 mobiles in different resource blocks.

18.2.3 Downlink Transmission and Feedback

To use eight layer spatial multiplexing [16–18], the base station starts by configuring the mobile into a new transmission mode, mode 9. This supports both single user and multiple user MIMO, so the base station can quickly switch between the two techniques without the need to change transmission mode.

The base station schedules the mobile using a new DCI format, 2C. In the scheduling command, it specifies the number of layers that it will use for the data transmission, between one and eight. It does not have to specify the precoding matrix, because that is transparent to the mobile. The base station then transmits the PDSCH on antenna ports 7 to $7 + n$, where n is the number of layers that the mobile is using. The maximum number of codewords is two, the same as in Release 8.

The mobile still has to feed back a precoding matrix indicator, which signals the discrepancy between the precoding that the base station is transparently providing and the precoding that the mobile would ideally like to use. Instead of using the PMI, however, the mobile feeds back two indices, i_1 and i_2. Both of these can vary from 0 to 15, which provides more finely-grained feedback than the PMI did and in turn improves the performance of the multiple user MIMO technique. The base station can then use these indices to reconstruct the requested precoding matrix.

18.3 Enhanced Uplink MIMO

18.3.1 Objectives

The only multiple antenna scheme supported by the Release 8 uplink was multiple user MIMO. This increased the cell capacity while only requiring the mobile to have a single transmit power amplifier and was far easier to implement than on the downlink. However, it did nothing for the peak data rate of a single mobile.

In LTE-Advanced, the uplink is enhanced to support single user MIMO, using up to four transmit antennas and four transmission layers. The mobile declares how many layers it supports as part of its uplink capabilities [19]. Release 10 only supports single user MIMO in TDD band 40 and in FDD bands 1, 3 and 7, but this support will be extended in future releases.

The peak uplink data rate in Release 10 is 600 Mbps. This is eight times greater than in Release 8, and results from the use of four transmission layers and two component carriers. Eventually, LTE should support a peak uplink data rate of 1500 Mbps, through the use of five component carriers.

Table 18.2 Uplink transmission modes in 3GPP Release 10

Mode	Purpose
1	Single antenna transmission
2	Closed loop spatial multiplexing

Table 18.3 Uplink antenna ports in 3GPP Release 10

Antenna port	Channels	Application
10		Single antenna transmission
20–21	PUSCH, SRS	2 antenna closed loop spatial multiplexing
40–43		4 antenna closed loop spatial multiplexing
100	PUCCH	Single antenna transmission
200–201		2 antenna open loop transmit diversity

18.3.2 Implementation

To support single user MIMO, the base station configures a Release 10 mobile into one of the transmission modes listed in Table 18.2 [20]. These are used in a similar way to the transmission modes on the downlink. Mode 1 corresponds to single antenna transmission, while mode 2 corresponds to single user MIMO, specifically the use of closed loop spatial multiplexing.

Once a mobile has been configured into mode 2, the base station sends it a scheduling grant for closed loop spatial multiplexing using a new DCI format, number 4 [21]. As part of the scheduling grant, the base station specifies the number of layers that the mobile should use for its transmission and the precoding matrix that it should apply. The maximum number of uplink codewords is increased to two, the same as on the downlink.

The PUSCH transmission process is then modified to include the additional steps of layer mapping and precoding [22], which work in the same way as the corresponding steps on the downlink. The antenna ports (Table 18.3) are numbered in an unexpected way. Port 10 is used for single antenna transmission of the PUSCH, ports 20 and 21 for dual antenna transmission and ports 40 to 43 for transmission on four antennas, while the same antenna ports are also used by the sounding reference signal. The PUCCH can be transmitted from a single antenna on port 100, or from two antennas using open loop diversity on ports 200 and 201.

18.4 Relays

18.4.1 Principles of Operation

Repeaters and *relays* (Figure 18.2) are devices that extend the coverage area of a cell. They are useful in sparsely populated areas, in which the performance of a network is limited by coverage rather than capacity. They can also increase the data rate at the edge of a cell, by improving the signal to interference plus noise ratio there.

LTE-Advanced and Release 10

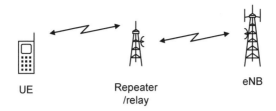

Figure 18.2 Operation of repeaters and relays.

A repeater receives a radio signal from the transmitter, and amplifies and rebroadcasts it, so appears to the receiver as an extra source of multipath. Unfortunately the repeater amplifies the incoming noise and interference as well as the received signal, which ultimately limits its performance. FDD repeaters were fully specified in Release 8, with the sole specification [23] referring to the radio performance requirements. TDD repeaters are harder to implement, because of the increased risk of interference between uplink and downlink, and have not yet been specified.

A relay takes things a step further, by decoding the received radio signal, before re-encoding and rebroadcasting it. By doing this, it removes the noise and interference from the retransmitted signal, so can achieve a higher performance than a repeater. Relays are first specified in Release 10, for both FDD and TDD modes.

18.4.2 Relaying Architecture

Figure 18.3 shows the architecture that is used for relaying [24, 25]. The relaying functions are implemented in the *relay node* (RN). This appears to the mobile as a perfectly normal base station: for example, it has one or more physical cell IDs of its own, broadcasts its own synchronization signals and system information, and is responsible for scheduling all

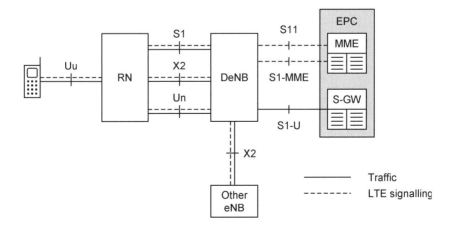

Figure 18.3 Relaying architecture in LTE.

the uplink and downlink transmissions on the Uu interface. The relay node is controlled by a *donor eNB* (DeNB), which is otherwise a normal base station that can control mobiles of its own.

The Un interface is the air interface between the relay node and the donor eNB. It is typically implemented as a point-to-point microwave link, so has a far greater range than the air interface to a mobile. Across this interface, the donor eNB acts like any other base station and the relay node acts like a mobile.

The Un and Uu interfaces can use either the same carrier frequency, or different ones. If the carrier frequencies are different, then the Un interface can be implemented in exactly the same way as a normal air interface. For example, the relay node acts like a base station on the Uu interface towards the mobile and independently acts like a mobile on the Un interface towards the donor eNB. If the carrier frequencies are the same, then the Un interface requires some extra functions, covered in the next section, to share the resources of the air interface with Uu.

There is a variant of the X2 interface between the donor eNB and the relay node, which supports handovers between the relay node and any other base station. The interface is implemented in the same way as a normal X2 interface, but transports the data and signalling messages using the functions of the Un interface instead of IP. A similar variant of the S1 interface allows the relay node to communicate directly with the serving gateway and the MME. Finally, a new instance of the S11 interface allows the MME to configure the S1 tunnelling functions inside the donor eNB, by treating it in the same way as a serving gateway.

There are a couple of restrictions on the use of relaying in Release 10. Relay nodes are assumed to be stationary, so a relay node cannot be handed over from one donor eNB to another. In addition, multi-hop relaying is not supported, so that one relay node cannot control another relay node. However, there is no impact on the mobile, which is completely unaware that it is being controlled by a relay. This implies that Release 8 mobiles support relaying in just the same way as Release 10 mobiles.

18.4.3 Enhancements to the Air Interface

If the Uu and Un interfaces use the same carrier frequency, then some enhancements are required to Un so that the resources of the air interface can be shared. The physical layer enhancements are ring-fenced in a single specification [26], while some extra RRC signalling messages are required as well [27]. There is no impact on the Uu interface, to ensure backwards compatibility with Release 8 mobiles, or on the layer 2 protocols.

The air interface resources are shared using time division multiplexing, with individual subframes allocated to either Un or Uu. This is implemented in two stages. Firstly, the donor eNB tells the relay node about the allocation using an RRC *RN Reconfiguration* message. Secondly, the relay node configures the Un subframes as MBSFN subframes on Uu, but does not transmit any downlink MBSFN data in them and does not schedule any data transmissions on the uplink. This is a little ugly, but is backwards compatible with mobiles that only support Release 8.

Unfortunately, the start of an MBSFN subframe is used by PDCCH transmissions on the Uu interface, typically scheduling grants for uplink transmissions that will occur a few subframes later. This prevents the use of the PDCCH on the Un interface. Instead, the specification introduces the *relay physical downlink control channel* (R-PDCCH), which takes the place of the PDCCH on Un. R-PDCCH transmissions look much the same as normal PDCCH transmissions, but occur in reserved resource element groups in the part of the subframe that is normally used by data. Transmissions in the first slot of a subframe are used for downlink scheduling commands, while transmissions in the second slot are used for uplink scheduling grants.

The Un interface does not use the physical hybrid ARQ indicator channel: instead, the donor eNB acknowledges the relay node's uplink transmissions implicitly, using scheduling grants on the R-PDCCH. In the absence of the PDCCH and PHICH, the physical control format indicator channel is not required either.

18.5 Release 11 and Beyond

18.5.1 Coordinated Multipoint Transmission and Reception

One of the issues being addressed beyond Release 10 is *coordinated multipoint* (CoMP) transmission and reception [28, 29]. This is a wide-ranging term, which refers to any type of coordination between the radio communications that are taking place in nearby cells. Its aim is to increase the data rate at the cell edge and the overall throughput of the cell.

There are two main varieties, which we will describe from the viewpoint of the downlink. (Similar issues apply on the uplink as well.) In *coordinated scheduling and beamforming* (CS/CB), a mobile receives data from one cell at a time, its serving cell. However, the serving cell can coordinate its scheduling and beamforming processes with those of cells nearby, so as to minimize the inter-cell interference. For example, a cell can configure its beamforming pattern on the sub-carriers that a mobile in a neighbouring cell is using, so as to place that mobile in a null.

In *joint processing* (JP), a mobile receives data from multiple cells. These cells can be controlled by one base station, which is not too hard to implement. Alternatively, the cells can be controlled by multiple base stations, which offers better performance but makes issues such as backhaul and synchronization far harder.

The cells used for joint processing can transmit the same data stream as each other, in which case they are operating as diversity transmitters. (The same technique is used for soft handover in UMTS.) Alternatively, they can transmit different data streams, in an implementation of spatial multiplexing that is known as *cooperative MIMO* (Figure 18.4). This has some similarities with multiple user MIMO, but instead of separating the mobile antennas onto two different devices, we separate the network's antennas onto two different cells.

18.5.2 Enhanced Carrier Aggregation

Carrier aggregation will also be enhanced in forthcoming releases of LTE. There are three main aspects to this. The specifications will be enhanced to support aggregation using more component carriers and resource blocks, so as to increase the mobile's peak data rate. They will also be enhanced to support aggregation in more FDD and TDD bands.

A final aspect will allow the mobile to use multiple values of the uplink timing advance, one for each component carrier. This is helpful when carrier aggregation is used in

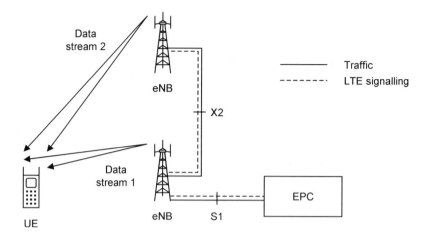

Figure 18.4 Cooperative MIMO on the LTE-Advanced downlink.

conjunction with relaying, as it allows the mobile to send one uplink component carrier to a donor eNB and another with a different timing advance to a relay node.

References

1. 3GPP TR 36.912 (March 2011) *Feasibility Study for Further Advancements for E-UTRA (LTE-Advanced)*, Release 10.
2. 3rd Generation Partnership Project (2011) *FTP directory*. Available at: ftp://ftp.3gpp.org/Information/WORK_PLAN/Description_Releases/ (accessed 12 December, 2011).
3. 4G Americas (February 2011) *4G Mobile Broadband Evolution: 3GPP Release 10 and Beyond*.
4. 3GPP TS 36.300 (October 2011) *Evolved Universal Terrestrial Radio Access (E-UTRA) and Evolved Universal Terrestrial Radio Access Network (E-UTRAN); Overall Description; Stage 2*, Release 10, sections 5.5, 6.4, 7.5.
5. 3GPP TS 36.101 (October 2011) *User Equipment (UE) Radio Transmission and Reception*, Release 10, sections 5.5A, 5.6A.
6. 3GPP TS 36.306 (October 2011) *User Equipment (UE) Radio Access Capabilities*, Release 10, sections 4.1, 4.3.5.2.
7. 3GPP TS 36.331 (October 2011) *Radio Resource Control (RRC); Protocol Specification*, Release 10, section 6.3.6 (*UE-EUTRA-Capability*).
8. 3GPP TS 36.212 (September 2011) *Multiplexing and Channel Coding*, Release 10, section 5.3.3.1.
9. 3GPP TS 36.213 (September 2011) *Physical Layer Procedures*, Release 10, sections 7.1, 8.0, 9.1.1.
10. 3GPP TS 36.306 (October 2011) *User Equipment (UE) Radio Access Capabilities*, Release 10, sections 4.3.4.12, 4.3.4.13, 4.3.4.14.
11. 3GPP TS 36.211 (September 2011) *Physical Channels and Modulation*, Release 10, section 5.4.2A.
12. 3GPP TS 36.212 (September 2011) *Multiplexing and Channel Coding*, Release 10, sections 5.2.2.6, 5.2.3.1.
13. 3GPP TS 36.213 (September 2011) *Physical Layer Procedures*, Release 10, sections 7.3, 10.
14. 3GPP TS 36.321 (October 2011) *Medium Access Control (MAC) Protocol Specification*, Release 10, sections 5.13, 6.1.3.8.
15. 3GPP TS 36.211 (September 2011) *Physical Channels and Modulation*, Release 10, sections 6.10.3, 6.10.5.
16. 3GPP TS 36.211 (September 2011) *Physical Channels and Modulation*, Release 10, section 6.3.4.2.3, 6.3.4.4.
17. 3GPP TS 36.212 (September 2011) *Multiplexing and Channel Coding*, Release 10, section 5.3.3.1.5C.
18. 3GPP TS 36.213 (September 2011) *Physical Layer Procedures*, Release 10, section 7.1.

19. 3GPP TS 36.306 (October 2011) *User Equipment (UE) Radio Access Capabilities*, Release 10, section 4.3.4.6.
20. 3GPP TS 36.213 (September 2011) *Physical Layer Procedures*, Release 10, section 8.0.
21. 3GPP TS 36.212 (September 2011) *Multiplexing and Channel Coding*, Release 10, section 5.3.3.1.8.
22. 3GPP TS 36.211 (September 2011) *Physical Channels and Modulation*, Release 10, sections 5.2.1, 5.3.2A, 5.3.3A.
23. 3GPP TS 36.106 (April 2011) *Evolved Universal Terrestrial Radio Access (E-UTRA); FDD Repeater Radio Transmission and Reception*, Release 10.
24. 3GPP TS 23.401 (September 2011) *General Packet Radio Service (GPRS) Enhancements for Evolved Universal Terrestrial Radio Access Network (E-UTRAN) Access*, Release 10, sections 4.3.20, 4.4.10.
25. 3GPP TS 36.300 (October 2011) *Evolved Universal Terrestrial Radio Access (E-UTRA) and Evolved Universal Terrestrial Radio Access Network (E-UTRAN); Overall Description; Stage 2*, Release 10, section 4.7.
26. 3GPP TS 36.216 (September 2011) *Evolved Universal Terrestrial Radio Access (E-UTRA); Physical Layer for Relaying Operation*, Release 10.
27. 3GPP TS 36.331 (October 2011) *Radio Resource Control (RRC); Protocol Specification*, Release 10, section 5.9.
28. 3GPP TR 36.814 (March 2011) *Evolved Universal Terrestrial Radio Access (E-UTRA); Further Advancements for E-UTRA Physical Layer Aspects*, Release 9, section 8.1.
29. 3GPP TR 36.819 (September 2011) *Coordinated Multi-Point Operation for LTE Physical Layer Aspects*, Release 11.

19
Self Optimizing Networks

In common with other mobile telecommunication technologies, an LTE network is controlled by a network management system. This has a wide range of functions: for example, it sets the parameters that the network elements are using, manages their software, and detects and corrects any faults in their operation. Using such a management system, an operator can remotely configure and optimize every base station in the radio access network and every component of the core network. However, the process requires manual intervention, which can make it time-consuming, expensive and prone to error. To deal with this issue, 3GPP has gradually introduced a technique known as *self optimizing* or *self organizing networks* (SON) into LTE.

In this chapter, we will cover the main self optimization features that have been added to LTE in each of its releases. Release 8 already includes a few capabilities, including self configuration of a base station, automatic establishment of communications with its neighbours, interference coordination and load balancing. These features were enhanced in Release 9, along with the new capabilities of mobility robustness optimization, random access channel optimization and energy saving. Release 10 led to further enhancements and another new capability, drive test minimization.

Self optimizing networks are summarized in TR 36.902 [1] and TS 36.300 [2]. Their main impact is on the radio access network's signalling procedures, notably the ones on the X2 interface [3]. For some more detailed accounts of the use of self optimizing networks in LTE, see References [4, 5, 6].

19.1 Self Optimizing Networks in Release 8

19.1.1 *Self Configuration of an eNB*

LTE has been designed so that a network operator can set up a new base station with minimal knowledge of the outside world, which might include the domain name of the network management system, and the domain names of its MMEs and serving gateways. The base station can acquire the other information it needs by a process of self configuration [7]. During this process, the base station contacts the management system and downloads the software it will require for its operation. It also downloads a set of configuration parameters [8], such as a tracking area code, a list of PLMN identities, and the global cell identity and maximum transmit power of each cell.

In the configuration parameters, the management system can explicitly assign a physical cell identity to each of the base station's cells. However, this places an unnecessary burden on the network planner, as every cell must have a different identity from any other cells that are nearby. It also causes difficulties in networks that contain home base stations, which can be sited without any knowledge of their neighbours at all.

As an alternative, the management system can simply give the base station a short list of allowed physical cell identities. From this short list, the base station rejects any identities that mobiles list in measurement reports and any that nearby base stations list during the X2 setup procedure described below. If the base station has a suitable downlink receiver, it can also reject any physical cell identities that it discovers by itself. The base station then chooses a physical cell identity at random from the ones that remain.

As part of the self configuration process, the base station also runs a procedure known as *S1 setup* [9], to establish communications with each of the MMEs that it is connected to. In this procedure, the base station tells the MME about its tracking area code and PLMN identities, as well as any closed subscriber groups that it belongs to. The MME replies with a message that indicates its globally unique identity and can now communicate with the base station over the S1 interface.

19.1.2 Automatic Neighbour Relations

During the configuration process described above, there is no need for a base station to find out anything about its neighbouring cells and no need for it to set up a neighbour list. This removes a large burden from the network operator and a large potential source of error. Instead, a mobile can identify a neighbouring cell by itself and can tell the base station about it later on using the RRC measurement reports that we covered in Chapter 14. The base station can then establish communications with its neighbour using the *automatic neighbour relation* procedure shown in Figure 19.1 [10].

The procedure is triggered when the base station receives a measurement report containing a physical cell identity that it was not previously aware of (1). The base station cannot contact the new cell right away, so it sends the mobile a second measurement configuration to ask for more information (2). In response, the mobile reads the neighbouring cell's system information and returns its global cell identity, tracking area code and PLMN list in a second measurement report (3). The base station now has enough information to initiate an S1 based handover to the new cell.

To support X2 based handovers, the base station sends the global cell ID to the MME and asks it to return an IP address that the neighbouring base station is using for communications over X2 (4). The MME is already communicating with the neighbouring base station over S1, so it can send the request onwards (5) and can return the neighbour's reply (6). The two base stations can now establish communications across the X2 interface (7), using a procedure known as *X2 setup* [11]. During this procedure, the base stations exchange information about all the cells they are controlling, including their global cell identities, physical cell identities and carrier frequencies. This last field might include frequencies that the original base station was not previously aware of, which it can use to populate the list of neighbouring frequencies in SIB 5.

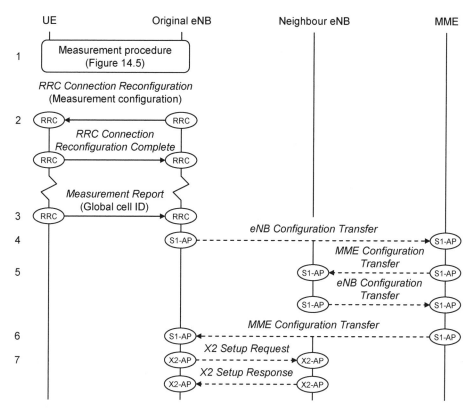

Figure 19.1 Automatic neighbour relation procedure.

A base station can also use steps 1 to 3 of this procedure to learn about neighbouring cells that are using UMTS, GSM or cdma2000. In the case of UMTS, for example, the mobile returns the neighbour's global cell identity, location area code, routing area code and PLMN identity list. This is enough information for the base station to initiate an inter-system handover over S1 and also allows the base station to populate the neighbour lists that we noted in Chapter 15.

19.1.3 Interference Coordination

The X2-AP *Load Indication* procedure [12] helps a network to minimize the interference between neighbouring base stations and to implement the fractional frequency re-use schemes that we introduced in Chapter 4. To use the procedure, a base station sends an X2-AP *Load Information* message to one of its neighbours. In the message, it can include three information elements for each cell that it is controlling. The first describes the transmitted power in every downlink resource block. The neighbour can use this information in its scheduling procedure, by avoiding downlink transmissions to distant mobiles in resource blocks that are subject to high levels of downlink interference.

The second information element describes the interference that the base station is receiving in every uplink resource block. The neighbour can use this in a similar way, so that it does not schedule uplink transmissions from distant mobiles in resource blocks that are subject to high uplink interference. The third is a list of uplink resource blocks in which the base station intends to schedule distant mobiles. Here, the second base station is expected to avoid scheduling uplink transmissions from distant mobiles in those resource blocks, so that it does not return high levels of uplink interference to the first.

19.1.4 Mobility Load Balancing

Figure 19.2 shows another procedure, known as *mobility load balancing* or *resource status reporting* [13]. Using this procedure, nearby base stations can cooperate to even out the load in the radio access network and to maximize the total capacity of the system.

Using an X2-AP *Resource Status Request* (1), a Release 8 base station can ask one of its neighbours to report three items of information. The first is the percentage of resource blocks that the neighbour is using in each of its cells, for both GBR and non GBR traffic. The second is the load on the S1 interface, while the third is the hardware load. The neighbour returns an acknowledgement (2) and then reports each item periodically for both the uplink and downlink, using an X2-AP *Resource Status Update* (3). As a result of this information, a congested base station can hand over a mobile to a neighbouring cell that has enough spare capacity and can even out the load in the radio access network.

Release 9 adds two enhancements. Firstly, the neighbour reports a fourth field in its Resource Status Update, the *composite available capacity group*, which indicates the capacity that it has available for load balancing purposes on the uplink and downlink. The original base station can use this information to assist its handover decision.

Secondly, there is a risk after such a handover that the new base station will hand the mobile straight back to the old one. To prevent this from happening, Release 9 introduces another X2 procedure, known as *mobility settings change* [14]. Using this procedure, a base station can ask a neighbour to adjust the thresholds that it is using for measurement reporting, by means of the cell specific offsets that we introduced in Chapter 14. After the adjustment, the mobile should stay in the target cell, instead of being handed back.

There is a further enhancement in Release 10. Using an S1 procedure known as a *direct information transfer* [15], a base station can initiate the exchange of *radio access*

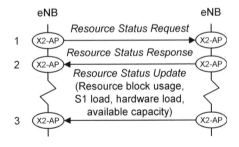

Figure 19.2 Resource status reporting procedure. Reproduced by permission of ETSI.

network information management (RIM) information with a UMTS or GSM neighbour. The information includes the composite available capacity group in the case of an LTE cell and similar information known as the *cell load information group* in the case of the other technologies. In turn, this information can trigger a load balancing handover to a UMTS or GSM neighbour.

19.2 New Features in Release 9

19.2.1 Mobility Robustness Optimization

Mobility robustness optimization [16] is a self optimization technique that first appears in Release 9. Using this technique, a base station can gather information about any problems that have arisen due to the use of unsuitable measurement reporting thresholds. It can then use the information to adjust the thresholds it is using and to correct the problem.

There are three main causes of trouble, the first of which is shown in Figure 19.3. Here, the base station has started a handover to a new cell (1) but it has done this too late, because its measurement reporting thresholds have been poorly set. Alternatively, it may not have started the handover at all. Before any handover is executed, the mobile's received signal power falls below a threshold and its radio link fails (2). In response, the mobile runs the cell selection procedure and discovers the cell that it should have been handed to. It contacts the new cell using the random access procedure (3) and a procedure known as *RRC connection reestablishment* (4, 5, 6), in which it identifies itself using the old cell's physical cell ID and its old C-RNTI. In step 6, the mobile can also indicate that it has measurements from immediately before the radio link failure of the power

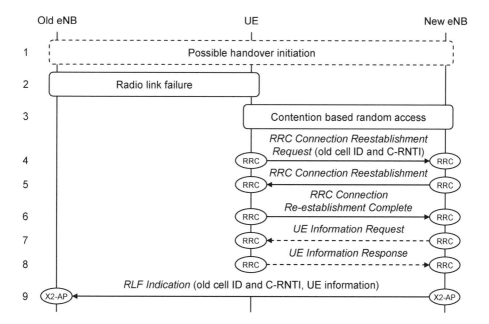

Figure 19.3 Mobility robustness optimization, triggered by a handover that was too late.

received from the old cell and its neighbours. If it does, then the base station retrieves this information using an RRC *UE Information* procedure (7, 8).

The new base station can now tell the old base station about the problem, using a Release 9 message known as an X2-AP *Radio Link Failure* (RLF) *Indication* (9). After a series of such reports, the old base station can take action by adjusting its measurement reporting thresholds, using proprietary optimization software that is outside the scope of the 3GPP specifications.

The next problem is in Figure 19.4. Here, the base station has carried out a handover too early (1), perhaps in an isolated area where the mobile is briefly receiving line-of-sight coverage from the new cell. The mobile completes the handover, but its radio link soon fails (2). On running the cell selection procedure, the mobile rediscovers the old cell, re-establishes an RRC connection (3), and identifies itself using its new physical cell ID and C-RNTI. The old base station notifies the new one, as before (4). However, the new base station notices that it had just received the mobile in a handover from the old one, so it tells the old base station using another Release 9 message, an X2-AP *Handover Report* (5). Once again, the old base station can use the information to adjust its measurement thresholds.

The final problem is in Figure 19.5. Here, the base station has handed the mobile over to the wrong cell, perhaps due to an incorrect cell specific measurement offset (1). The mobile's radio link fails as before (2) and it re-establishes an RRC connection with a third cell, the one it should have been handed to in the first place (3). In response, the third base station sends a radio link failure indication to the second (4), which notifies the original base station using a handover report as before (5).

Mobility robustness optimization is enhanced in Release 10, to let the system detect unnecessary handovers to another radio access technology. After a handover to UMTS or GSM, the new radio access network can ask the mobile to continue measuring the signal power that it is receiving from nearby LTE cells. If the signal power is sufficiently high, then the network can tell the LTE base station that it triggered the handover unnecessarily, using the S1-AP direct information transfer procedure that we saw earlier. As before, the base station can use a series of such reports to adjust its measurement reporting thresholds.

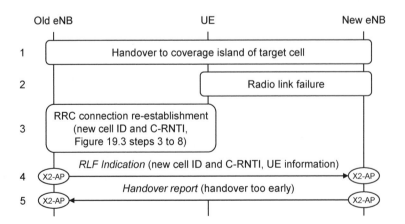

Figure 19.4 Mobility robustness optimization, triggered by a handover that was too early.

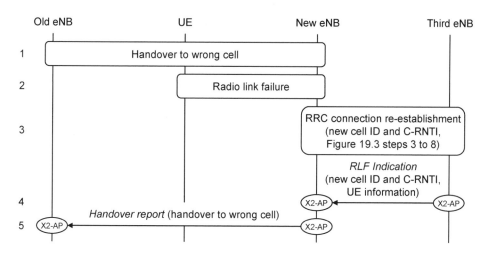

Figure 19.5 Mobility robustness optimization, triggered by a handover to the wrong cell.

19.2.2 Random Access Channel Optimization

A base station can gather two types of information to help it optimize the random access channel. Firstly, the base station can use the RRC UE information procedure to retrieve information about a mobile's last successful random access attempt. The information includes the number of preambles that the mobile sent before receiving a reply and an indication of whether the contention resolution procedure failed at any stage. Using this information, the base station can adjust the random access channel's power settings and resource block allocations, so as to minimize the load that the channel makes on the air interface.

Secondly, neighbouring base stations can exchange information about the parameters that they are using for the random access channel, during the X2 setup procedure that we saw earlier. The information includes the PRACH frequency offset and PRACH configuration index, which determine the resource blocks that the channel is using, and the root sequence index, which determines the cell's choice of random access preambles. Using this information, the base stations can minimize the interference between random access transmissions in nearby cells, by allocating them different sets of resource blocks and different preambles.

19.2.3 Energy Saving

The aim of the final procedure [17] is to save energy by switching off cells that are not being used. A typical situation is the use of picocells in a shopping centre, in which a cell can be switched off outside shopping hours if it only contributes to the network's capacity, but not to its coverage.

If a base station supports this feature, then it can decide to switch the cell off after a long period of low load. To do this, it hands any remaining mobiles over to cells that have overlapping coverage, tells them about the change using an X2-AP *eNB Configuration*

Update and switches the cell off. The base station itself remains switched on, so, at a later time, a neighbour can ask the base station to switch the cell on again using an X2-AP *Cell Activation Request*.

19.3 Drive Test Minimization in Release 10

Network operators have traditionally assessed the coverage of a radio access network by transporting measurement devices around its intended coverage area, in a technique known as drive testing. As well as being time-consuming and expensive, this technique provides coverage data that are limited to the route of the drive test and supplies little or no information about coverage indoors. Network operators do, however, have another ready supply of measurement devices in the form of the users' mobiles. In a technique known as *minimization of drive tests* (MDT) [18–20], an operator can ask its mobiles to return measurements that supplement or even replace the ones obtained from traditional drive testing.

As part of the customer care process, the operator is obliged to obtain the users' consent for using their mobiles in drive test minimization. The network stores the relevant information in the home subscriber server and checks it before measurement activation.

If the user does consent, then two measurement modes are available: immediate measurements for mobiles in RRC_CONNECTED state and logged measurements for mobiles in RRC_IDLE. Immediate measurements follow the same reporting procedure that we saw in Chapter 14. The mobile measures the downlink RSRP or RSRQ and reports these quantities to the base station along with any location data that it has available. The base station can then return the information to the management system, using the existing network management procedures for trace reporting.

A base station can also send an RRC message known as *Logged Measurement Configuration* to an active mobile, to configure it for logged measurements once it enters RRC_IDLE. In idle mode, the mobile makes its measurements with a period that is a multiple of the discontinuous reception cycle. It then stores the information in a log, along with time stamps and any location data that it has available. When the mobile next establishes an RRC connection, it can signal the availability of its measurement log using a field in the message RRC Connection Setup Complete. The base station can then retrieve the logged measurements from the mobile using the RRC UE Information procedure and can forward them to the management system as before.

References

1. 3GPP TR 36.902 (April 2011) *Evolved Universal Terrestrial Radio Access Network (E-UTRAN); Self-Configuring and Self-Optimizing Network (SON) Use Cases and Solutions*, Release 9.
2. 3GPP TS 36.300 (October 2011) *Evolved Universal Terrestrial Radio Access (E-UTRA) and Evolved Universal Terrestrial Radio Access Network (E-UTRAN); Overall Description; Stage 2*, Release 10, section 22.
3. 3GPP TS 36.423 (September 2011) *Evolved Universal Terrestrial Radio Access Network (E-UTRAN); X2 Application Protocol (X2AP)*, Release 10.
4. Hämäläinen, S., Sanneck, H. and Sartori, C. (2011) *LTE Self-Organizing Networks: Network Management Automation for Operational Efficiency*, John Wiley & Sons, Ltd, Chichester.
5. Ramiro, J. and Hamied, K. (2011) *Self-Organizing Networks: Self Planning, Self Optimization and Self Healing for GSM, UMTS and LTE*, John Wiley & Sons, Ltd, Chichester.
6. 4G Americas (July 2011) *Self-Optimizing Networks: The Benefits of SON in LTE*.

7. 3GPP TS 32.501 (April 2011) *Telecommunication Management; Self-Configuration of Network Elements; Concepts and Requirements*, Release 10, section 6.4.2.
8. 3GPP TS 32.762 (September 2011) *Telecommunication Management; Evolved Universal Terrestrial Radio Access Network (E-UTRAN) Network Resource Model (NRM) Integration Reference Point (IRP); Information Service (IS)*, Release 10, section 6.
9. 3GPP TS 36.413 (September 2011) *Evolved Universal Terrestrial Radio Access Network (E-UTRAN); S1 Application Protocol (S1AP)'*, Release 10, section 8.7.3.
10. 3GPP TS 36.300 (October 2011) *Evolved Universal Terrestrial Radio Access (E-UTRA) and Evolved Universal Terrestrial Radio Access Network (E-UTRAN); Overall Description; Stage 2*, Release 10, section 22.3.3.
11. 3GPP TS 36.423 (September 2011) *Evolved Universal Terrestrial Radio Access Network (E-UTRAN); X2 Application Protocol (X2AP)*, Release 10, section 8.3.3.
12. 3GPP TS 36.423 (September 2011) *Evolved Universal Terrestrial Radio Access Network (E-UTRAN); X2 Application Protocol (X2AP)*, Release 10, section 8.3.1.
13. 3GPP TS 36.423 (September 2011) *Evolved Universal Terrestrial Radio Access Network (E-UTRAN); X2 Application Protocol (X2AP)*, Release 10, section 8.3.6, 8.3.7.
14. 3GPP TS 36.423 (September 2011) *Evolved Universal Terrestrial Radio Access Network (E-UTRAN); X2 Application Protocol (X2AP)*, Release 10, sections 8.3.8.
15. 3GPP TS 36.413 (September 2011) *Evolved Universal Terrestrial Radio Access Network (E-UTRAN); S1 Application Protocol (S1AP)*, sections 8.13, 8.14, Release 10, annex B.
16. 3GPP TS 36.423 (September 2011) *Evolved Universal Terrestrial Radio Access Network (E-UTRAN); X2 Application Protocol (X2AP)*, Release 10, sections 8.3.9, 8.3.10.
17. 3GPP TS 36.423 (September 2011) *Evolved Universal Terrestrial Radio Access Network (E-UTRAN); X2 Application Protocol (X2AP)*, Release 10, sections 8.3.5, 8.3.11.
18. 3GPP TS 37.320 (October 2011) *Universal Terrestrial Radio Access (UTRA) and Evolved Universal Terrestrial Radio Access (E-UTRA); Radio Measurement Collection for Minimization of Drive Tests (MDT); Overall Description; Stage 2*, Release 10.
19. 3GPP TS 36.331 (October 2011) *Radio Resource Control (RRC); Protocol Specification*, Release 10, sections 5.6.6, 5.6.7, 5.6.8.
20. 3GPP TS 32.422 (September 2011) *Telecommunication Management; Subscriber and Equipment Trace; Trace Control and Configuration Management*, Release 10, sections 4.2.8, 6.

20

Performance of LTE and LTE-Advanced

In a mobile telecommunication system, two main factors limit a cell's performance: coverage and capacity. Coverage is more important in rural areas, since a mobile far from the base station may not receive a signal that is strong enough for it to recover the transmitted information. Capacity is more important in urban areas, since every cell is limited by a maximum data rate. We consider these issues in this final chapter, by reviewing the issues that determine the coverage of a cell in LTE, and by examining the peak data rate of a mobile and the typical capacity of a cell.

20.1 Coverage Estimation

In Chapter 3, we defined the *propagation loss* or *path loss*, PL, as the ratio of the transmitted signal power P_T to the received signal power P_R:

$$\text{PL} = \frac{P_T}{P_R} \qquad (20.1)$$

In a *link budget*, we estimate the largest value of P_T that the transmitter can send and the smallest value of P_R at which the receiver can recover the original information. We can then use the equation above to estimate the greatest propagation loss that the system can handle. There are some example link budgets for LTE and discussions of the various parameters in [1] and [2], while other sources of information include [3–6].

Propagation models relate the propagation loss to the distance between the transmitter and the receiver. Several propagation models exist and vary greatly in their complexity. A simple and frequently used example is the Okumura-Hata model, which predicts the coverage of macrocells in the frequency range 150 to 1500 MHz [7]. The model was later extended to the range 1500 to 2000 MHz, as part of a project in the *European Cooperation in Science and Technology* (COST) framework known as COST 231 [8]. As another example, the *Wireless World Initiative New Radio* (WINNER) consortium has developed propagation models for several different propagation scenarios in the frequency range 2

An Introduction to LTE: LTE, LTE-Advanced, SAE and 4G Mobile Communications, First Edition. Christopher Cox.
© 2012 John Wiley & Sons, Ltd. Published 2012 by John Wiley & Sons, Ltd.

to 6 GHz [9]. By combining the link budget with a suitable propagation model, we can estimate the maximum distance between the base station and the mobile, which is the maximum size of the cell.

We do, however, need to issue a warning about the use of propagation models. The parameters are estimated by fitting the predictions to a large number of measurements, but, in practice, the actual propagation loss can vary greatly from one environment to another. To deal with this, it is important for a network operator to measure the actual radio propagation in the region of interest by means of drive tests and to adjust the parameters in the chosen model using the results. If we use a propagation model without making these adjustments, then we get a rough estimate of the coverage of a cell, but no more.

With this warning issued, we can run through the main results from coverage estimation in LTE. The first result is that coverage depends on data rate. A high data rate requires a fast modulation scheme, a high coding rate and possibly the use of spatial multiplexing, all of which increase the receiver's susceptibility to noise and interference. This implies that high data rates can only be achieved if the mobile is close to the base station, where the received signal power is high and the interference from neighbouring transmitters is low. We will return to this point when we discuss the peak data rate of LTE later on.

The second result is that coverage depends on carrier frequency: low carrier frequencies such as 800 MHz are associated with a high coverage, while at high carrier frequencies such as 2600 MHz, the coverage is less. The main reason is that the receive antenna has an effective collecting area proportional to λ^2, where λ is the wavelength of the incoming radio waves. As the carrier frequency increases, so the wavelength falls and the power collected by the receive antenna becomes progressively less. Because of this, operators typically prefer low carrier frequencies for wide area networks, while reserving high carrier frequencies for boosting network capacity in urban areas.

In addition, indoor coverage can be badly degraded by penetration losses through the walls of buildings. If the base station is outdoors but the mobile is indoors, then penetration losses typically reduce the received signal power by 10 to 20 decibels (a factor of 10 to 100), which can greatly reduce the indoor coverage area. This is one of the motivations behind the progressive introduction of femtocells.

If we keep these other factors equal, by comparing systems with the same data rate, carrier frequency and penetration loss, then we find that the maximum ranges of LTE and 3G systems are actually very similar. This implies that network operators can often re-use their existing 3G cell plans for LTE, with relatively few adjustments to the operating parameters. The absolute cell sizes typically range from 1 km or less for indoor coverage at high carrier frequencies, up to 10 km or more for low frequency outdoor coverage. Wide variations are possible, however.

20.2 Peak Data Rates of LTE and LTE-Advanced

20.2.1 Increase of the Peak Data Rate

Figure 20.1 shows how the peak data rate of LTE has increased since its introduction in Release 8 and compares it with the peak data rate of WCDMA from Release 99. The data are taken from the most powerful UE capabilities available in FDD mode at each release [10–12]. The vertical axis is logarithmic, which is consistent with the large increases in peak data rate that have been achieved since the introduction of 3G systems.

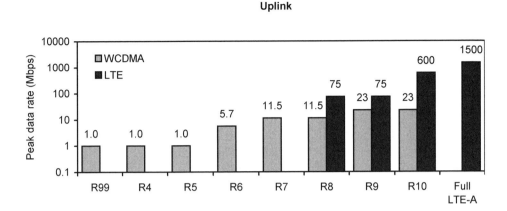

Figure 20.1 Evolution of the peak data rates of WCDMA and LTE in FDD mode.

In Release 99, WCDMA had a peak data rate of 2 Mbps on the downlink and 1 Mbps on the uplink. The introduction of high speed downlink packet access in Release 5 increased the peak downlink data rate to 14.4 Mbps, by the use of a faster coding rate and a new modulation scheme, 16-QAM. There was a similar increase for the uplink in Release 6, through the introduction of high speed uplink packet access. Later releases have increased the peak data rate further, through the introduction of 64-QAM and spatial multiplexing, and the use of multiple carriers.

The peak data rate in LTE Release 8 is 300 Mbps in the downlink and 75 Mbps in the uplink. These figures are easily understood. In the uplink, the mobile's greatest likely allocation is 96 resource blocks, because we need to reserve some resource blocks for the PUCCH and because we need an allocation with prime factors of 2, 3 or 5 only. Each resource block lasts for 0.5 ms and carries 72 PUSCH symbols (Figure 8.8), so it supports a symbol rate of 144 ksps. With a modulation scheme of 64-QAM, the resulting bit rate on the PUSCH is 82.9 Mbps. We can therefore support an information rate of 75 Mbps using a coding rate of 0.90, which is a reasonable maximum figure.

In the downlink, the appropriate symbol rate per resource block is 136 ksps, assuming the use of four transmit antennas (Figure 7.2) and a control region that lasts for one symbol. Using four layers, 100 resource blocks and a modulation scheme of 64-QAM, the PUSCH bit rate is 326.4 Mbps, so we can support an information rate of 300 Mbps using a coding rate of 0.92. This figure is almost identical to the coding rate used for a channel quality indicator of 15 (Table 8.4).

The peak data rates in release 10 are about 1200 Mbps on the downlink and 600 Mbps on the uplink. These arise through the use of two component carriers, eight layers on the downlink and four layers on the uplink, and meet the peak data rate requirements of IMT-Advanced. The peak data rates will eventually increase to 3000 and 1500 Mbps respectively, through the use of five component carriers.

20.2.2 Limitations on the Peak Data Rate

As we noted in Chapter 1, we can only reach the peak data rates shown above in special circumstances. There are five main criteria. Firstly, the cell must be transmitting and receiving in its maximum bandwidth of 20 MHz. This is likely to be an unusually large allocation, at least in the early days of LTE, with a value of 10 or 5 MHz being more common. In these lower bandwidths, the peak data rate will be a factor of 2 or 4 less.

Secondly, the mobile must have the most powerful UE capabilities that are available at each release. In Release 8, for example, category 5 mobiles are the only ones that support four layer spatial multiplexing on the downlink, or the use of 64-QAM on the uplink. If we switch to a category 2 mobile, then the peak downlink data rate falls by another factor of 6 and the peak uplink data rate falls by another factor of 3.

Thirdly, the mobile should be close to the base station. If it is not, then the received signal to interference plus noise ratio may be low, and the receiver may be unable to handle the fast modulation schemes and coding rates that are required for a high data rate.

Fourthly, the cell should be well isolated from other nearby cells. This condition can often be achieved in femtocells and picocells, which are usually indoors and are isolated by the surrounding walls. A similar but weaker result applies to microcells, which are partially isolated from each other by the intervening buildings. In macrocells there is little such isolation, so the receiver may pick up significant interference from nearby cells. This reduces the SINR, and prevents the receiver from handling a fast modulation scheme or a high coding rate.

The final condition is that the mobile must be the only active mobile in the cell. If it is not, then the cell's capacity will be shared amongst all its mobiles, resulting in a large drop in the peak data rate that is available to each one.

20.3 Typical Data Rates of LTE and LTE-Advanced

20.3.1 Total Cell Capacity

The typical data rates of LTE and LTE-Advanced are more important than the peak values, because they give a realistic measure of how a network will actually perform and because they can be used to estimate the required capacities of the S1-U backhauls. They are normally estimated using simulations. To illustrate them, we will look at some simulations that have been carried out by 3GPP during system design [13, 14].

Performance of LTE and LTE-Advanced   305

The simulations we will describe are known as 3GPP case 1 and case 3. They are both carried out for macrocell geometries, in which the receiver can pick up significant interference from nearby cells. The coverage requirements are demanding, as the base station is outdoors but the mobile phone is indoors, with a penetration loss of 20 decibels through the walls of the building. Furthermore, the carrier frequency is 2 GHz, a higher-than-average value that is associated with worse-than-average propagation losses. To compensate for these issues, the distance between the base stations is rather low for a macrocell geometry, at 500 metres in case 1 and 1732 metres in case 3. There are three sectors per site: we will show the capacity of each sector throughout.

Figure 20.2 shows the total sector capacity in FDD mode from a number of different simulations in the uplink and downlink, which cover baseline results from WCDMA Release 6, early results from LTE, later results from LTE and early results from LTE-Advanced. The figure shows the results for a bandwidth of 10 MHz, which is the same

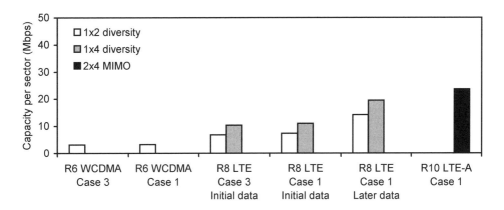

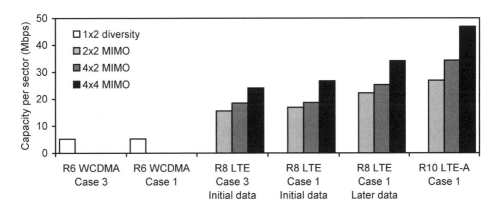

Figure 20.2 Capacity per sector of a typical LTE macrocell in a 10 MHz bandwidth.

bandwidth that was actually used for the simulations of LTE, but implies the use of two carriers per sector for the case of WCDMA. The vertical axis is linear, which is different from Figure 20.1, but is consistent with the smaller increase in capacity since the introduction of 3G systems.

There are several points to note. Firstly, the cell size makes little difference to the results, at least in the range we are considering, since there is little difference between cases 1 and 3. Instead, the data rates are limited by the interference from neighbouring cells, which depends on the amount of overlap between them. To increase the cell capacity, we would have to move to a microcell or femtocell geometry, in which the greater isolation would reduce the interference and permit a higher data rate.

Secondly, the capacity increases significantly as we move from the baseline WCDMA results to the early results for LTE. There are several reasons for this, particularly the more effective treatment of fading and inter-symbol interference through the use of OFDMA and SC-FDMA, and the introduction of higher order diversity on the uplink and spatial multiplexing on the downlink.

Thirdly, there is another increase in capacity as we move from the initial simulations of LTE to the later ones. Possible reasons include improvements in the receiver software, which allow the receiver to support a high data rate at lower signal to interference ratios than before. There is another capacity increase as we move from LTE to LTE-Advanced. This is primarily due to the introduction of single user MIMO in the uplink and a proper implementation of multiple user MIMO in the downlink.

Estimates of sector capacity should be treated with caution, since they are sensitive to issues such as the bandwidth, the antenna geometry and the amount of incoming interference. Roughly speaking, however, we can say that the capacity of a 10 MHz LTE macrocell in these simulations is about 25 Mbps per sector in the downlink and 15 Mbps per sector in the uplink. The introduction of LTE-Advanced increases these figures to about 35 and 20 Mbps per sector respectively. These figures are far less than the peak data rates quoted in the previous section and have to be shared amongst all the mobiles in the cell.

There is a useful analogy here with the motor industry. In Release 8 LTE, we have seen that the maximum data rate on the downlink is 300 Mbps, while the typical capacity of a macrocell is somewhere around 25 Mbps per sector. Coincidentally, the top speed of a Ferrari is around 300 kph (180 mph), while the average traffic speed across Greater London is somewhere around 25 kph (15 mph) [15]. So, at the risk of stretching the analogy too far, we are about as likely to achieve 300 Mbps in a Release 8 macrocell, as we are to drive a Ferrari at 300 kph on the streets of London. A similar situation applies in Release 10, except that we can exchange the Ferrari for the Thrust SSC [16].

20.3.2 Data Rate at the Cell Edge

Simulations have also been carried out of the data rate at the cell edge. 3GPP's definition of this quantity is the data rate exceeded by 95% of mobiles, from a number of simulations in which 10 mobiles at a time are dropped at random within the sector [17]. The results are shown in Figure 20.3, once again for a 10 MHz bandwidth. With 10 mobiles per

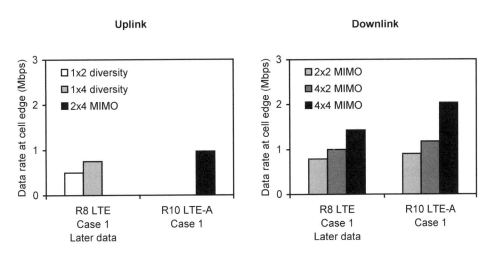

Figure 20.3 Data rate at the edge of a typical LTE macrocell, with 10 users per sector in a 10 MHz bandwidth.

sector, the results suggest that an LTE mobile can expect a data rate exceeding about 1 Mbps in the downlink and 0.6 Mbps in the uplink, 95% of the time.

Now, the earlier results from Figure 20.2 suggest that, with 10 mobiles per sector, an LTE mobile can expect an average data rate of about 2.5 Mbps in the downlink and 1.5 Mbps in the uplink. By combining the two sets of results, we can infer that the data rate at the cell edge is roughly 40% of the average.

References

1. Penttinen, J. (2011) *The LTE/SAE Deployment Handbook*, John Wiley & Sons, Ltd, Chichester.
2. Holma, H. and Toskala, A. (2011) *LTE for UMTS: Evolution to LTE-Advanced*, Ch. 10, John Wiley & Sons, Ltd, Chichester.
3. Sesia, S., Toufik, I. and Baker, M. (2009) *LTE – The UMTS Long Term Evolution*, Ch 18, John Wiley & Sons, Ltd, Chichester.
4. Salo, J., Nur-Alam, M. and Chang, K. (2010) *Practical Introduction to LTE Radio Planning*. Available at: http://www.eceltd.com/lte_rf_wp_02Nov2010.pdf (accessed 12 December, 2011).
5. 3GPP TR 25.814 (October 2006) *Physical Layer Aspect for Evolved Universal Terrestrial Radio Access (UTRA)*, Release 7, annex A.
6. 3GPP TR 36.814 (March 2010) *Further Advancements for E-UTRA Physical Layer Aspects*, Release 9, annex A.
7. Hata, M. (1980) Empirical formula for propagation loss in land mobile radio services, *IEEE Transactions on Vehicular Technology*, **29**, 317–325.
8. European Cooperation in Science and Technology (1999) *Digital Mobile Radio Towards Future Generation System*, COST 231 final report. Available at: http://www.lx.it.pt/cost231/final_report.htm (accessed 12 December, 2011).
9. Wireless World Initiative New Radio (2008) *WINNER II Channel Models*, WINNER II deliverable D1.1.2, version 1.2. Available at: http://www.ist-winner.org/WINNER2-Deliverables/D1.1.2.zip (accessed 12 December, 2011).
10. 3GPP TS 25.306 (October 2011) *UE Radio Access Capabilities*, Release 10, section 5.

11. 3GPP TS 36.306 (October 2011) *User Equipment (UE) Radio Access Capabilities*, Release 10, section 4.1.
12. 3GPP TS 36.101 (October 2011) *User Equipment (UE) Radio Transmission and Reception*, Release 10, section 5.6A.
13. 3GPP TR 25.912 (April 2011) *Feasibility Study for Evolved Universal Terrestrial Radio Access (UTRA) and Universal Terrestrial Radio Access Network (UTRAN)*, Release 10, section 13.5.
14. 3GPP TR 36.814 (March 2010) *Further Advancements for E-UTRA Physical Layer Aspects*, 3rd Generation Partnership Project, Release 9, section 10.
15. Department for Transport, UK (2008) *Road Statistics 2008: Traffic, Speeds and Congestion*. Available at: http://data.gov.uk/dataset/road_statistics_-_traffic_speeds_and_congestion (accessed 12 December, 2011).
16. Coventry Transport Museum (2008) *Thrust SSC*. Available at: http://www.thrustssc.com/thrustssc.html (accessed 12 December, 2011).
17. 3GPP TR 36.913 (April 2011) *Requirements for Further Advancements for Evolved Universal Terrestrial Radio Access (E-UTRA) (LTE-Advanced)*, Release 10, section 8.1.3.

Bibliography

LTE Air Interface

Dahlman, E., Parkvall, S. and Sköld, J. (2011) *4G: LTE/LTE-Advanced for Mobile Broadband*, Academic Press.
Ghosh, A., Zhang, J., Andrews, J. G. and Muhamed, R. (2010) *Fundamentals of LTE*, Prentice Hall.
Holma, H. and Toskala, A. (2011) *LTE for UMTS: Evolution to LTE-Advanced*, John Wiley & Sons, Ltd, Chichester.
Johnson, C. (2010) *Long Term Evolution in Bullets*, Createspace.
Khan, F. (2009) *LTE for 4G Mobile Broadband: Air Interface Technologies and Performance*, Cambridge.
Rumney, M. (2012) *LTE and the Evolution to 4G Wireless: Design and Measurement Challenges*, 2nd edn, John Wiley & Sons, Ltd, Chichester.
Sesia, S., Toufik, I. and Baker, M. (2011) *LTE: The UMTS Long Term Evolution: From Theory to Practice*, 2nd edn, John Wiley & Sons, Ltd, Chichester.

Signalling and System Operation

Kreher, R. and Gaenger, K. (2010) *LTE Signaling: Troubleshooting and Optimisation*, John Wiley & Sons, Ltd, Chichester.
Olsson, M., Sultana, S., Rommer, S., Frid, L. and Mulligan, C. (2009) *SAE and the Evolved Packet Core: Driving the Mobile Broadband Revolution*, Academic Press.

White Papers

4G Americas (2011) *4G Mobile Broadband Evolution: 3GPP Release 10 and Beyond – HSPA+, SAE/LTE and LTE-Advanced* (updated annually in February).
4G Americas (2011) *Mobile Broadband Explosion: 3GPP Broadband Evolution to IMT-Advanced* (updated annually in September).

Index

Access point name (APN), 24, 181, 184–5, 210
Access stratum (AS), 29, 33, 41
Account balance management function (ABMF), 212
ACK/NACK bundling, 137
ACK/NACK multiplexing, 137, 142
Acquisition procedure, 113–14, 175, 219
Active time, 147
Adaptive multi rate (AMR) codec, 126, 168
Adaptive re-transmission, 125
Advanced encryption standard (AES), 197
Aggregate maximum bit rate (AMBR)
 per APN (APN-AMBR), 184, 202
 per UE (UE-AMBR), 184, 202
Air interface, 2, 29–30, 95–6
Alamouti's technique, 79–80
Allocation and retention priority (ARP), 202–3
Analogue processing, 47, 64, 95
Antenna ports, downlink, 107, 113, 118
 port 0–3, 107, 116
 port 4, 267
 port 5, 107, 117
 port 6, 270
 port 7–8, 273, 282
 port 9–14, 282
 port 15–22, 283
Antenna ports, uplink, 284
APN configuration, 183–4
Application function (AF), 204, 207–8, 252, 254
Application layer, 5–6

Application server (AS), 255
Attach procedure, 174, 179–87, 234, 241–3, 257
Authentication and key agreement (AKA) procedure, 182, 191, 193–5, 218, 222
Authentication centre (AuC), 4
Authentication token (AUTN), 194
Authentication vector, 193–4
Authentication, authorization and accounting (AAA) server, 239
Automatic neighbour relation procedure, 292–3
Automatic repeat request (ARQ), 57–8, 60

Bandwidth, 3, 10, 11, 304
 acquisition, 113, 117
 of cdma2000, 8
 of LTE, 13, 106–7
 of LTE-Advanced, 15, 277
 of TD-SCDMA, 7
 of WCDMA, 7
Base station, 2–4, 23
 see also Evolved Node B, Node B
Base station controller (BSC), 4
Base transceiver station (BTS), 4
Beamforming, 90–4, 108, 117, 128
 dual layer, 92–3, 108, 273, 282
Bearer, 36–40, 233–4
 dedicated, 36, 206, 208–9
 default, 36, 40, 180, 184–5, 205, 210
 EPS, 36–7, 41
 evolved radio access (E-RAB), 37

Bearer (*continued*)
  GBR, 36, 202
  non GBR, 36, 202
  radio, 37, 41, 215–18
  S1, 37, 41, 215–18
  S5/S8, 37, 41, 216
Bearer binding and event reporting function (BBERF), 204, 240
Bearer modification procedure, 210
Billing domain, 211
Binary phase shift keying (BPSK), 48, 81, 136, 143
Broadcast channel (BCH), 98
Broadcast control channel (BCCH), 97, 165
Broadcast services, 265
Broadcast/multicast service centre (BM-SC), 268
Buffer status reporting, 161

Call session control function (CSCF), 254
Call setup procedure, 255, 258–60
Camping, 173
Capacity
  cell, 11, 13, 15, 304–7
  channel, 10
Carrier aggregation (CA), 277–81, 287–8
Carrier aggregation bandwidth class, 279
Carrier frequency, 3, 42–4, 47, 302
Carrier indicator field (CIF), 280
cdma2000, 7–8, 16, 107, 120
  evolution data optimized (EV-DO), 8, 244–9
  high rate packet data (HRPD), 8, 244–9
  1xRTT, 7, 261–2
cdmaOne, 7
Cell, 3, 23, 28
  suitable, 176, 220–21
Cell broadcast centre (CBC), 25, 120
Cell broadcast service (CBS), 25
Cell identity, 28, 113, 114
Cell identity group, 114, 116
Cell reselection procedure, 42, 120, 219–22, 234–5, 243–4, 246
Cell selection procedure, 173, 176–7, 234

Channel estimation, 50–1, 68–9, 81–2, 100–1
Channel matrix, 86–7
Channel quality indicator (CQI), 99, 137–8, 139–40, 163
Channel state information (CSI), 100, 142
Charging, 202, 210–12
Charging data function (CDF), 211
Charging data record (CDR), 211
Charging event, 211
Charging gateway function (CGF), 211
Charging key, 202
Charging trigger function (CTF), 211
Ciphering, 168, 192, 196–7
Circuit switched fallback, 256–62
Circuit switching, 2, 12, 256
Closed subscriber group (CSG), 24, 120
Closed subscriber group reselection procedure, 220, 221
Closed subscriber group selection procedure, 173, 175–6
Code division multiple access (CDMA), 52, 137
Codeword, 56, 133
Coding rate, 56, 132, 138
Coherence bandwidth, 55
Coherence time, 54
Combined EPS/IMSI attach, 257
Commercial mobile alert system (CMAS), 273–4
Common control channel (CCCH), 97, 165, 177
Complex number, 48, 81
Component carrier (CC), 277–8
Control channel element (CCE), 131
Control format indicator (CFI), 100, 113, 118–19, 132
Control information, 96, 99, 100
Control plane, 28
Control region, downlink, 109, 118–19, 131–2, 136, 267–8
Cooperative MIMO, 287
Coordinated multipoint (CoMP) transmission, 287
Coordinated scheduling and beamforming (CS/CB), 287

Core network, 1, 4, 24–5
Coverage, 13, 105, 301–2
Credit request, 205, 207–8, 212
Cyclic prefix, 69–70, 103
   acquisition, 113, 116
   extended, 103, 267–8
   insertion, 69–70, 134
   normal, 70, 103
   PRACH, 151
Cyclic redundancy check (CRC), 58–9, 131–3
Cyclic shift, 115, 136, 142–4
Cyclic shift hopping, 144

Data link layer, 6
Data rate
   cell edge, 306–7
   peak, 13, 15, 282–3, 302–4
Dedicated bearer establishment procedure, 208–9, 210, 234, 255
Dedicated control channel (DCCH), 97, 166
Dedicated traffic channel (DTCH), 97, 166
Delay spread, 55, 103
Demodulation reference signal (DRS), 100, 109, 135, 142–4
Detach procedure, 187–8
Diameter protocol, 32, 204, 232, 239, 271
   AA-Request, 208
   Authentication Information Request, 193
   Cancel Location Request, 183
   CC-Request, 205–6
   LCS Routing Info Request, 272
   ME Identity Check Request, 182
   Notify Request, 186
   Provide Location Request, 272
   Re-Auth-Request, 207–8
   Update Location Request, 183
Direct information transfer procedure, 294, 296
Direct tunnelling, 232
Discontinuous reception (DRX), 146–8
   DRX command, 148, 160
   in RRC_CONNECTED, 147–8, 163
   in RRC_IDLE, 146–7, 217, 219
Discrete Fourier transform spread OFDMA (DFT-S-OFDMA), 280
Diversity, 77–80, 83, 88
   closed loop transmit, 78–9, 108, 128, 139
   open loop transmit, 79–80, 108, 128
   receive, 77–8
Doppler shift, 54, 72
Downlink (DL), 2
Downlink assignment index, 130, 137
Downlink control information (DCI), 100, 123, 127–8, 131–2
   DCI formats, 127–8, 280
      format 0, 128, 146, 281
      format 1, 128–30
      format 1A, 128, 155
      format 1B, 128
      format 1C, 128, 155, 270
      format 1D, 128
      format 2, 128
      format 2A, 128
      format 2B, 273
      format 2C, 283
      format 3, 128, 146, 281
      format 3A, 128, 146, 281
      format 4, 284
Downlink shared channel (DL-SCH), 98, 123, 132–3
Drive testing, 298, 302
Dual stack mobile IP (DSMIP), 241
Dynamic host configuration protocol (DHCP), 179, 186

Earthquake and tsunami warning system (ETWS), 25, 120, 175, 274
Eigenvalue decomposition, 86–7
Emergency call, 175, 177, 197, 256, 270
Encapsulating security payload (ESP), 198
Encryption, 192
End marker packet, 228, 239
Energy saving, 297–8
Enhanced Data Rates for GSM Evolution (EDGE), 7

EPS connection management (ECM) state
    diagram, 40–1, 173–4, 215–19
  ECM-CONNECTED state, 40
  ECM-IDLE state, 41
EPS encryption algorithm (EEA), 196–7
EPS integrity algorithm (EIA), 197
EPS mobility management (EMM)
    protocol, 32–5
  Attach Accept, 185
  Attach Request, 181, 257
  Authentication Request, 194
  Detach Request, 187
  Extended Service Request, 258
  GUTI Reallocation Command, 34
  Identity Request, 182
  Security Mode Command, 195
  Service Request, 217
  Tracking Area Update Request, 222
  Uplink NAS Transport, 261
EPS mobility management (EMM) state
    diagram, 40, 173–4
  EMM-DEREGISTERED state, 40
  EMM-REGISTERED state, 40
EPS session management (ESM)
    protocol, 32
  Activate Dedicated EPS Bearer
      Context Request, 208
  Activate Default EPS Bearer Context
      Request, 185
  Bearer Resource Allocation Request,
      206
  ESM Information Request, 183
  PDN Connectivity Request, 180
Equalizer, 56
Equipment identity register (EIR), 25, 183
Ethernet, 6, 30
European Cooperation in Science and
    Technology (COST), 301
E-UTRAN cell global identifier (ECGI),
    28, 292
E-UTRAN cell identity (ECI), 28
Evolved access network/evolved packet
    control function (eAN/ePCF), 244
Evolved HRPD (eHRPD), 244
Evolved Node B (eNB), 23–4
  donor, 286

home, 24, 175–6, 200
  serving, 23, 41, 177
Evolved packet core (EPC), 12, 24–5
Evolved packet data gateway (ePDG),
    240
Evolved packet system (EPS), 12
Evolved serving mobile location centre
    (E-SMLC), 271
Evolved UMTS terrestrial radio access
    network (E-UTRAN), 12, 23–4

Fading, 53–5, 78, 146
  frequency dependent, 55–6, 66, 128,
      138
Fast moving mobiles, 79, 88, 222
Federal Communications Commission
    (FCC), 44, 270, 273
Femtocell, 3, 24, 304
File transfer protocol (FTP), 6
First generation (1G), 6
Forward error correction, 56–7, 133
Forward link, 2
Fourier transform, 64
  fast (FFT), 65, 107
  forward, 66, 73, 134
  inverse, 64, 66, 74, 134
Fourth generation (4G), 15–16
Fractional frequency re-use, 67–8, 293
Fractional power control, 145
Frame, 103–4, 113, 116
Frame structure type 1, 103
Frame structure type 2, 104
Frequency division duplex (FDD) mode,
    3, 42–4, 52–3
  acquisition, 113, 116
  and beamforming, 92
  and carrier aggregation, 278
  frame structure, 103
  half duplex, 53
  and hybrid ARQ acknowledgements,
      137
  and random access, 153
  transmission timing, 124, 126
Frequency division multiple access
    (FDMA), 51, 70
Frequency hopping, 129

Gateway control and QoS rules provision procedure, 210, 243
Gateway control session establishment procedure, 210, 242, 245
Gateway GPRS support node (GGSN), 4, 25, 234
Gateway mobile location centre (GMLC), 271
Gating, 202
General packet radio service (GPRS), 7
Generic routing encapsulation (GRE), 31, 32, 38–9, 240
Global navigation satellite system (GNSS), 270
Global Positioning System (GPS), 270
Global System for Mobile Communications (GSM), 1–4, 6, 67, 120, 231–9
Globally unique MME identity (GUMMEI), 28, 178, 181
Globally unique temporary identity (GUTI), 28, 34, 181, 185, 191, 223
Gn interface, 233
Gold sequence, 116
Gp interface, 233
GPRS tunnelling protocol control part (GTP-C), 32, 37–8, 232–3
  Bearer Resource Command, 206
  Create Bearer Request, 208
  Create Indirect Data Forwarding Tunnel Request, 238, 246
  Create Session Request, 184
  Delete Indirect Data Forwarding Tunnel Request, 239
  Delete Session Request, 187, 249
  Downlink Data Notification, 216
  Forward Relocation Complete Notification, 239
  Forward Relocation Request, 238
  Identification Request, 182
  Modify Bearer Request, 186, 219, 228, 239
  Release Access Bearers Request, 216
GPRS tunnelling protocol user part (GTP-U), 31, 37–8
Gr interface, 233

GSM Association (GSMA), 254
GSM EDGE radio access network (GERAN), 2, 4, 232–3
Guaranteed bit rate (GBR), 202
Guard band, 51, 71, 106, 153
Guard period, 69, 104–5, 152
GUTI reallocation procedure, 33–4
Gx interface, 204

Handover, 4, 42
  too early, 296
  inter-system, 4, 15
  too late, 295–6
  optimized, 236, 243
  soft, 4, 23
  to wrong cell, 296
Handover procedure
  inter-system, 236–9, 243–4, 246–9, 256, 258–9
  S1-based, 230, 292
  X2-based, 227–9, 292
Hard decision, 49
Header compression, 168–9
High speed downlink packet access (HSDPA), 7
High speed packet access (HSPA), 7
High speed uplink packet access (HSUPA), 7
Home location register (HLR), 4
Home network, 2, 25–6, 204, 211–12
Home routed traffic, 25
Home subscriber server (HSS), 4, 24, 183, 232, 239, 255
HRPD serving gateway (HSGW), 244
Hybrid ARQ, 58–60, 123–6
  asynchronous, 124
  synchronous, 125
Hybrid ARQ acknowledgement, uplink, 99, 137, 142, 163, 280
Hybrid ARQ indicator (HI), 100, 135–7
Hybrid ARQ process, 59, 124, 125, 129
Hybrid ARQ re-ordering function, 59, 166
Hypertext transfer protocol (HTTP), 5–6
Hysteresis, 220, 222, 226, 236

Idle mode signalling reduction (ISR), 233, 235, 246

IMS media gateway (IM-MGW), 255
IMT 2000, 15
IMT-Advanced, 15, 277
Indirect data forwarding, 236
In-phase (I) component, 48, 63
Institute of Electrical and Electronics
    Engineers (IEEE), 8
Integrity protection, 168, 181–3, 192,
    197
Interference coordination, 67–8, 163,
    293–4
International mobile equipment identity
    (IMEI), 28, 183
International mobile subscriber identity
    (IMSI), 28, 147, 182
International Telecommunication Union
    (ITU), 11, 15, 42
Internet Engineering Task Force (IETF),
    5, 30, 32, 33, 197, 255
Internet key exchange (IKE), 198
Internet protocol (IP), 6, 14, 30
    version 4, 14, 22, 30–1
    version 6, 14, 22, 30–1
Internet protocol security (IPSec), 198,
    240
Interrogating call session control function
    (I-CSCF), 254
Inter-symbol interference (ISI), 55–6,
    61–2, 69–70, 103
IP address, 6, 22, 30, 40
    allocation, 36, 179, 184–5, 243
    dynamic, 179, 184
    during inter-system handovers, 234,
        243–4
    use by location services, 271
    use by policy and charging control,
        205, 207
    static, 179, 184
    version 4, 22, 36, 179, 186, 234
    version 6, 22, 36, 179, 186, 234
IP multimedia services identity module
    (ISIM), 255
IP multimedia subsystem (IMS), 14, 24,
    253–6
IP short message gateway (IM-SM-GW),
    256

IP-CAN session establishment procedure,
    184, 205–6, 243
IP-CAN session modification procedure,
    206–8
IP-CAN session termination procedure,
    187, 210
IS-95, 7

Joint processing (JP), 287

Latency, 11, 13–14, 23, 179, 260
Layer mapping, 83, 134, 284
LCS application protocol (LCS-AP), 271
    Location Request, 273
Link budget, 301
Load indication procedure, 163, 293–4
Local breakout, 26, 204
Location services (LCS), 270–3
Location update procedure, 257
Logical channel, 96, 97
    prioritization function, 159, 163
    priority, 163, 203
Long Term Evolution (LTE), 1, 11–13
LTE positioning protocol (LPP), 271
    Request Location Information, 273
LTE-Advanced, 15–17, 277–88

M temporary mobile subscriber identity
    (M-TMSI), 28
M1 interface, 269
M2 interface, 268
M3 interface, 268
MAC control elements, 148, 156, 160–2,
    269, 281
Macrocell, 3, 220, 304, 305
Market research data, 8–10, 251–2
Master information block (MIB), 97, 113,
    117–18, 119
Maximum bit rate (MBR), 202
MBMS gateway (MBMS-GW), 268
MBSFN area, 266–7, 269
MBSFN subframe, 267–8, 286
MCH scheduling information, 269
MCH scheduling period, 269
Measurement, 219–22, 224–7, 235–6
    logged, 298

Measurement configuration, 224
Measurement event, 225, 235, 281
Measurement gap, 226–7
Measurement identity, 224
Measurement object, 224
Measurement reporting configuration, 224
Measurement reporting procedure, 224–5
  for circuit switched fallback, 258
  for inter-system handover, 236, 246
  for self optimizing networks, 292, 298
  for X2-based handover, 227
Media gateway (MGW), 4, 252, 255
Media gateway control function (MGCF), 255
Medium access control (MAC) protocol, 30, 95, 159–64
Microcell, 3, 220, 304
Minimization of drive tests (MDT), 298
Minimum mean square error (MMSE) detector, 86
MME code (MMEC), 27
MME group identity (MMEGI), 27
MME identifier (MMEI), 27
MME pool area, 26–7, 181, 223, 230
Mobile application part (MAP), 233
Mobile country code (MCC), 27
Mobile equipment (ME), 21
Mobile IP (MIP), 240
Mobile network code (MNC), 27
Mobile switching centre (MSC), 4
Mobile termination (MT), 22
Mobility load balancing procedure, 294
Mobility management (MM) protocol, 259
Mobility management entity (MME), 25, 27–8
  selection, 181
  serving, 25, 40, 180
Mobility robustness optimization procedure, 295–6
Mobility settings change procedure, 294
Modulation, 47, 48, 134
  adaptive, 50, 138
Modulation and coding scheme, 129
MSC server, 4, 257
Multicast channel (MCH), 98, 267

Multicast control channel (MCCH), 97, 267
Multicast services, 265
Multicast traffic channel (MTCH), 97, 267
Multicast/broadcast over a single frequency network (MBSFN), 266–8
Multicell/multicast coordination entity (MCE), 268
Multimedia broadcast/multicast service (MBMS), 97, 265–70
Multipath, 53–5, 70, 88
Multiple access, 51–2, 66–7
Multiple input multiple output (MIMO), 77
  multiple user (MU-MIMO), 88–90, 93–4, 108, 128, 139, 281–3
  single user (SU-MIMO), 89, 282, 283–4
  see also Spatial multiplexing
Multiplexing, 52

Neighbour list
  inter-system, 234–5, 246, 293
  intra-LTE, 220, 224, 292
Network layer, 6, 30
Network management system, 291–2
Network operator, 1
Network planning, 114, 292
Network reselection procedure, 224
Network selection procedure, 173, 175, 234
New data indicator, 129
Next hop (NH), 196
Node B, 2, 23, 238
Non access stratum (NAS), 29, 33–5, 40
Non adaptive re-transmission, 125

Observed time difference of arrival (OTDOA), 270
Offline charging system (OFCS), 210–211
Offset
  cell-specific, 220, 225, 226
  frequency-specific, 221, 225–6, 236
Okumura-Hata model, 301

One Voice, 254
Online charging function (OCF), 212
Online charging system (OCS), 210, 212
Open service architecture (OSA), 271
Open systems interconnection (OSI) model, 5–6, 30
Orthogonal frequency division multiple access (OFDMA), 51, 61–72, 134
 orthogonality, 71–2
 power variations, 72–3
 and precoding, 79, 91
Orthogonal frequency division multiplexing (OFDM), 61–6
Orthogonal sequence index
 PHICH, 136
 PUCCH, 142–4

Packet data convergence protocol (PDCP), 30, 95, 167–70
Packet data network (PDN), 1, 24, 36
Packet data network gateway (P-GW), 24–5, 179, 184, 204, 232, 239
Packet data protocol (PDP) context, 233–4
Packet delay budget, 202
Packet error/loss rate, 202
Packet filter, 38
Packet flow, 36
Packet forwarding, 170, 227, 239, 243, 246
Packet switching, 2, 12
Paging channel (PCH), 98
Paging control channel (PCCH), 97, 165
Paging procedure, 208, 216–17, 235, 260, 261, 273
Path loss, 49, 145, 301
PDCP status reporting procedure, 169–70, 228
PDN gateway initiated bearer deactivation procedure, 249
Physical broadcast channel (PBCH), 98, 109, 113, 117–18
Physical cell identity, 28, 114
 use by access stratum, 116, 118–19, 133

 acquisition, 113, 116
 use by non access stratum, 225
 self configuration, 292
Physical channel, 96, 98–9, 100
Physical channel processing, 95, 133–5, 280, 283, 284
Physical control channel, 96
Physical control format indicator channel (PCFICH), 100, 109, 113, 118–19
Physical data channel, 96, 98–9
Physical downlink control channel (PDCCH), 100, 109, 123, 125, 131–2
 candidate, 132
 order, 155
Physical downlink shared channel (PDSCH), 98, 109, 123, 133–4
Physical hybrid ARQ indicator channel (PHICH), 100, 109, 125, 135–7, 281
 configuration, 113, 117, 119, 135
 duration, 135–6
 group, 136
 resource, 136
Physical layer
 air interface, 29, 95
 OSI model, 6
Physical multicast channel (PMCH), 99, 267
Physical random access channel (PRACH), 98, 112, 151–4, 297
 preamble format, 152
 preamble sequence, 151, 153–5, 227
 root sequence index, 153
Physical signal, 96, 100–101
Physical uplink control channel (PUCCH), 100, 109, 123, 140–3, 280
 formats, 140–3
  format 1/1a/1b, 110, 141–3
  format 2/2a/2b, 110, 141–3
  format 3, 280
 resource, 142, 161, 177
Physical uplink shared channel (PUSCH), 98, 112, 123, 125, 134–5, 280
Picocell, 3, 304

Policy, 202
Policy and charging control (PCC) rule, 202, 203–5, 207, 208
Policy and charging enforcement function (PCEF), 204, 240
Policy and charging rules function (PCRF), 203, 240, 252, 254
Polling bit, 166
Power control procedure, 145–6, 154, 281
Power headroom reporting, 162
Power-on procedure, 173–4, 234, 241–3
Precoding, 78–9, 84–8, 91–4, 134, 139, 284
Precoding matrix indicator (PMI), 78–9, 85, 93, 99, 139, 283
Pre-emption capability, 203
Pre-emption vulnerability, 203
Preregistration with cdma2000 HRPD procedure, 244–5
Primary cell (PCell), 278
Primary synchronization signal (PSS), 101, 109, 113, 115–16
Prioritized bit rate (PBR), 163, 203
Priority
   carrier frequency, 220–1, 234
   logical channel, 163, 203
   network, 175, 224
Privacy profile register (PPR), 271
Propagation loss, 49, 145, 301
Propagation model, 301–2
Protocol, 4–6, 28–33
   signalling, 29, 31–3
   transport, 29–31
   user plane, 29, 31
Protocol data unit (PDU), 6
Proxy call session control function (P-CSCF), 254
Proxy mobile IP (PMIP), 33, 38–9, 204, 210, 233, 240
Pseudonym mediation device (PMD), 271
Public land mobile network (PLMN), 1
Public land mobile network identity (PLMN-ID), 27, 119–20, 175, 178, 181

Public switched telephone network (PSTN), 1, 255
Public warning system (PWS), 274
Puncturing, 56

QoS class identifier (QCI), 202, 255
Quadrature (Q) component, 48, 63
Quadrature amplitude modulation (QAM), 48, 50, 134, 138
Quadrature phase shift keying (QPSK), 47–8, 50, 134, 138
Quality of service (QoS), 36, 201–3
   mobile originated request, 206–7, 210
   and scheduling, 163
   server originated request, 207–8, 210, 252, 254–5
   in UMTS and GSM, 233

Radio access network, 2–4, 23–4
Radio access network application part (RANAP), 238
Radio interface, 2
Radio link control (RLC) protocol, 30, 95, 124–5, 164–7
   acknowledged mode (AM), 166–7, 169
   transparent mode (TM), 165
   unacknowledged mode (UM), 165–6
Radio network controller (RNC), 4, 23
Radio network layer, 29
Radio network temporary identifier (RNTI), 130–2, 191
   cell (C-RNTI), 130, 156
   MBMS (M-RNTI), 131, 270
   paging (P-RNTI), 130, 147
   random access (RA-RNTI), 131, 155
   semi persistent scheduling cell (SPS C-RNTI), 130
   system information (SI-RNTI), 130
   temporary cell, 131, 155, 156
   TPC-PUCCH-RNTI, 131
   TPC-PUSCH-RNTI, 131, 146, 281
Radio resource control (RRC) protocol, 32–33, 95, 238
   DL Information Transfer, 34
   Handover From EUTRA Preparation Request, 246

Radio resource control (RRC) protocol (*continued*)
  Logged Measurement Configuration, 298
  MBSFN Area Configuration, 269
  Measurement Report, 225, 258, 292
  Mobility From EUTRA Command, 239, 246
  Paging, 42, 97, 121, 128, 147, 217
  RN Reconfiguration, 286
  RRC Connection Reconfiguration, 186, 209, 219, 224, 227, 258, 281
  RRC Connection Reestablishment Request, 295
  RRC Connection Release, 216, 223
  RRC Connection Request, 177–8
  RRC Connection Setup, 177–8
  RRC Connection Setup Complete, 177–8, 181, 218, 222, 258
  Security Mode Command, 196
  System Information, 97, 119, 121, 128
  UE Capability Enquiry, 33
  UE Information Request, 296
  UL Handover Preparation Transfer, 246
  UL Information Transfer, 34
Radio resource control (RRC) state diagram, 41–2, 173–4, 177–8
  RRC_CONNECTED state, 41, 97, 147, 224–30
  RRC_IDLE state, 41, 97, 146, 219–24
Random access channel (RACH), 98, 297
Random access procedure, 140, 154–6, 161
  contention based, 155–6, 177–8, 228
  non contention based, 154–5, 228
Random access response, 128, 155
Rank indication (RI), 83, 87–8, 99, 138–9, 163
Rate matching, 56, 132
Rating function (RF), 212
Real time protocol (RTP), 168
Received signal code power (RSCP), 236
Received signal strength indicator (RSSI), 236
Redundancy version, 130, 133

Reference signal (RS), downlink, 101
  cell specific, 101, 109, 113, 116–17
  CSI, 283
  MBSFN, 267
  positioning, 270, 273
  UE specific, 101, 117, 273, 282
Reference signal, uplink
  demodulation (DRS), 100, 109, 135, 142, 143–4
  sounding (SRS), 101, 112, 143–5
Reference signal received power (RSRP), 176, 219, 221, 226
Reference signal received quality (RSRQ), 176, 220, 222, 226
Reference symbol, 50–1, 69, 81–2
Relay, 285
Relay node (RN), 285
Relay physical downlink control channel (R-PDCCH), 100, 287
Relaying, 100, 284–8
Release, 16–17
Release 10, 17, 277–87
  carrier aggregation, 277–81
  maximum bit rate, 202
  multiple antenna techniques, 89, 94, 101, 281–4
  relaying, 100, 284–7
  self optimizing networks, 294, 296, 298
Release 11, 17, 287–8
Release 5, 253
Release 6, 265
Release 8, 17, 291–5
  DCI formats, 129, 280
  MBMS, 98–9, 101, 266
  multiple antenna techniques, 88–9, 94, 132–5
Release 9, 17, 265–74
  cell reselection, 220, 222, 274
  cell selection, 176, 224, 274
  dual layer beamforming, 93, 273
  emergency calls, 197, 256, 274
  location services, 101, 270–3
  MBMS, 97–9, 101, 265–70
  self optimizing networks, 294–8
Release 99, 22, 193

Remote authentication dial in user service (RADIUS), 32
Repeater, 285
Resource allocation, 128–9, 163, 280
Resource allocation header, 129
Resource block, 106
  physical, 128
  virtual, 128
Resource block assignment, 129
Resource block group (RBG), 128
Resource element, 106
Resource element group (REG), 119, 131, 136
Resource element mapping
  downlink, 109, 115–16, 118–19, 136
  functional block, 66, 134
  uplink, 109–12, 135, 141, 144, 151
Resource grid, 106
Resource status reporting procedure, 294–5
Reverse link, 2
Roaming, 2, 25–6, 204, 211–12
Robust header compression (ROHC), 168–9
Router, 2, 6
Routing area, 235
Routing area update procedure, 235, 239
RRC connection establishment procedure, 42, 177–9, 180, 188, 217, 222, 258
RRC connection reestablishment procedure, 295–6
Rx interface, 204, 252, 254

S temporary mobile subscriber identity (S-TMSI), 28, 147, 177, 191, 217
S1 application protocol (S1-AP), 32
  Downlink NAS Transport, 34
  Downlink S1 CDMA2000 Tunneling, 246
  eNB Configuration Transfer, 292
  E-RAB Setup Request, 209
  Handover Command, 238
  Handover Required, 237
  Initial Context Setup Request, 186, 195, 218, 258
  Initial UE Message, 181, 218, 222, 258
  MME Configuration Transfer, 292
  Paging, 41, 217
  Path Switch Request, 228
  UE Capability Info Indication, 186
  UE Context Release Command, 187
  UE Context Release Request, 216
  Uplink NAS Transport, 34
  Uplink S1 CDMA2000 Tunneling, 246
S1 interface, 21, 24–5, 199, 200, 304
S1 release procedure, 174, 215–16
S1 setup procedure, 292
S2 interface, 240–1
S3 interface, 232
S4 interface, 232
S5/S8 interface, 26, 199
  GTP option, 37–8, 180, 204, 233
  PMIP option, 38–9, 204, 210, 233
S6 interface, 24, 232
S9 interface, 204
S10 interface, 24
S11 interface, 24
S12 interface, 232
S16 interface, 232
S101 application protocol (S101-AP), 244
  Direct Transfer Request, 246
  Notification Request, 248
S101 interface, 244
S102 interface, 261
S103 interface, 244
Sampling rate, 63, 101
Scheduling, 103, 123–7, 163–4, 279–80
Scheduling command, 100, 123, 127–8
Scheduling grant, 100, 125, 127–8, 140, 155, 161
Scheduling request (SR), 100, 140, 142, 161
Scrambling, 133
Search space, 131–2, 280
Second generation (2G), 6–7
Secondary cell (SCell), 278
Secondary synchronization signal (SSS), 101, 109, 113, 116
Sector, 3, 305–6

Secure gateway (SEG), 198
Security, 191–200
   keys, 192–3, 198
   network access, 191–7, 239, 254
   network domain, 197–200
Security activation procedure
   access stratum, 186, 195–6, 218, 258
   non access stratum, 182, 195, 218, 222
Security domain, 198
Self optimizing networks, 291–8
Semi persistent scheduling (SPS), 126–7, 130
Sequence group, 143
Sequence hopping, 144
Service data flow, 36, 202, 206–7, 210
Service data unit (SDU), 6
Service request procedure, 41, 208, 217–19, 273
Serving call session control function (S-CSCF), 254
Serving gateway (S-GW), 25, 232, 244
Serving gateway service area, 26–7, 223, 229
Serving GPRS support node (SGSN), 4, 25, 232–3
Session description protocol (SDP), 255
Session initiation protocol (SIP), 255
SGi interface, 21, 24
SGi-mb interface, 269
SGmb interface, 268
SGs application protocol (SGsAP), 257
   Location Update Request, 257
   Paging Request, 260–1
   Uplink Unitdata, 261
SGs interface, 257
Shannon, Claude, 10
Short message service (SMS), 6, 251, 256, 260–1, 262
Signal to interference plus noise ratio (SINR), 10, 49–50, 56–7, 137, 304
Signalling connection, 40
Signalling radio bearer (SRB), 39–40
   SRB 0, 39, 97, 177
   SRB 1, 40, 97, 174, 177, 215, 217
   SRB 2, 40, 97, 174, 186, 215, 217
Simple mail transfer protocol (SMTP), 6

Single carrier frequency division multiple access (SC-FDMA), 72–6, 134, 280
Single radio voice call continuity (SRVCC), 256
Singular value decomposition, 88
16 quadrature amplitude modulation (16-QAM), 48, 50, 134, 138
64 quadrature amplitude modulation (64-QAM), 48, 50, 134, 138
Skype, 252
Slot, 103, 106, 268
Sm interface, 268
SMS gateway MSC (SMS-GWMSC), 256
SMS interworking MSC (SMS-IWMSC), 256, 261
SMS over generic IP access, 256
SMS over SGs, 257, 260–1
SMS service centre (SC), 256, 261
Soft decision, 49, 58
Sounding procedure, 144–5, 164
Sounding reference signal (SRS), 101, 112, 143, 144–5
Sp interface, 204
Spatial multiplexing, 80–90, 138, 281–4
   closed loop, 84–5, 108, 128, 139
   open loop, 82–4, 108, 128
   see also Multiple input multiple output (MIMO)
Special subframe, 104, 144, 152
Special subframe configuration, 104
Specifications, 16–18
   releases, 16–17
   series, 17–18
   stages, 18
Spectral efficiency, 13, 15, 71
State diagrams, 40–42
Stateless address auto-configuration, 179, 186
Status PDU, 166
Stream control transmission protocol (SCTP), 31
Sub-carrier, 61–6, 106
Sub-carrier spacing, 62, 71–2, 153, 268
Subframe, 103, 104, 268
Subscriber identity module (SIM), 22
Subscription data, 183–4, 205, 242

Subscription profile repository (SPR), 204
Suitable cell, 176, 220, 221
Supplementary services (SS), 271
   LCS Location Notification, 273
Symbol, 47, 62, 103
Symbol duration, 62, 71–2, 103
System architecture evolution (SAE), 12–15
System frame number (SFN), 103, 113, 117–18, 121
System information, 97, 114, 119–21, 228
System information block (SIB), 97, 119–21
   SIB 1, 119, 121, 175, 176
   SIB 2, 120, 141, 146, 153–4, 177, 267
   SIB 3, 120, 219–22
   SIB 4, 120, 220
   SIB 5, 120, 220–21, 292
   SIB 6, 120, 234
   SIB 7, 120, 234
   SIB 8, 120, 245–6
   SIB 9, 120, 175
   SIB 10, 120
   SIB 11, 120
   SIB 12, 274
   SIB 13, 269

Technical report (TR), 18
Technical specification (TS), 18
Terminal equipment (TE), 22
Third generation (3G), 7–8
Third Generation Partnership Project (3GPP), 1, 16
Third Generation Partnership Project 2 (3GPP2), 8
Time alignment timer, 161
Time division duplex (TDD) mode, 3, 42–4, 52–3
   acquisition, 113, 116
   and beamforming, 92
   and carrier aggregation, 278
   frame structure, 104
   and hybrid ARQ acknowledgements, 137
   and random access, 153
   and sounding, 144
   TDD configuration, 104
   transmission timing, 124, 126
Time division multiple access (TDMA), 51
Time division synchronous code division multiple access (TD-SCDMA), 7, 107
Timing advance, 105, 155, 160–1, 287
Timing synchronization, 155–6, 161
Tracking area (TA), 27, 41
Tracking area code (TAC), 28, 222
Tracking area identity (TAI), 28, 181
Tracking area list, 185, 223
Tracking area update procedure, 41, 222–3, 229, 235
Traffic flow template (TFT), 38, 206, 208
Transmission control protocol (TCP), 6, 31
Transmission mode
   downlink, 107–8, 127–8, 273, 283
   uplink, 284
Transmission procedure, 123–6
Transmission time interval (TTI), 123, 132
Transmit power control (TPC) command, 100, 127–8, 130, 146, 281
Transport block, 123, 129, 132, 159
Transport channel, 96–8
Transport channel processing, 95, 132–3
Transport layer, 6, 31
Transport network, 24, 30
Transport network layer, 29
Trusted access network, 239, 244
Tunnel endpoint identifier (TEID), 37–8, 184–6
Tunnelling, 31, 36–9
Turbo coding, 56, 132

UE capability transfer procedure, 33, 186
UE category, 22, 279, 304
UE contention resolution identity, 156, 160
UE information procedure, 296
UE requested PDN connectivity procedure, 210, 255

Ultra Mobile Broadband (UMB), 16
UMTS terrestrial radio access network (UTRAN), 2–4, 232–3
Un interface, 286–7
Universal integrated circuit card (UICC), 22, 192, 255
Universal Mobile Telecommunication System (UMTS), 1–4, 7, 67, 120, 231–9
Universal subscriber identity module (USIM), 22, 24, 175, 193–4
Untrusted access network, 239
Uplink (UL), 2
Uplink control information (UCI), 99–100, 133, 137–40, 280
Uplink shared channel (UL-SCH), 98, 125, 132–3
User datagram protocol (UDP), 6, 31
User equipment (UE), 2, 21–2
   non access stratum capabilities, 181, 273
   radio access capabilities, 22, 279–80, 283, 304
   speed, 13, 72, 79, 88, 222
User plane, 28
Uu interface, 21

Value tag, 121
Visited network, 2, 25–6, 204, 211–12
Visitor location register (VLR), 4
Voice calls, 14, 251–62
Voice over IP (VoIP), 11, 126, 168, 252–6
Voice over LTE (VoLTE), 254
Voice over LTE via generic access (VoLGA), 262
VoLGA access network controller (VANC), 262
VoLGA forum, 262

Wideband code division multiple access (WCDMA), 7, 107, 302–3
Wireless local area network, 61, 239–44, 262
Wireless World Initiative New Radio (WINNER), 301
Worldwide Interoperability for Microwave Access (WiMAX), 8, 16, 61, 239–44

X2 application protocol (X2-AP), 32
   Cell Activation Request, 298
   eNB Configuration Update, 297–8
   Handover Report, 296
   Handover Request, 227
   Load Information, 293–4
   Resource Status Request, 294
   RLF Indication, 296
   SN Status Transfer, 170, 227
   UE Context Release, 229
   X2 Setup Request, 292
X2 interface, 24, 68, 199–200
X2 setup procedure, 292, 297

Za interface, 198
Zadoff-Chu sequence, 115–16, 143, 153
Zb interface, 198
Zero-forcing detector, 82, 86